Modern Histo

Modern Historical Geographies

edited by

Brian Graham and

Catherine Nash

Longman

Pearson Education Limited
Edinburgh Gate, Harlow
Essex CM20 2JE
England
and Associated Companies throughout the World

First published 2000

ISBN 0-582-35779-9

British Library Cataloguing-in-Publication Data

A catalogue record for this book is available from the British Library.

Library of Congress Cataloging-in-Publication Data

A catalog entry for this title is available from the Library of Congress.

Set by 35 in 11/12pt Adobe Garamond
Produced by Addison Wesley Longman Singapore Pte Ltd.
Printed in Singapore

Contents

List of figures

List of tables

Contributors

Dr Richard Dennis, Department of Geography, University College London, 26 Bedford Way, London WC1H 0AP, UK

Professor Brian Graham, School of Environmental Studies, University of Ulster, Coleraine, Northern Ireland BT52 1SA, UK

Dr Nuala C. Johnson, School of Geography, The Queen's University of Belfast, Belfast BT7 1NN, Northern Ireland, UK

Dr Alan Lester, Geography Department, St Mary's University College, University of Surrey, Waldegrave Road, Twickenham TW1 4SX, UK

Dr Catherine Nash, Department of Geography, Royal Holloway, University of London, Egham, Surrey TW20 0EX, UK

Dr Miles Ogborn, Department of Geography, Queen Mary and Westfield College, Mile End Road, London E1 4NS, UK

Dr Teresa Ploszajska, Department of Environmental and Biological Studies, Liverpool Hope University College, Hope Park, Liverpool L16 9JH, UK

Professor J. M. Powell, Department of Geography and Environmental Science, Monash University, Clayton, Victoria 3168, Australia

Dr Susanne Seymour, School of Geography, University of Nottingham, Nottingham NG7 RD, UK

Dr Brenda S. A. Yeoh, Department of Geography, National University of Singapore, Kent Ridge, Singapore 0511

Preface

The approaches, themes and theories of historical geography have changed significantly over the last twenty years. This reflects in part the ways in which the divisions between the sub-disciplines of cultural and historical geography have become less distinct. As cultural geography has grown, so too has scholarship that considers the historical dimensions of cultural, economic and political geography. Contemporary historical geography also reflects theoretical shifts in the social sciences and humanities away from positivist and disinterested models of research and knowledge production. Much recent work acknowledges the situated nature of research and openly addresses the political nature of interpretation. In this, contemporary historical geography also draws on long established traditions of considering the material and symbolic geographies of hierarchical social relations and extends the concern with class to other axes of domination and identity such as race and gender.

This book provides a comprehensive account of the substantial areas of research in historical geography over the past two decades. It is organised thematically around the historical geographies of key processes and issues which have shaped the modern world in the post-medieval period, including patterns of globalisation and processes of state formation, systems of imperialism and environmental change. Above all, it seeks to demonstrate the interconnections between places, which ensured that modernisation was experienced and enacted in a multiplicity of ways throughout the world, thereby producing significantly different historical geographies of modernity. In turn, these have manifest implications for the present and the book is concerned too with the ways in which the past is remembered and represented to the contemporary era in both formal or official senses and within popular forms.

We are indebted to the co-operation and intellectual commitment of the contributors, which made editing the book to be such a profoundly stimulating and thought-provoking experience. Our thanks, also, to Kilian McDaid and Nigel McDowell of the University of Ulster for preparing many of the figures and to Luciana de Lima Marten for her careful indexing. Above all, we would like to acknowledge our gratitude to Matthew Smith (who first suggested the book) and his colleagues at Addison Wesley Longman, most especially Kate Henderson, for their enthusiastic and ever-encouraging support in seeing the project through to completion.

Brian Graham and Catherine Nash
Belfast and London
October 1998

Acknowledgements

We are grateful to the following for permission to reproduce copyright material:

Bristol Museums and Art Gallery for Figures 1.2 and 1.4; Annie Lovejoy for Figure 1.3; Houghton Library, Harvard University for Figure 2.5; British Library for Figure 2.7; National Archives of Singapore for Figures 6.1 and 6.2; Marianne North Gallery, Royal Botanic Gardens, Kew for Figure 8.1; Trustees of the Croome Estate for Figure 8.2; The National Gallery, London for Figure 8.3; Holkham Estate for Figure 8.4; The British Museum for Figure 8.5; City of Westminster Archives for Figure 9.9; Founder's Gallery, Royal Holloway, University of London for Figure 9.11; Strang Print Room, University College, London for Figure 9.13.

We also acknowledge the support of: the Historical Geography Research Group of the Royal Geographical Society with the Institute of British Geographers; the Social and Cultural Group, Department of Geography, Royal Holloway, University of London; Ian Cook, David Atkinson and Eric Laurier who involved Catherine Nash in the collective research project on 'Bristol and the Sea: Maritime Heritage, Memory and Empire', funded by the Department of Geography, University of Wales, Lampeter, which inspired and informed part of Chapter 1.

Every effort has been made to obtain permission to reproduce material contained in this book. If proper acknowledgement has not been made, we would like to take this opportunity to apologise to any copyright holders whose rights we may have inadvertently infringed.

Introduction

The making of modern historical geographies

Catherine Nash and Brian Graham

Historical geographies of the modern world

The 'modern' in the title of this book is meant in two ways. First, it denotes our concern with contemporary or new approaches, perspectives and themes within historical geography at the turn of the twenty-first century. Secondly, it refers to the specific portion of the past on which the book focuses. While historical geographers have explored – and continue to investigate – the medieval and more distant pasts, the title here alludes to the time period of the 'modern', usually divided into the early modern of the sixteenth and seventeenth centuries, the modern eighteenth and nineteenth centuries, and the (post)modern twentieth century. Instead of being an exhaustive survey of all work in historical geography, the themes of the book reflect one important aspect of the field and the interests that coalesce around the study of 'modernity' (Harris, 1991; Ogborn, 1998). The text is an introduction to the historical geographies of processes and developments that have shaped the modern world.

Pursuing both an interdisciplinary perspective and the long traditions of historical geography in exploring the material geographies shaped by agricultural change, urbanisation, industrialisation, transport and the lives, deaths and migrations recorded in the statistics of demography, the central function of this book is to explore the practices, institutions and ideologies that such geographies entailed. Bringing together questions of material change, culture, and power, we address the meanings, implications and material geographies of 'modernity' through three key themes: interconnected historical geographies, differential historical geographies of modernity and the politics of the past in the present.

Interconnected historical geographies

Historical geography has a long tradition of locating local studies within broader processes operating at wider spatial scales, of paying attention to both the specificity of the local and the wider economic, cultural and political processes and institutional structures. *Modern Historical Geographies* extends this tradition of considering the interconnections between places – for example, within modern networks of communication, transport or trade that connected Europe and its imperial sphere of influence – by analysing the kinds of geographical discourses and practices which have structured the ways in which the relationships between places or regions are understood. These include, for instance, the idea of the nation-state and the complex web of interconnections defined by imperialism. Again, considering the connections between places involves stressing the ways in which the historical geographies of colonial countries themselves were shaped by colonialism. This sense of hierarchical but reciprocal networks of influence and power avoids focusing exclusively on Western case studies, or naturalising Western models of modernity and development. Consequently, the book explores global networks and flows of information, capital, people, culture, plants, animals and objects that linked the world in the past and rebound on its present.

Differential historical geographies of modernity

Despite this emphasis on interconnected historical geographies, the broad processes of modernisation were experienced differently by people in different social groups and places, and were locally shaped in different contexts. Thinking about modernity means paying attention to the spatially and socially differentiated nature of its processes, so that our accounts do not repeat the marginalisation of certain people or places from the narratives of human history. For example, much of the work on the historical geography of modernity has focused on metropolitan centres such as London, Paris or New York. While there is increasing sensitivity to the different ways in which modernity was experienced through hierarchical social relations based on gender, race and class, less attention has been paid to the different historical geographies of modernity beyond the metropolis, in the margins of Europe or in the non-European world. Again this focus on the differential geographies of modernisation attempts to de-centre the West – or, at a different scale, the national core – in narratives of progress and development.

The politics of the past in the present

Historical geography is also increasingly informed by an interest in the ways in which the past is remembered and represented in both formal or official senses and within popular forms, and the implications which these have for the present. The stories and places of contemporary heritage are often the most overt examples of the contemporary politics of the past, but other historical forms, including the writing of historical geography, need to consider the implications of the ways in which the past is retold. Much of the stimulus for this critical reflection upon the practice of Anglo-American history comes from challenges to local, national, or global historical narratives, which have excluded or marginalised women, the working class, minority ethnic groups, indigenous people or the histories and cultures of the non-Western world. Thus instead of simply attempting to uncover or reconstruct the geographies of the past, historical research involves acknowledging the ways in which interpretation is context bound and power laden. Consequently, the ways in which the historical geographies of nationalism, colonialism, or the Atlantic world, are written, for example, have crucial implications for contemporary understandings of social relations or cultural identity.

Modern historical geographies

The first sense of the 'modern' in this book's title is worth exploring further since it both suggests a specific approach to contemporary historical geography, and new 'modern' departures in the discipline. As a specific academic perspective within European and Anglo-American universities, historical geography developed in partnership with the work of reconstructing past environments and understanding processes of landscape change within physical geography. Current research in the discipline, as evident in the content of the *Journal of Historical Geography*, recent conferences (Holdsworth and Kobayashi, 1996; Bowen, 1997) and progress reports (Ogborn, 1996, 1997, 1999), builds on these long-standing empirical traditions and historical geography's focus on industrialisation, rural land ownership, agriculture and agrarian change, migration and population patterns, urbanisation, and transport networks. Recent work, however, is also moving in new directions. Many contemporary historical geographers, informed by feminism, post-structuralism, anti-racism and post-colonial perspectives, share concerns about questions of power and meaning with other researchers more readily located within the traditional sub-disciplines of economic, cultural, political and social geography. These interests are also held in common with those outside the discipline altogether but who espouse a shared concern with the historical and geographical, material and cultural shaping of the world.

The key themes evident in much recent work include: a more overtly theorised exploration of the material, political and symbolic dimensions of travel and exploration; imperialism and colonialism; nationhood and state formation; and ideas of nature and environmental change. In these regards, contemporary historical geography draws on the long-established tradition of considering the material and symbolic geographies of hierarchical social relations, and extends its concern with questions of class and capital to other axes of domination and identity such as race and gender. Historical geographers share interests in the formal and informal knowledges, practices, and forms of representation, which have attempted to make sense of the world and our place within it, from systems of mapping and measurement to the classification of things, animals and people. This entails both considering the historiography of geography as a discipline itself, as well as the popular representation and memorialisation of the past through commemorative events, in film, performance or writing, and in museums or heritage sites.

Considerable energy has already been expended on exploring the nature and philosophy of historical geography and establishing the nature of its distinctive perspective (Langton, 1988; Meinig, 1989; Butlin, 1993; Philo, 1994; Earle, 1995; Wishart, 1997). This introspection, sometimes anxious, sometimes confident, has a long history (for example, Darby, 1953, 1983; Harris, 1971, 1978; Baker, 1972; Guelke, 1997). Nevertheless, our concern here is not with establishing precise boundaries between geography and history, or between historical geography and other sub-disciplinary areas, but in the arguably more interesting and fruitful task of exploring work that is both geographical and historical, and is being produced outside as well as within geography. In addition, the tendency to bemoan the erosion of historical geography by other sub-disciplines within the subject, or its dilution through the influences of other disciplines, belies the way in which historical geography has always been in some respects an interdisciplinary endeavour, drawing on economic and social history, ecology, social theory, just as it draws on feminism, environmental history, post-colonialism and cultural history today. Thus those who berate the contemporary 'colonisation' of historical geography by other theoretical frameworks forget the always hybrid and interdisciplinary richness of its past, one shaped, unsurprisingly, by influences that were simply different from those which apply today. In 1997, Michael Heffernan, as the new editor of the *Journal of Historical Geography*, reiterated the insistence of his predecessor, Alan Baker, that historical geography should be 'eclectic and liberal: no particular dogma about the nature of historical geography was to be promoted' nor should the journal be concerned with 'the precise definition and rigorous policing of historical geography's borders' (Baker, 1987: 1–2). Heffernan powerfully reinforced this perspective:

> To those who worry that eclecticism breeds conceptual chaos and threatens the intellectual integrity of historical geography as an academic project, I would offer an alternative and more optimistic prospectus for the future. Historical geography is, above all, a hybrid discipline and is therefore, likely to benefit from a widespread tendency, equally discernible in North America and in Europe, to question the conventional intellectual categories through which the modern world has been interpreted and conceptualized. (Heffernan, 1997: 2)

Locating and explaining *Modern Historical Geographies*

Its location

If this book, therefore, is not an attempt to define the limits of historical geography against the encroachments of other areas of research, neither is it a dramatic call for a 'new historical geography', a label which would diminish what historical geography is and has been. An instructive analogy is provided by the limitations of the term, 'new cultural geography', which became popular in the early 1990s. While marking a sense of departure and new direction, this has also relied on over-simplifying the nature of the 'old' while overlooking continuities between these traditions. In naming the 'fathers' to be killed, an exclusive genealogy of key figures, institutions and their academic descendants is reinforced, a process which edits out other trajectories of influence and other traditions (for example, Que, 1995; Kinda, 1997). By setting up the key figures or approaches that are being left behind, an exclusive and hierarchical disciplinary historiography is created. *Modern Historical Geographies* is meant to be less an agenda, or a programmatic statement for a 'new historical geography', than a reflection of already current interests, new developments and continuities of concern within historical geography. It does, nevertheless, insist upon the importance of its three key themes.

Historical geography has also widened its interests alongside history itself. As historians increasingly explore popular forms of history today as well as cultural sources for the past, historical geographers have also extended what counts as source material in researching contemporary imaginative historical geographies. These include the objects, practices and discourses of popular culture as well as the past itself. If this makes historical geography more cultural, is does not make it any less concerned with the material shaping of places, spaces, landscapes and lives. These current intersections reflect the ways in which cultural and historical geography have overlapped in the past – in the study of landscape, for example – just as they do in the present, whatever the labels academics chose to define their research interests. The opposition between an empirically grounded historical geography and a purely qualitative cultural geography is clearly reductive and redundant. Similarly, the modern emphasis on the ways in which the past features in the present does not make our work any less concerned with happenings of history. Rather, historical geography is about the interplay between the past and the present, between symbolic and material worlds.

An explanation of its structure

Modern Historical Geographies is clearly a product of its particular time and place, reflecting the interests and expertise of its authors. Like previous books on historical geography, it is aimed at an undergraduate audience while acting as a showcase of current work for researchers outside as well as within the subdiscipline. However, unlike other collections, which are often organised both temporally and regionally, working through themes and time periods and on a clearly defined region, country, or countries (for example, Ireland, Graham and Proudfoot, 1993; or England and Wales, Dodgshon and Butlin, 1990), this present text inevitably is harder to place in terms of its spatial and temporal focus. Its thematic organisation around the concept of the 'modern' means that neither its time-scale, nor its regional focus, can be easily or strictly delimited. Instead of precisely defining the founding moments or paradigmatic spaces of modernity, the book traces the spatially, socially and historically differentiated processes of modernisation as well as the history and geography of the concept of the 'modern'. In so doing, it is divided into four parts, which are given continuity and coherence of purpose by the three themes outlined above: the interconnected nature of historical geographies; the differential experiences of modernity; and the politics of the past in the present.

Part I: The context

In the single chapter in Part I, Catherine Nash provides the conceptual context for the book. She explores the concept of modernity as a European discourse, which has prioritised certain kinds of places and people over others in its equation of modernity, civilisation and European metropolitan geographies. It is argued that modernity needs to be spatially and socially differentiated to avoid some of its totalising implications. A dual focus on the asymmetrical relations of power between places and their reciprocal interconnections, and on the ways in which gender has been both symbolically central to ideas of the modern and relatively neglected within the study of modernity, serves, she suggests, both to 'provincialise Europe' and differentiate modernity.

Part II: Modernity and its consequences

The chapters in Part II deal with modernity and its consequences, dwelling on the complex interconnections between places created by the networks and routes of people, power, objects, culture, capital and commodities that linked places around the world and help structure their present circumstances. The analysis begins in Chapter 2 with Miles Ogborn's study of the historical geographies of globalisation, which takes as its spatial frame an early modern Atlantic world linked by trade, slavery and exploration. His discussion inevitably focuses on certain kinds of connections and networks that linked Britain and Europe to other places, but does not centre on British or European history. Rather, Ogborn critically explores both the asymmetries of power and the reciprocal exchanges that have shaped contemporary political and cultural geographies.

In Chapter 3, the focus switches to historical geographies of identity and the ways in which these were shaped by the impulse of modernisation. Adopting a largely European perspective, Brian Graham stresses the contingent and situated nature of nationalist identities, vested in particular readings of the past and tropes of exclusivity, but inevitably being subjected to processes of continuous renegotiation to face the demands of changing historical circumstances. Graham argues that any assessment of the continuing importance of nationalist tropes of identity is bound up with the parallel debate on the importance of the territorial state, which draws its legitimacy – itself a modernist notion – from linear narratives that connect representations of the past to the present nexus of power.

This concern with the differential experiences of nationalism in the modern era further emphasises the ways in which the politics of the past help shape contemporary political and cultural geographies. Chapters 4–6, which take three different but complementary approaches to the historical geographies of imperialism, colonialism and the colonised world, and the relationship between imperialism and the history of geography, also reflect this perspective, albeit from a very different scale. In Chapter 4, Alan Lester argues that historical geographies of imperialism within the English-speaking world have largely focused on how imperialism was understood in the European metropoles, exploring the kinds of knowledge about other 'races', climates and landscapes produced, for example, by European geographers as well as within informal or popular circuits of knowledge. Focusing on the eastern Cape Colony in South Africa in the nineteenth century, Lester traces the movements of people and goods and information between Europe, Africa, the Far East, the Caribbean and the Antipodes, as well as the localised conditions of material struggle that conditioned the discourses and practices of imperialism in these contexts and in the imperial 'centre'.

Similarly, in Chapter 5, Teresa Ploszajska both contributes to and extends this study of the intersection of imperialism and the historiography of geography. She explores the kinds of knowledge transmitted and produced within the texts and practices of geography teaching in English schools. Using descriptions of late nineteenth- and early twentieth-century textbook representations of Australia and Australians, this analysis demonstrates above all that the interconnected histories of geography and imperialism were complex, variable and multiple, as are their legacies. Thus imperialist discourses of 'racial others' were not simply constructed in the metropolis and exported with explorers, settlers and travellers, but as both Lester and Ploszajska argue, were produced through the circuits of information and knowledge, and their representations, which connected different colonial contexts to each other and to the imperial metropoles.

In Chapter 6, Brenda Yeoh suggests that a 'third way' for historical geographies of colonialism is to shift the focus away from the ideologies and practices of colonialism to consider the ideas, actions, and spaces of the colonised world. Even critical approaches to colonialism, she argues, can erase the subjects of colonisation and empire from history, presenting the colonised as homogenous victims lacking in agency. By focusing on three cases of resistance amongst the Chinese community in nineteenth-century Singapore, Yeoh challenges the absence of the colonised from the historical geographies of colonialism and theorises these spaces of resistance within the constrained world of colonial authority and domination.

Part III: Spatial contexts

In Part III, particular geographical contexts are used to further explore the book's three themes. In Chapter 7, Joe Powell chronicles the history of environmental concerns within geography from early work on the relationship between the activities of prehistoric and pre-modern communities and vegetation change to the exploration of the massive modern global environmental changes. Powell locates these concerns in the complex and changing interplay between historical geography and environmental history. Historical geographies of the environment, he argues, have always been deeply connected to their context, informing, reflecting and challenging environmental imaginations, practices, management policies and institutions.

In Chapter 8, Susanne Seymour extends the consideration of the aesthetics, politics and ideologies of rural landscape to contexts beyond Britain by following the transport and circulation of aesthetic categories, capital and landscaping practices between Britain and the Caribbean, and the social relations they entailed in the eighteenth and nineteenth centuries. Her analysis adds a further dimension to the kinds of connections that linked Britain and Europe to other places and stresses that these cannot be analysed in economic and political terms alone.

Continuing this emphasis on the aesthetic into the very different environment of historical geographies of urbanisation, Richard Dennis argues in Chapter 9 that modernisation in nineteenth-century cities was both taking place 'on the ground' and 'in the mind'. On one hand, it led to the creation of new spaces, new scales and new patterns of segregation and specialisation, and new forms of technology. On the other, modernisation also changed 'how cities were spoken about, how they were visualised, mapped, painted and photographed'. His chapter is a rich illustration of the ways in which historical geographers are increasingly eclectic in their sources and approaches, using qualitative and quantitative sources to mutually inform each other.

Part IV: Past and present

Finally, in the single chapter in Part IV, Nuala Johnson addresses the explicit theoretical, moral and practical questions raised by bringing the past into the present. In exploring the contrasting examples of the creation of landscapes of remembrance in the aftermath of the First World War and heritage tourism, she suggests that historical geographies of the present are fundamentally conjugated in the present tense and calibrated by shifting sands of interpretation, which evoke and provoke diverse responses in our readings of our relationships with the past. Her interpretation effectively concludes a book which in its making, and through its themes, fully demonstrates the vibrancy and relevance of modern historical geography in an era in which as Heffernan (1997: 2) argues:

> Traditional disciplinary allegiances, like traditional political ideologies and economic structures, are collapsing into more fluid (and potentially liberating) kaleidoscope of reformulations, reconfigurations and deconstructions. In these circumstances, the intellectual necessity of a hybrid disciplinary arena in which spaces, places, environments and landscapes are considered and analysed historically, not as an exercise in antiquarianism, but as a commentary with direct bearing on the contemporary scene, becomes all the more urgent.

It is that intellectual urgency which has informed the selection of themes and the necessarily hybrid and eclectic content of *Modern Historical Geographies*. Ultimately, however, these are not academic games. The metaphors of place created by the interconnected and differential experiences of the sort discussed here must be grounded in the concrete spaces of everyday life (Lefebvre, 1991), and the geographies of advantage and disadvantage, of life and death, which they, in turn, reflect.

References

Baker, A. R. H. (ed.) (1972) *Progress in Historical Geography*, David and Charles, Newton Abbot.

Baker, A. R. H. (1987) Editorial: the practice of historical geography. *Journal of Historical Geography*, **13**, 1–2.

Bowen, D. S. (1997) Historical geography at the annual meeting of the AAG North Carolina, April, 1996. *Journal of Historical Geography*, **23**, 76–7.

Butlin, R. A. (1993) *Historical Geography: Through the Gates of Space and Time*, Arnold, London.

Darby, H. C. (1953) On the relations of geography and history. *Transactions of the Institute of British Geographers*, **19**, 1–11.

Darby, H. C. (1983) Historical geography in Britain, 1920–1980: continuity and change. *Transactions of the Institute of British Geographers*, **8**, 421–8.

Dodgshon, R. A. and Butlin, R. A. (eds) (1990) *An Historical Geography of England and Wales*, Academic Press, London.

Earle, C. (1995) Historical geography in extremis? splitting personalities on the postmodern turn. *Journal of Historical Geography*, **21**, 455–9.

Graham, B. J. and Proudfoot, L. (eds) (1993) *An Historical Geography of Ireland*, Academic Press, London.

Guelke, L. (1997) The relations between geography and history reconsidered. *History and Theory*, **36**, 216–34.

Harris, R. C. (1971) Theory and Synthesis in Historical Geography. *Canadian Geographer*, **15**, 157–72.

Harris, R. C. (1978) The historical mind and the practice of geography. In Ley, D. and Samuels, M. S. (eds) *Humanistic Geography: Prospects and Problems*, Croom Helm, London, 123–37.

Harris, R. C. (1991) Power, modernity and historical geography. *Annals of the Association of American Geographers*, **81**, 67–83.

Heffernan, M. (1997) Editorial: the future of historical geography. *Journal of Historical Geography*, **23**, 1–2.

Holdsworth, D. W. and Kobayashi, A. (1996) Historical geography in a post-colonial world – multiple voices (not) in search of theories: the Ninth International Conference of Historical Geographers, 1995. *Journal of Historical Geography*, **22**, 198–201.

Kinda, A. (1997) Some traditions and methodologies of Japanese historical geography. *Journal of Historical Geography*, **23**, 62–75.

Langton, J. (1988) The two traditions of geography: historical geography and the study of landscape. *Geografiska Annaler*, **70B**, 17–26.

Lefebvre, H. (1991) *The Production of Space*. trans. Donald Nicholson-Smith, Basil Blackwell, Oxford.

Meinig, D. W. (1989) The historical geography imperative. *Annals of the Association of American Geographers*, **79**, 79–87.

Ogborn, M. (1996) History, memory and the politics of landscape and space: work in historical geography from autumn 1994 to autumn 1995. *Progress in Human Geography*, **20**, 222–9.

Ogborn, M. (1997) (Clock)work in historical geography: autumn 1995 to winter 1996. *Progress in Human Geography*, **21**, 414–23.

Ogborn, M. (1998) *Spaces of Modernity: London's Geographies, 1680–1780*, Guilford Press, New York.

Ogborn, M. (1999) The relations between geography and history: work in historical geography in 1997, *Progress in Human Geography*, 23, 95–106.

Philo, C. (1994) History, geography and the 'still greater mystery' of historical geography. In Gregory, D., Martin, R. and Smith, G. (eds) *Human Geography: Society, Space and Social Science*, Macmillan, London, 252–81.

Que, W. (1995) Historical Geography in China. *Journal of Historical Geography*, **21**, 361–70.

Wishart, D. (1997) The selectivity of historical representation. *Journal of Historical Geography*, **23**, 111–18.

PART I

THE CONTEXT

Chapter 1

Historical geographies of modernity

Catherine Nash

Introduction

Modernity is a key but contested concept in the humanities and social sciences. While the meaning of the term and its timing as a period is debated, most writers on modernity explore the broad and far reaching transformations of spaces, lives, institutions, experiences and subjectivities through a whole series of interconnected processes. The development and expansion of capitalist economies and their new modes and relations of production, distribution and consumption, have been central concerns within theories of modernity. But these modernising developments also included urbanisation, colonialism, and state formation, and their attendant new technologies of production, communication and transport, bureaucratic organisation, and new concepts of rational knowledge. More recently scholars are paying attention to those class based, gendered, national, sexual and racialised identities that emerged from and made modernity, and the discourses of 'newness' which articulated a sense of a shift from traditional society governed by communal custom and religion to a secular state of autonomous individuals. Modernity, it seems, is a matter of multiple processes and meanings. While many aspects of these processes of change can be found in Europe in the medieval period, modernity is usually distinguished by both a greater rate of change, and by deeper and more extended degrees of interconnection between places at a variety of scales. These modes of spatial integration could occur, for example, as local or regional forms of governance were subordinated to a centralised state and local economies more integrated, or as networks of communication, trade and influence deepened and expanded on a global scale, 'transforming the intimate geographies of everyday life and animating grand transformation on a national or global scale' (Ogborn, 1998: 19). The historical geography of modernity involves not only tracing these material geographies but also the history and geography of the whole concept of the 'modern'.

This chapter navigates through these complexities by considering what kinds of historical geographies emerge from rethinking modernity in two specific ways. The first involves challenging the generalising and exclusive tendencies of some versions of modernity by considering how the experiences of modernisation have been differentiated across space and through the social divisions of class, gender and race. The second approach explores the interconnections between places shaped by and in turn shaping modernity. These two themes are explored by considering the interwoven geographies of citizenship, race and gender that connected the Caribbean, India and England in the nineteenth century. Both of these perspectives are also based on a critical sense of the implications of the ways in which the past, and specifically modernity, has been and is understood not only in formal, official and academic knowledges but also in the 'living history' of popular culture (Samuel, 1994). This discussion of modernity is therefore framed by the example of a recent festival which used a specific history to shape an image and identity for a city in the west of England. While much of the material in this chapter inevitably comes from British, European or North American contexts, the aim is to reconsider rather than reiterate the conventional ways of telling their histories.

Port stories: maritime and modern historical geographies

Over a bank holiday weekend in May 1996 the International Festival of the Sea took place in Bristol (Figure 1.1). During four days approximately 35,000 people moved around the harbour area and its several hundred maritime exhibitions and events – including Newfoundland sea rescue dogs, model boats, fish smoke houses, shipping containers, canal boats, and Tudor maritime archery – amongst the 1,000 or so performers, the crews of 800 ships of all sizes from inflatable tenders to a replica of a seventeenth-century naval vessel. The occasion of the festival was the 500th anniversary of a particular transatlantic voyage,

Figure 1.1 A Voyage from the Past into the Future, 1996 (*source*: promotional postcard for the International Festival of the Sea, Bristol, 24–27 May 1996).

whose history and geography was used in the Festival to promote a maritime heritage for the city and help redevelop the city's old dockside areas (Atkinson and Laurier, 1998). With the support of the king and a royal edict to 'discover and claim a new world' and funded by the merchant adventurers of Bristol, an Italian explorer, John Cabot, set sail from Bristol in 1496 to discover a new trade route to the East Indies. He sailed westwards and landed in Newfoundland in 1497. The central event of the Festival was the dedication of the *Matthew*, a carefully reconstructed replica of his ship, which later left Bristol in May 1997 to sail to Newfoundland and arrive exactly 500 years after the first *Matthew*, as a focal point of the 'Cabot 500' commemorative celebrations. Cabot's venture was part of the growth in European travel, trade and exploration in the early modern period, as places previously unknown to Europeans were 'discovered', named the 'New World', and increasingly caught up in growing networks of power and culture. These early modern processes of European expansion were tied to the development of the nation state system, imperialism, capitalism, industrial development and urbanisation.

The geographical processes which have shaped the modern world and the ways in which the past is understood and culturally represented in the present are central concerns within historical geography. As this example shows, exploring modernity involves not only trying to understand the past but considering how both ideas about the present and past are shaped by the ways historical geographies appear in popular culture – in film, heritage trails, or exhibitions, for example – as well as formal or academic accounts. Thinking about the historical geographies of Bristol, and historical geography in general, means thinking about the transformation of material landscapes, but also about questions of meaning, power and identity. European exploration, travel and trade were cultural and political as well as economic processes which were shaped by and in turn shaped European imaginative geographies of the 'new' and 'old' world; how people thought of themselves and were thought of by others. They were also deeply implicated in unequal power relations between people and groups in Europe and between Europe and other parts of the world. The movements of people, plants, animals, ideas, cultures, raw materials, commodities, and capital that were set in train by the twined processes of European capitalist development and overseas expansion transformed social relations as they dramatically altered environments.

Environmental historians have chronicled the massive environmental changes resulting from capitalist and colonial settlement and agricultural systems in white settler contexts in North America or the 'New World' (see Chapter 7), and have been highly critical of capitalist agriculture and 'its commodification of land, its drastic ecological simplification, its affection for dangerously vulnerable monocultures, its promotion of divisions of labor that in the long run can do great damage to nature and human community' (Cronon, 1990: 1,129). These new forms of agrarian capitalism and plantation agriculture depended upon global circuits of plants and people. By the mid-nineteenth century, this process was highly organised as a series of British imperial botanical gardens were established in colonial capitals to facilitate economic botany. Rubber seeds, for example, smuggled out of Brazil in 1876 by an English botanical collector, were sent to be propagated in London's Botanical Garden at Kew. Of the 1,900 resulting young rubber plants sent to the Botanical Garden in Sri Lanka, 22 were sent on to the Botanical Garden in Singapore to be distributed in Malaysia as indentured Chinese labourers worked to clear and plant rubber plantations (Hoyles, 1991). But ecologies were also affected by the introduction of species for more cultural reasons, as overseas species of plants were introduced to English greenhouses and gardens and as 'English' species were 'naturalised' in colonial contexts to recreate a sense of 'home' (Wynn, 1997). As the International Festival of the Sea showed, remaking the history of Bristol around its ports, involved tracing historical geographies which linked the city to other places. These complex connections across space – in this case though trade with mainland Europe, Africa and the Caribbean, settlement in Newfoundland, the Caribbean and the Southern States of America, or the 'improvement' of agricultural productivity in the landed estates in Bristol's rural hinterland and in the West Indies (see Chapter 8) – exemplify the interconnected historical geographies that constitute one theme of this book.

One way of conceptualising the histories of these networks and the geographies they shaped as places were transformed though these flows between them is through the concept of *globalisation* – the economic, cultural and political processes which have increasingly linked together formerly relatively separate areas of the world. Within the extensive literature on globalisation most accounts focus on the rapid flows and complex networks of information, finance and cultural media around the late twentieth-century world and emphasise the novelty of these global links (Allen, 1995). While the degree of integration and speed of change has increased enormously in the later part of the twentieth century, patterns of trade, communication and migration today are deeply shaped by their historical precedents over centuries. From at least the fifteenth century, global interconnections were developing and deepening in tandem with modern European industrial and urban development. As Chapter 2 shows, these European networks variously displaced, altered or were grafted onto pre-existing circuits of trade and travel. The historical geographies of globalisation involve, according to Hall (1995: 190),

> exploration by the West of hitherto 'unknown' parts of the globe (unknown to Europe, that is); expansion of world trade and the early stages of the construction of a 'world market'; movements of capital investment and the transfer of profits and resources between metropolis and periphery; large-scale production of raw materials, food, minerals and commodities for industries and markets elsewhere; the process of conquest and colonisation which imposed systems of rule and other cultural norms and practices on subordinated cultures; the migrations which were set in motion and the settlements and colonised outposts which were established; and the establishment, even where direct colonisation was avoided, of powerful imperial spheres of influence: Britain, France, the Netherlands and Portugal in the Middle and Far East; the British, Spanish, Portuguese and Dutch in Latin America; the colonisation's by the Dutch, British and French of the Pacific; the scramble by the great powers for colonies in Africa.

These deeply geographical processes shaped places, spaces, and landscapes in new ways while they shaped and were shaped by new social relations, not just in the 'New World' but also in the 'Old'. In turn, they involved new ways of understanding, representing, ordering and imagining the world including measurements of space, forms of classification (Withers, 1995), metaphors of 'progress' and 'development', and the whole concept of a world divided into the 'Old' and the 'New'. As Doreen Massey (1994, 1995) has argued, this long-term historical perspective can be a vital antidote to the fear that the old, supposedly 'better' and secure world of the past, especially in Europe, is being eroded by the 'invasion' of people and cultures from other places. In turn, recent work on the socially and spatially differentiated nature of contemporary globalisation offers a guide to considering the interconnected historical geographies of the past. Despite the prominence of images of a 'smaller' more mobile world of simultaneous communication and increasingly homogenous culture, many authors argue that these images obscure the specific patterns of global economic or political networks as well as the different ways people are linked or marginal to these flows of capital and information. Globalisation is a highly uneven as well as unequal process in the present as different places and people are tied into these processes in different ways (Allen, 1995). This was also the case in the past. The rest of this chapter takes up the themes and issues introduced through the example of Bristol and the sea, firstly by exploring the historical geographies of modernity and secondly, returning to the Festival, to consider the politics of historical representation in the present.

Historical geographies of modernity

As Miles Ogborn (1998: 2) has written, the 'literature on modernity is full, paradoxically, of both ambiguity and totalisation'. A grand narrative of modernity as a total history of sweeping changes, overlooks the social, spatial and historical differences in the nature and experiences of the dramatic transformations it highlights. At the same time accounts of modernity emphasise the contradictory combination of dynamic change and the increasingly routinised timetable of work, of the sense of liberation from the constraints of tradition and the new disciplinary powers of the state. Despite the diversity of processes encompassed within theories of modernity and their overarching tendencies, Ogborn argues that modernity can remain a useful framework, if it can be refined through a more careful sense of its complex geographies. This serves to further fracture and fragment the idea of a singular and totalised modernity without denying the relevance of the concept altogether. Contextual and historical geographies of modernity, he argues, allow for a multiplicity of modernities while retaining a sense of large-scale and far-reaching changes. As globalisation has been a spatially and socially differentiated process, there have also been different modernities in different places. But if there is no one geography of modernity, the concept of the 'modern' can be located within European discourses which have used concepts of the modern, traditional, and primitive to make distinctions between huge areas of the world, countries, regions and between people. Thinking about modernity then, means considering the substantive transformations that have shaped the modern world, their differentiated geographies and the implications of particular Western models of the modern world. This means critically addressing the place of Europe in narratives of world history.

Post-colonial theories and approaches pose at least two key challenges to any attempt to write the historical geographies of modernity. The first is to trace the destructive impact of European colonialism in its violent, coercive and insidious cultural practices, whose results are still lived out in the contemporary global geographies of poverty and death. The second is to challenge eurocentric versions of history. Western models of modernisation and development, that crystallised and hardened in the nineteenth century, mapped a hierarchy of social development onto an imaginative geography of the 'centre' and 'periphery', defining European superiority, modernity and civilisation and centrality against the supposed inferiority, lack of development and primitive character of the 'margins'. Despite the eclipse of formal imperialism, this eurocentrism lingers on in the persistent assumption that world history can be summarised by accounts of the origins and development of capitalism, democracy, and the nation state in Europe and tracing their benign but haphazard diffusion to the rest of the world. As Ella Shotat and Robert Stam (1994: 2) argue,

> Eurocentric discourse projects a linear historical trajectory leading from classical Greece (constructed as 'pure', 'Western', and 'democratic') to imperial Rome and then to the metropolitan capitals of Europe and the US. [. . .] In all cases, Europe, alone and unaided, is seen as the 'motor' for progressive historical change: it invents class society, feudalism, capitalism and the industrial revolution.

To be modern, is to be European and to be European is to be at the pinnacle of cultural achievement and social evolution. Despite the complex regional and cultural diversity in Europe and the complex exchanges which have shaped its cultures, Eurocentrism supports an idealised notion of the purity, isolation, authority and superiority of the 'centre'. An undifferentiated global model of modernity, progress and development performs a form of cultural imperialism in which all other histories are irrelevant or subordinated. A post-colonial project of considering these historical geographies has to mediate this tension between, on the one hand, paying attention to the problematic impact of European modernities outside Europe, and, on the other, challenging the eurocentrism endemic in the Western practice of history. Critical understandings of modernity involve making sense of the powerful and influential impact of European expansion, without reinforcing Eurocentric versions of the modern. This can be done by exploring a more spatially and socially differentiated sense of European modernity, and the inseparability of economic, social, cultural, political change in Europe from the complex encounters and mutual flows of culture, capital, objects and people between Europe and the colonised world. Though burdened by the language of the 'centre' and 'margins', differentiated and interconnected historical geographies of modernity can help erode this loaded discourse of 'periphery' and 'centre'. A critical but progressive sense of the asymmetrical but reciprocal relationships between 'Europe and its others' challenges the simple spatial dualism that has been central to concepts of modernity, development and Europe itself. What I want to consider here are two, in reality inseparable, perspectives on the 'modern': firstly differentiated, and secondly interconnected historical geographies of modernity. These approaches are illustrated by tracing the geographies of campaigns surrounding citizenship, women's rights and for social reform that linked India, Britain and the West Indies in the nineteenth century.

Differentiating modernity

The focus of much work on modernity has been on the dramatically changing spaces and social relations of metropolitan centres of Europe, especially Paris, London, Berlin, Stockholm and Vienna (Berman, 1982, Kern, 1983, Pred, 1995), especially in the 'high modernist' decades around the turn of the twentieth century. This has been the iconic modern period for social theory and cultural studies, but while justifiably attending to massive changes taking place in modern cities, the continued focus on urban modernities relegates the non-metropolitan to the margins of history, as inconsequential, overlooked and unimportant. The alternative is to differentiate historical geographies of modernity spatially – by considering those places and spaces beyond the metropolitan, especially rural modernities, and socially – by exploring the social divisions of class, gender and 'race'. The model of centre and margins is disrupted by a more differentiated sense of the 'centre' and 'periphery' in which larger scale external asymmetries of power and privilege intersected with internal hierarchical social relations based on gender, race and class. By returning to historical geography's older traditions as well as considering its more recent turn to questions of culture and power, this can be illustrated by two brief examples: the historical geography of agrarian change, and the historical geography of the city.

Despite the dramatic nature of late nineteenth-century European urbanisation, the origins of modernisation can be traced back to the equally fundamental transformations that took place in the fifteenth and sixteenth century transition in England from feudalism to capitalist modes and relations of production. This involved the shift to new forms of wage labour and the consolidation of rural land ownership through enclosure beginning in the fifteenth and sixteenth century, and later backed by government, in the Parliamentary enclosures of the late eighteenth and early nineteenth centuries (Walton, 1990; Yelling, 1990). Along with schemes of improving land and stock, in new crop rotations or by the draining of wetlands – the classic case studies of rural historical geography – this practice of converting former communally worked open field systems to individually worked, separate plots, transformed the landscape as it transformed the nature of agricultural labour. Rural landscapes of modernity were newly ordered, productive, capitalised and commodified. Though general accounts of enclosure have been challenged by more regionally and temporally differentiated historical geographies, and its impact on the class structure of rural society has been central to the social histories of the countryside, enclosure was also part of a larger process of agrarian capitalism which impacted in specific ways on the family economies of the rural poor and on the productive activity of women. Jane Humphries (1990) argues that the loss of traditional rights of grazing, gleaning, wood and turf gathering as common land was enclosed and a 'moral economy' of paternalism eroded, had particular impact on women whose unwaged labour was vital to the income of poor cottagers, increasing their dependence on wages and wages earners. In addition, new patterns of disciplined and regulated labour, as waged work 'was intensified, effort was concentrated, and the workday and workplace were formalised' (Humphries, 1990: 36) created new difficulties for combining pregnancy, breast feeding and child care with waged work in agriculture or industry. Thus new capitalist modes of production entailed significant shifts in the local geographies of the gendered division of labour, women's property rights, and regional differences in women's activity in paid work (Burt and Archer, 1994). In addition, the drama of modernity identified in massive urban transformations, can be traced in the new rural industrial complexes such as Coalbrookdale (Daniels, 1992), and the regulatory devices of capital and the modern state can be found in the harsh laws against vagrancy and poaching that attempted to discipline the movement and activity of the rural poor. These rural modernities entailed changes in rural landscapes, patterns of ownership and relations of production but they also involved the articulation of new modern subjectivities.

As historical geographers and cultural historians have shown, the aesthetics of landscape in the England of the eighteenth and early nineteenth century were inseparable from the social relations and material transformation of agrarian capitalism. They were also enmeshed in debates about the qualities of the ideal civilised modern individual: an autonomous 'self', freed from the constraints of tradition, capable of abstract and rational thought and distinguished from nature. As feminist historians have argued, and as John Barrell (1988, 1990) has explored in relation to the political discourses of picturesque landscape aesthetics, this version of modern subjectivity was dependent on the construction of the inferiority and difference of women, other 'races' and the working class, all defined as pre-modern, primitive and still located in the immanent world of nature. As Barrell (1988: 117–18) argues, the 'metaphor by which the progress of civilisation could be represented as a process of "cultivation" helped in that construction, for it could represent the normal and proper process of "human development" as a matter of differentiation from, and control of nature'. This exclusive individualism was 'essential to the progress of commercial capitalism in general, and to agricultural revolution in particular'. So as the economies and landscapes of rural agrarian capitalism in eighteenth- and nineteenth-century England were increasingly tied to economic activity overseas (see Chapter 8), the aesthetics of landscapes were bound to the construction of modern subjectivities defined through ideas of internal difference and external 'others'.

Though modernity has been largely defined through the examples of European metropoles, a more differentiated sense of the modern can be achieved by attending to these rural modernities, themselves fractured by class and gender, and by returning to the city to trace there the geographies of power and difference in the material transformation of the city and in the idea of its 'modern' status (see Chapter 9). Its material and imaginative geographies and social structures (slums and suburbs, new commercial centres, new divisions of labour and space, new social relations) were also shaped by and experienced through gender and class. As Griselda Pollock (1988) has analysed with regard to the visual culture of late nineteenth-century Paris, the modern metropolitan culture of consumption, spectacle and excitement was based on the central but unacknowledged spaces of commercial and sexual cross-class exchange between middle-class men and working-class women and the construction of bourgeois and privatised womanhood in difference from both men and working-class women. The classical figure of the *flâneur*, strolling freely through the city observing its spectacle has been thoroughly deconstructed as a specifically male and middle class, and of course white, figure. Despite the presence of women on the streets – out working, shopping, or walking – the gendered associations of freedom, movement and the public that the *flâneur* epitomises still persist. However, not only were modern identities and geographies structured through gender and class, but gender was also central to discourses which tried to make sense of the 'modern'.

Rita Felski (1994) explores the gendered metaphors within the ambivalent and contradictory discourses of modernity as both freedom and loss, individualism and alienation, opportunity and insecurity, in the cultures of capitalist, industrialised, urban societies. In the culture of nineteenth-century Europe, she argues, two gendered versions of modernity circulated. The first 'equates the modern with a male-directed logic of rationalisation, objectification and developmental progress, the second emphasises the "feminine" qualities of artificiality and decadence, irrationality and desire' (Felski, 1994: 149). In contrast to both ideas of 'masculine' progress and 'feminine' frivolity, the romantic image of a redeeming, authentic and pre-modern femininity was constructed as an alternative to the apparent alienation and fragmentation of modern life. Growing numbers of historical, scientific and anthropological texts, Felski argues, 'sought to demonstrate women's greater continuity with organic processes and natural rhythms of pre-industrial society'. This pre-modern 'woman' was 'seen to be less specialised and differentiated than man, located within the household and intimate web of familial relations, more closely linked to nature through her reproductive capacity' (Felski, 1994: 146). This version of femininity could be denigrated as of limited value and a constraint on progress, or romantically celebrated as an antidote to the superficiality and meaningless nature of modern life. In turn, femininity as the emblem of pleasure, desire, irrationality could be denounced as the source of excessive and defiant women's consumerism or fetishised as the icon of exhilarating illusion, spectacle and liberation from tradition. When male artists travelled to Brittany from Paris in the late nineteenth and early twentieth century in search of the pre-modern, they sought in the rural women an alternative to both the idealised purity of middle-class women and the commercialised sexuality they enjoyed in the cities' 'spaces of modernity', but found there a region already shaped by modern agricultural economies which linked together the city and its rural environs (Orton and Pollock, 1980). These searches for authenticity in the margins of Europe were precursors to the radicalised as well as sexualised discourses of European 'cultural primitivism' (Foster, 1985; Varnedoe, 1990; Hiller, 1991; Perry, 1993) in which the cultures of non-European people were appropriated, in European ethnographic museums or by artists travelling in search of an often eroticised, exotic (Pollock, 1992). 'Primitive' cultures and their apparently unmediated, authentic relationship to nature were seen as a means of regenerating jaded modern cultures in ways which confirmed a hierarchical model of world development. So called 'pre-modern' non-European cultures were both desired and denigrated as 'primitive'. European discourses of gender and the 'primitive' were intermeshed in complex ways, gender on the one hand, structuring these asymmetrical power relations and imaginative geographies of primitivism, and on the other deployed to reinforce European superiority. European discourses of modern and gendered domesticity were central to the ways in which Europeans defined themselves as 'modern' in contrast to the 'primitivism' of non-European people, and set the privatised, patriarchal, bourgeois nuclear family and culturally specific concepts of reason, individual freedom, citizenship and the nation state as global registers of modernity, and narratives of 'progress' as a universal target of development (Chakrabarty, 1992).

Interconnected geographies of modernity

Historical geographies of modernity map complex and specific interconnections between places and between different processes. But thinking about interconnected historical geographies of modernity also means paying attention to how these connections between places are understood. In one sense, tracing lines of European influence and endeavours is a familiar feature of historical scholarship. Europe is centred in a world made global through European trade, settlement and colonialism, in which lines of contact are dominated by narratives of 'discovery' and settlement. Crucially, the theme of interconnection is deployed in this chapter not to reinforce models of European influence, but as a way of highlighting hierarchical but reciprocal networks of influence and power. Rather than linear and unidirectional flows of European culture, long histories of international circuits of people and cultures have shaped Europe as much as colonised countries. As Ogborn (1998: 19) argues:

> [m]odernity's geographies are not [. . .] place-specific in any singular sense. These differentiated geographies are made in the relationships *between* places and *across* spaces. Again, this has tended to be understood as the 'exportation' of modernity from centre to periphery both for metropole and empire and for city and country. This conceptualisation, however, ignores the crucial ways in which these geographies of connection are moments in the making of modernities rather than being matters of their transfer or imposition.

Stories of interconnection that link the local and the global can challenge neat versions of European modernity which, by failing to announce their specificity, subsume all the world within a story of Europe. Anti-eurocentric perspectives do not demand the end of all work on Europe but call for new ways of thinking about the historical geographies of Europe, or Britain, or Bristol. Critical historical geographies can take different but complementary directions in following through the challenges of modernist nationalism (see Chapter 3) and of anti-eurocentrism, in exploring the networks of colonial power and colonial discourses (see Chapter 4) or the historical geographies of the colonised world (see Chapter 6). Anti-eurocenticism can also involve exploring the construction of Europe's sense of itself as 'central' thereby deconstructing its 'natural' status. New critical work on the 'centre', 'whiteness' or 'Englishness' or 'masculinity' explores the construction of categories, previously naturalised as both superior and immutable, through ideas of difference. In the case of Englishness in the nineteenth century, identities were constructed by contrasts 'with the troublesome margins of the United Kingdom, with the dispossessed and unrepresented within the society, and with the subject peoples of the empire, with the world outside' (Hall, 1993: 216). Catherine Hall's (1994) work on the complex lines of connection between Birmingham, England and Jamaica within the discourses of racialised, gendered, national and class subjectivities and citizenship surrounding the British Reform Act of 1867, is a powerful example of this critical reading which de-centres the 'centre' in the process of rewriting its history. The rest of this section follows the geographies of citizenship, feminism and social reform as individuals and ideas travelled between the West Indies, India, Birmingham and London in the latter half of the nineteenth century.

Votes for men, votes for women: citizenship, feminism and social reform

The development of democratic government through the extension of the franchise is one of the classic markers of modernisation. The extension of male suffrage to most adult working-class men in Britain in 1867 could be read as a moment in this process. Yet its act of inclusion, Hall argues, depended upon discourses of difference within Britain and in British colonies which positioned some as subject to rather than citizens of the state, and that drew up distinctions between the 'civilised' and 'barbarous'. Focusing on Birmingham and its radical tradition in the anti-slavery campaigns since the 1820s and reform movements of the 1860s, Hall traces the ways in which the arguments surrounding the Reform Bill were articulated though the ideas of race, class and gender and informed by attitudes to the political organisation of the colonies and the social character of the colonised. The debates about political reform in Britain were framed by empire though the construction of imagined others in Australia, New Zealand, Canada, Ireland, the post-independence United States, and, following the black rebellion and mass executions of Morant Bay in 1865, Jamaica. As Hall illustrates, both those who fought against and those who supported the Reform Bill based their arguments on the beneficial or pernicious results of extending suffrage in the Empire. The extension of the vote to British men who were household heads and in permanent residence, constructed the model of the respectable, independent, industrious working man against the supposed unfitness of black people and British women for citizenship. On this basis, a degree of self government for Jamaica was replaced by direct rule and a petition to grant suffrage to British women on the same basis as men was defeated. In the debates leading up to the Reform Act, questions of women's status, male suffrage in Britain, the system of colonial rule in Jamaica and the abolition of slavery, Hall argues, were inseparable. Following 1867, suffrage was no longer based on property, 'but "race", gender, labour and level of civilisation now determined who was included in and excluded from the political nation' (Hall, 1994: 29). The historical geographies of British feminism also inscribe an interconnected and differentiated modernity.

Though legal rights, education and employment were more important than the vote to the British feminist campaigners of the 1860s, the rise of a feminist consciousness and feminism's 'sustained attack on male bourgeois society' (Pollock, 1995: 14) was a key, but often overlooked, aspect of the development of modern subjectivities, shaped in tandem with as well as in opposition to modern male subjectivity. The historical geographies of feminism both differentiate European modernity and link women across colonial space in contradictory and unequal relationships of affinity and difference. Though imperial histories have chronicled the activities and achievements of the men as explorers, statesmen, and soldiers, or more critically explored the gendered discourses of discovery, control and domination, only recently have the roles of women in the empire as missionaries, educationalists, reformers, nurses, wives of colonial soldiers and officials, travellers and permanent settlers been addressed (Chaudhuri and Strobel, 1992; Bush, 1994). Not only were women actively involved in imperialism, and central to the cultural exchanges that have shaped English culture, bringing home shawls and recipes from India for example (Chaudhuri, 1992), but ideas of Englishness were structured through the supposed contrast between white, pure, chaste and noble English women and their colonial counterparts. While feminists drew on an analogy between the condition of women and the condition of slaves, their discourses of emancipation for women in Britain and for colonial women were both complicit with and resistant to white, European, bourgeois and masculine structures of power. As Vron Ware (1992) and Catherine Hall (1992) have argued, British feminism was shaped by imperialism. English feminists drew on the language of slavery and civil rights in the 1830s and as middle-class women articulated their concerns in the context of popular imperialism later in the nineteenth century.

In the 1860s significant numbers of upper- and middle-class English women became increasingly concerned about the condition of women in India and many travelled there as reformers and missionaries to provide education and medical aid to Indian women. Though motivated by a sense of shared female subjectivity, many saw themselves as 'maternal imperialists' in a benevolent but authoritative mothering role over India and Indians (Ramusack, 1992). Believing the condition of Indian women – and especially the practices of sati, purdah and child marriage – resulted from the 'tyranny' of 'native' customs, their sense of gender alliance was based on a model of racial superiority in which Western women had a duty to bring civilisation to the uncivilised in Britain and in 'heathen' lands. Drawing on the ideal of Victorian femininity and women's moral superiority, middle-class liberal feminists used the 'condition' of colonial women, especially Indian women, to justify their own role in the public sphere, thus challenging their subordination, and as examples against which to gauge their own progress, racial superiority and civilised status (Burton, 1992). According to Burton (1992: 150), the Woman of the East was a

> pivotal reference in arguments for female emancipation and she became the embodiment of personal, social and political subjugation in a decaying civilisation – the very symbol in short of what British feminists were struggling to progress away from in their own struggle for liberation.

In turn, the 'unnatural' despotism of some English men could be reformed by Westernising them through the moral influence of English women. This 'orientalist feminism' challenged patriarchal power but through the racist trope of the despotic Oriental man (Meyer, 1990; Zonana, 1993). In the 1880s, for example, British women in India were deeply opposed to the relatively progressive Ibert Bill which would have allowed legal cases involving Europeans to be tried by Indian judges, fearful of the loss of dignity and liberty at the hands of native judges and their 'barbaric' views of women (Ware, 1992: 122). Thus, both dominant and resistant forms of modern gendered subjectivity in Europe were deeply structured through 'race' and imperialism.

Though late nineteenth-century British feminism was deeply compromised by its imperialist discourses of the suffering of Indian women as the subjects of 'barbarous' colonial men, its reforming efforts found support amongst some Indian men as well as women. This was especially the case amongst those, such as the wealthy Parsi of Bombay, whose adoption of European concepts of women's liberty and support for British reform initiatives could be used to distinguish themselves from 'less civilised' Hindus. The story of these tense alliances takes us back to London. In 1890 Behramji Malabari, a Bombay Parsi, journalist and reformer, visited London to seek support for campaigns against 'child marriage' and prohibitions on the re-marriage of widows in India. As Antoinette Burton (1996: 190) so importantly shows, the travel of Malabari, like other 'well-known people of colour – M. K. Gandhi, Mary Prince, Frederick Douglass, Mary Seacole, Pandita Ramabai, C. L. R. James' to the European metropole,

> reminds us that the flow of ideas, commerce and people was not just from West to East. Either because they were part of permanent communities with long histories and traditions in the British Isles, or because they were travellers or temporary residents in various metropoles and regions throughout the United Kingdom, a variety of colonial 'others' circulated at the very heart of the British Empire before 1945. (Burton, 1996: 176)

Malabari's account of his visit, *The Indian Eye on English Life*, like the accounts of other 'national reformers from Rammohun Roy in the 1820s through

Rabindranath Tagor in the 1870s to Gandhi in the 1880s', as Burton (1996: 188) argues, inverts the tradition of ethnographies of Europe's 'others'. But his account also, as Burton shows, narrates his uneasy negotiation of status and masculinity as an Indian gentleman *flâneur* critically observing the character of English society, especially English women, while he seeks, as self-styled 'mother' of Indian's daughters, support from English men for reform for Indian women. His racialised encounters on the street deprived his account of the 'sense of possession or confident familiarity' (Burton, 1996: 188) characteristic of the *flâneur*. His writing, Burton suggests, challenges the equation of civilisation, Englishness and modernity in deploring England's secular materialism, while at least ambivalently advocating its 'liberation' of women. It also, she argues, illustrates the constant mediation of masculinity by culture, class and race in a

> context of British imperialism, where Indian men competed with British men for terrain, legitimacy, and authority over colonised women, [when] the performance of colonial masculinity varied enormously and was contingent on the ever-shifting ground of imperial hegemony and nationalist/indigenous challenges to it. (Burton, 1996: 177)

These gendered geographies provide a more differentiated and interconnected sense of European modernity. They can also be seen as part of a project what Dipesh Chakrabarty (1992: 21) calls 'provincialising "Europe"'. This means deconstructing the image of an homogenous, geographically isolated entity; recognising that 'Europe's acquisition of the adjective *modern* for itself is a piece of global history of which an integral part is the story of imperialism' and despite their anti-imperial direction, the reproduction of this equation of the modern and European within third-world nationalisms. 'Provincialising Europe', Chakrabarty argues, is

> to see the modern as inevitably contested; it involves writing into the history of modernity the ambivalences, contradictions, the use of force, and the tragedies and the ironies that attend it. (Chakrabarty, 1992: 21)

Critical historical geographies which rewrite the histories of modernity as de-centred, interdependent, differentiated and power-laden perform this 'provincialising' work. This is central to a post-colonial perspective which attends to the colonial 'histories of exploitation and the evolution of strategies of resistance' and which 'bears witness to those countries and communities – in the North and the South, urban and rural – constituted,' as Homi Bhabha describes them, 'other wise than modernity'.

> Such cultures of a post-colonial *contra-modernity* may be contingent to modernity, discontinuous or in contention with it, resistant to its oppressive, assimilationist technologies; but they also deploy the cultural hybridity of their borderline conditions to 'translate', and therefore reinscribe, the social imaginary of both metropolis and modernity. (Bhabha, 1994: 6)

Catherine Hall's call for a new way of belonging involves re-thinking ideas of history, geography and identity in a provincialised and post-colonial Britain. It means rejecting myths of homogeneity, recognising the power laden and painful discourses of national inclusion and exclusion, and understanding interconnections, interdependence and power relations differentiated through gender and race:

> In re-imagining the past and re-evaluating the relations of Empire 'we' can begin to understand the ties which bind the different peoples of contemporary Britain in a web of connections which have been mediated through power, the power of the coloniser over colonised, that have never moved only from 'centre' to 'periphery', but rather have criss-crossed the globe. (Hall, 1996: 76)

The politics of the past in the present

The ways in which the historical geographies of colonialism or the Atlantic world are written have crucial implications for contemporary understandings of social relations or cultural identity. The representation of the past is deeply tied to contemporary conflicts surrounding how collective identities are constituted (see Chapter 10). As the character of communities, cities, regions or nation states are constructed through the stories of the past, the versions of history narrated in academic and popular forms, and how the past is understood – What is included? What is left out? How are these histories made sense of? – have profound implications for the ways in which individuals and groups are located within, granted significance and or excluded from membership of states and nations, and local, family or community histories. Not only have the modern transformations of the past shaped the present, but a historical perspective is vital to any project to challenge the apparently natural and immutable organisation of societies, economies and politics today, and the romanticised versions of the past which are used to reinforce conservative versions of the present and future. Historical geography can illuminate the historical, cultural and geographical variability of concepts of nationhood, race, gender and sexuality and their social construction, undermining essentialist versions of their natural status.

As the representation of the past is so bound up in questions of power and identity, what counts as the past, and what kinds of histories are valued, are often deeply contested. As feminist and postcolonial historians have shown, attending to these overlooked histories does not just mean including them in dominant narratives but challenging conventional notions of history itself. Over the past thirty years, traditional national histories of kings, queens and statesmen have undergone a sustained and far-reaching challenge from socialist, feminist and post-colonial historians exploring the neglected and alternative histories of the working class, women and indigenous peoples. The controversies which arise when authoritative and public accounts of the past put on display in museums are challenged, or the extremely sensitive and controversial debates surrounding the commemoration or memorialisation of famine, slavery, or the Holocaust (Young, 1983) are powerful reminders of the politics of history.

While the historical dimensions of museums are self-evident, they are also institutions whose internal geographies of display are linked to imaginative geographies of communities, regions, nations and human history on a global scale. 'New museology' (Vergo, 1989) and its critical perspectives and innovative curatorial practices has grown from an awareness of the ways in which the traditional authority of the museum has been harnessed in the production of élite, racist and imperial knowledges. The historical geographies of their exhibition practices and their representation of the past today make museums key sites for exploring the politics of historical representation and their, often contested, construction of concepts of the native, the nation, the community, the past and the present (Karp and Lavine, 1991; Karp *et al.*, 1992; Pointon, 1994). As Carol Duncan (1995: 3) has observed:

> the issue of what western museums do to other cultures, including minority cultures within their own societies has become especially urgent as post-colonial nations attempt to define and redefine their cultural identities and as minority cultures in the West seek cultural recognition.

Much has been written on the problematic histories of the museum 'exhibitionary complex' (Bennett, 1996), Western practices of collecting (Breckenridge, 1989), display and exhibition of other places and peoples (Greenhalagh, 1988, 1993; Mitchell, 1989) in the past and their legacies in the present (Lidchi, 1997). Ethnographic museums and their forms of classification have been interpreted as powerful ways in which European societies defined themselves through the apparent 'primitivism' of non-European people (Chapman, 1985; Dias, 1994). With non-European objects classified as artefacts rather than art (Clifford, 1988) and often ordered through an evolutionary framework, the collections put on display an imaginative geography of the 'centre' as authoritative, scholarly, cultured and custodian institutions and individuals, and the 'periphery' as a general source of artefacts and the 'exotic'. Yet, recent attempts to rework the ideologies and practices of museums provide vivid examples of efforts to produce progressive, inclusive and dialogic versions of the historical geographies of places and communities. The three examples briefly discussed here – Canadian First Nations' museums, the China Town history project in New York, and an ethnographic exhibition in Birmingham – offer some answers to the questions Felix Driver and Raphael Samuel ask about the ways the histories of communities and local places can be understood and represented in the present 'without falling prey to introverted (and ultimately exclusionary) visions of the essence or spirit of places':

> If conventional notions of place have been destabilised, what are the alternatives? Can we understand the identity of places in less bounded, more open-ended ways? Can we write local histories which acknowledge that places are not so much singular points as constellations, the product of all sorts of social relations which cut across particular locations in a multiplicity of ways? What are the ways of telling the story of places that might be appropriate to such a perspective? How are we to reconcile radically different senses of place? Such questions arise not simply within the projects of local history, but within all those varieties of writing concerned with places and their pasts. (Driver and Samuel, 1995: vi)

In 1992 a Canadian Task Force established in collaboration between the Canadian Museums Association and the Assembly of First Nations produced a report on 'Museums and First Peoples' (Hurle, 1994). The Task Force was initiated following protest and debate surrounding the exhibition 'The Spirit Sings' at Glenbow Museum in Calgary. This exhibition of 'native' culture was organised to coincide with the 1988 Winter Olympics, but was boycotted by the Lubicon Cree in part because of its sponsorship by Shell Oil, then extracting oil from land claimed by the Cree. The resulting ethical framework highlighted the need for First Nations people to be empowered to represent their own histories in partnership with mainstream museums in light of the histories of inequality and of the continued religious and ceremonial significance of 'native' objects to First Peoples. This issue of empowering and alternative and self-determined forms of museum representation was central to James Clifford's (1991) analysis of two mainstream 'majority' and two community 'tribal' museums on the north-west coast of Canada, all of which display ceremonial masks, rattles, robes and sculpture. In contrast to majority museums broadly characterised by

> (1) the search for the 'best' art or most 'authentic' cultural forms; (2) the interest in exemplary or representative objects; (3) the sense of owning a collection that is a treasure for the city, for the national patrimony, and for humanity; and (4) the tendency to separate (fine) art from (ethnographic) culture. (Clifford, 1991: 225)

the tribal museum, is informed by different objectives. For Clifford (1991: 225–6)

> (1) its stance is to some degree oppositional, with exhibits reflecting excluded experiences, colonial pasts and current struggles; (2) the art/culture distinction is often irrelevant, or positively subverted; (3) the notion of unified or linear History (whether of the nation, of humanity or of art) is challenged by local, community histories; (4) the collections do not aspire to be included in the patrimony (of nation, of great art, etc.) but to be inscribed within different traditions and practices, free of national, cosmopolitan patrimonies

commemorating instead local meanings and memories and a living culture and ceremonial tradition of which the objects are a part. At the same time this distinction is fractured by the efforts of mainstream museums to ethnically represent native peoples, by involving native artists and craft makers and changing the ways in which their culture is displayed. It is also disrupted by the ways in which tribal museums re-appropriate the exhibitionary practices of metropolitan museums as well as their material cultures in national ethnographic collections, to recuperate a sense of a history that is shared by native peoples as well as varying amongst different tribal groups. Both metropolitan and minority culture museums, Clifford (1997: 192) suggests, can be thought of as 'contact zones' – places of cultural interaction, exchange, appropriation, albeit in asymmetrical relations of power – at a whole variety of scales from global international exhibitions to neighbourhood museums and cultural centres. This sense of both cultural specificity and cultural interconnection has also informed other community history projects.

In the China Town History Museum Experiment in New York, the aim was to reconstruct a neglected history of the oldest Chinese settlement in the

United States and mutually explore the meaning of Chinatown's past, in order to reclaim a marginalised history and counter the cultural damage of decades of institutionalised racisim. In this recent attempt to create an archive of this community as well as explore the meaning of history and memory, the organisers brought together formal methods of historical research with popular historical practices, using media productions, public and voluntary programmes, tapping community histories through reunions, school projects, family history and genealogy workshops, or salvaging discarded material from rubbish bins. The stories, photographs, documents and other objects gathered together are used to facilitate an ongoing exploration of neighbourhood memories, cultural representations, and strategies of survival. The intention has been to both reconstruct these community histories and resist a fixed, narrow, homogenous version of China Town as an isolated enclave of pure and stereotypical Chinese culture (Tchen, 1992: 294). Instead of over-emphasising Chineseness as an 'essentialist and quasi-genetic characteristic', or the uniqueness of the 'local history', the aim was to 'examine the roles of Chinese New Yorkers, non-Chinese New Yorkers, and tourists in the creation of New York's China Town whose history is part of, rather than isolated from, the development and identity of the Lower East Side, New York, and the United States as a whole. In this way a 'local' history project is also an exploration of the histories of the nation. While some community history projects resist the institutional format of a fixed collection in creating dispersed and dynamic archives (Fuller, 1992), in other cases curators and museum professionals deploy their existing collections to interrogate the ideologies through which they were shaped.

A recent project to reorganise an ethnographic collection in the Birmingham City Museum and Art Gallery, formed in the early years of the twentieth century by wealthy Birmingham industrialists, colonial soldiers, missionaries and administrators, is an attempt to critically explore the historical geographies of ethnographic collection in order to engage with the present (Peirson Jones, 1992). As emblem of civic pride and culture, the museum and gallery, like many others founded from the mid-nineteenth century in Britain, was opened in 1885 to 'extend the knowledge, refine the taste, instruct the judgement and strengthen the faculty of those who are engaged in Birmingham industries' (quoted in Peirson Jones, 1992: 222–3), whose respectable working-class men had been granted suffrage in 1867. Museums like this exhibited the nation and the world through the imaginative geographies of empire and were part of the project of 'social imperialism' which aimed to convince the working classes that all the country's ills could be cured and prosperity secured by the continued support for imperialism, rather than radical reform at home (Coombes, 1991).

Against this history and in the current context of a multi-ethnic city shaped by the post-war immigration from Britain's former colonies, especially India, the Caribbean, Africa, Pakistan, Bangladesh and Ireland, the exhibition 'Gallery 33: A Meeting Ground of Cultures', which opened in 1990, used part of the museum's ethnography collection to critically explore the process of collection in the past and the meaning of culture, identity and history in the late twentieth century. As the curator, Jane Peirson Jones, describes, the strongly thematic display brings together objects and images that are normally separated by the traditional division between the ethnographic material from 'foreign' cultures and Western art. The section on the 'The Decorated Body' included British Sikh turbans, African hairstyling, Italian plastic surgery and Japanese, Maori and British tattoos. In contrast to the conventions of classification, Gallery 33 declassifies and mixes 'the familiar with the unfamiliar, the past with the present, and the majority with the minority' to challenge visitors' 'sense of order, their sense of the "other", and thus their sense of themselves' (Peirson Jones, 1992: 227–8) through these juxtapositions. Following interactive exhibits visitors can trace the historical geographies of collectors active in the 1920s, their lives and their motives, or objects as they travelled from the site of acquisition to the site of display. By choosing from twenty-two categories such as geographical location, collectors' name or topic, visitors can explore for themselves the different ways objects can be understood. This reworking of ethnographic display practices, forms of interpretation and their assumed authority, locates the objects and the collection both in the context of colonial and post-colonial cultural politics, 'deconstructing colonialism, recontextualising twentieth-century migrations, and integrating the histories of white Britons and ethnic minorities' (Peirson Jones, 1992: 240). These critical historical geographies of interdependence, in academic and popular forms, deconstruct the 'centre' and metropolitan models of Western modernity. While heritage in the form of civic monuments 'redraws the boundaries of inclusion and exclusion for contemporary Birmingham' as its statues of the 'founding fathers' trace the city's 'white male line of descent' (Hall, 1994: 6), at the same time museum heritage is being deployed here to question these boundaries and this history. The public bodies which are often the target of criticism for their social conservatism in some respects are often also committed to progressive developments in others. While I have concentrated on museums in this section, as particularly pertinent cases, questions of power, memory, identity and the past can be explored in other cultural forms through which people shape 'living history' – re-enactments, historical novels, television drama, film, festivals and commemorative events. Bristol's contested maritime heritage at the International Festival of the Sea is one vivid example.

Alternative maritime historical geographies

The celebration of Cabot's story in Bristol, here pictured at the moment of departure and hanging in Bristol's Museum and Art Gallery (Figure 1.2), speaks of a longing for the security of the later age of Elizabethan exploration that Cabot's voyage foretold. When this painting was produced in 1907, the model of Elizabethan expansion, commerce and rural life provided an ideal for Edwardians anxious about threats to the empire abroad and social stability at home, and an antidote to the extremes of nineteenth-century industry and commerce (Howkins, 1986). This turn to a romanticised Elizabethan era of continuity, community, harmony and natural social hierarchies of religious and civil power as well as expansion and discovery offered an imaginative

Figure 1.2 *The Departure of John and Sebastian Cabot from Bristol on their First Voyage of Discovery in 1497*, Ernest Board (1906) (published with the permission of Bristol Museums and Art Gallery).

escape from anxieties about social change and challenges to old authorities and ideologies. In late twentieth-century, post-industrial and post-colonial British society, it also seems to evoke an implicit imperialist nostalgia, or at least an untroubled celebration of heroic endeavour. Though the most prominent official Festival narrative was of heroic exploration and discovery, its events also suggested other ways of thinking about this maritime history.

The Festival did not simply feature explorers. It also focused on those who crewed the vessels which crossed the sea and the skills of seamanship they required to survive its storms and trials. Yet this romantic image concentrating on the struggle between individuals, usually men, and nature deflects attention away from relationships between people at sea – among the crew and between the sailors and their masters. As Marcus Rediker (1987: 7–8) has argued,

> the history of seafaring people can and must be more than a chronicle of admirals, captains and military battles at sea. It must be made to speak of larger historical problems and processes.

These processes included the changing social, cultural, economic and political relations within early modern capitalist development. Its problems included the suffering of those who worked as wage labours and as slaves in the networks of capital and power that crossed the ocean and connected ports, palaces and plantations (see Chapter 2). The words and rhythms of the sea shanties and maritime folk music at the festival spoke of the hardship of life on fragile craft in a dangerous environment, but their refrains also echoed with a sense of the sailor, as Rediker (1987: 5) has written,

> caught between the devil and the deep blue sea: On one side stood his captain, who was backed by the merchant and royal official, and who held near-dictatorial powers that served a capitalist system rapidly covering the globe; on the other side stood the relentlessly dangerous natural world.

Many of the work songs at the Festival expressed a culture of communal and co-operative effort under the harsh conditions of wage labour, when in the eighteenth century the 'merchant's ideal of increased work and productivity usually meant increased exploitation of seamen' (Rediker, 1987: 75). The sea shanties at the Festival sang of the culture of a maritime working class and their resistance to the press gang, which sought to make up for the shortages of those who would willingly take to sea and resistance to ill treatment by the captains they worked under.

Alongside the celebration of the sea as lubricant of British maritime capital at the Festival, were the songs of an international working class with its alliances across race and ethnicity – pirates freeing indentured servants, seamen helping runaway slaves – songs of a port-side proletariat of many 'different nations, races, ethnicities and degrees of freedom' (Linebaugh and Rediker, 1990: 226). The movement of working people from port to port meant that the port cities of the Atlantic world in the early modern period were cosmopolitan places. As Peter Linebaugh and Marcus Rediker have argued, sailors shared experiences of oppression often led to interracial co-operation and rebellion in seaport strikes and riots and to a multi-cultural, anti-authoritarian and egalitarian tradition. At the Festival, the shanty group Bosun's Call sang songs of mills, mines, factories, sailors and fishermen. These songs re-figured the sea and port not only as places of working-class solidarity but also as multicultural. The Larry Brown Shanty Men drew from a repertoire of 700 songs from around the world and the songs of Stormalong John were described as 'typically powerful, vigorous, often with intricate rhythms derived from the international origins of the crews' (*Official Programme*, International Festival of the Sea, 1996: 58). This specifically class-based cross-cultural experience, goes some way to differentiating the experience of globalisation.

Yet, the Festival almost exclusively presented the sea as an arena of men's experiences, skills and stories. The heroic, independent and challenging life at sea tended to be contrasted to the communal, inconsequential and safe life of women and children left behind. Clearly maritime trade and exploration have been predominantly but not exclusively male. In the early modern period women pirates, slaves, passengers and indentured servants crossed the Atlantic. But those on land were not simply isolated from the processes of change. The development of the modern world was gendered not only though the changing sexual division of labour and women's role in the development of capitalism through waged and unwaged labour but through the construction, performance and negotiation of modern and class-based forms of masculinity and femininity. Women were not simply located outside social, cultural and economic change, but would have been present, for example, as rowdy prostitutes, as subversive messengers, as insurgents, as receivers of stolen goods in the brothels, dancing houses and taverns that are central to Linebaugh and Rediker's case for an international history of the maritime working class. This maritime historical geography, like historical geography in general needs to consider gendered discourses as well as gendered experiences (Rose and Ogborn, 1988). This might involve understanding the gendering of heroic voyage; the sea and its feminised mythology of dangers and attractions (of sirens and mermaids); the negotiation of homo-social and hierarchical masculinities and male sexualities on board ship; and highlighting more explicitly the specific role of women in the dockside encounters so fundamental to radical maritime history (Mahony, 1987; Creighton and Norling, 1996).

The story of the sea also needs to be thought about through questions of race. By the 1740s, traders, mostly from Bristol and Liverpool, ports which dominated the slave trade, had made Great Britain the world's leader in carrying human cargoes. At the Festival, in different sites and through different media, artists, writers and activists criticised the way in which problematic histories of oppression in general, through imperialism and colonialism, and in particular slavery were largely elided in the event. Annie Lovejoy's artistic intervention entitled *stirring@the international festival of the sea* (Figure 1.3) was a reminder of the human costs of mercantile success in slavery in the infamous triangular trade in which goods, slaves and sugar circulated around the Atlantic between Britain, Africa and the West Indies. From the 1640s, English settlers in the West Indies began to produce sugar and imported slaves to work their plantations and British manufactured goods. On Lovejoy's sugar packets – which were found and used by visitors in the cafés within the festival site – Bristol was located within these circuits of sugar, tobacco, cocoa, tea, spices, rum, slaves and sugar. She also produced a postcard which mapped the sites where the sugar packets were available on to the plan of the Festival and so made this dockside geography speak of the overlooked but central issue of oppression within Bristol's history of imperial and commercial success. The issue of slavery was also foregrounded by anonymous protesters who pasted stickers featuring the plan devised by a prominent Bristol slave merchant of how to restrain and pack a cargo of slaves, onto prominent signs around the festival site. It is this sense of modernity's power-laden circuits and interconnections that Paul Gilroy's (1993) concept of the 'Black Atlantic' encapsulates.

However, the most explicit expression of the problems of uncritically celebrating Bristol's past took place when the Black Writers Group, in a short slot in the Saturday programme of events, explored the centrality of colonialism and slavery to Bristol's past through poetry, short stories and song and memorialised a history that was largely obscured at the Festival. Yet they did not simply construct a history of Bristol as one of only wealthy exploitative merchants.

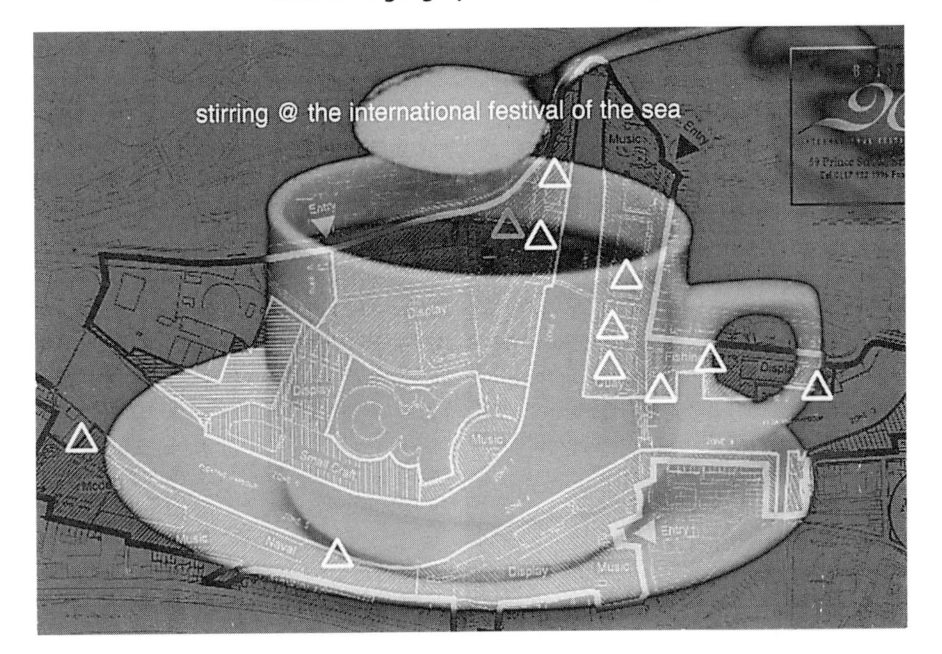

sugar, tobacco, cocoa, tea,
spices, rum, slaves, sugar,
tobacco, cocoa, tea, spices,
rum, slaves, sugar, tobacco,
cocoa, tea, spices, rum,
slaves sugar, tobacco, coc
oa, tea, spices, rum, slaves,
sugar, tobacco, cocoa, tea,
spices, rum, slaves, sugar,
tobacco, cocoa, tea, spices,
rum, slaves, sugar, tobacco,
cocoa, tea, spices, rum,

Bristol

"..Bristol's maritime heritage is a multi-layered construct. We remind ourselves of historical realities when we begin to peel away layers and look closely at seemingly innocuous things like packets of sugar. Creatively this idea encourages us to acknowledge the tensions & discomfort that has been so much a part of the historical trade in sugar." Eddie Chambers.

"It is an idea which connects history with the present by turning a familiar commodity into a symbol" Martin Lister.

postcard production assisted by.....Watershed & Easton Community Centre

sugar sponsored by.....

GWR fm

packet printing...
Good Morning
Foods & Disposables

a negotiation made possible thanks to the interest & support from...Eddie Chambers, Simon Cooper, Rupert Daniels, Mac Dunlop, Andrew Kelly, Philippa Goodall, Tessa Jackson, Martin Lister, Nigel Locker, (Venue magazine) & John Summers.

Annie Lovejoy '96. Tel: 0117 9515029

Figure 1.3 *stirring@ the international festival of the sea*, Annie Lovejoy (1996) (published with the permission of the artist).

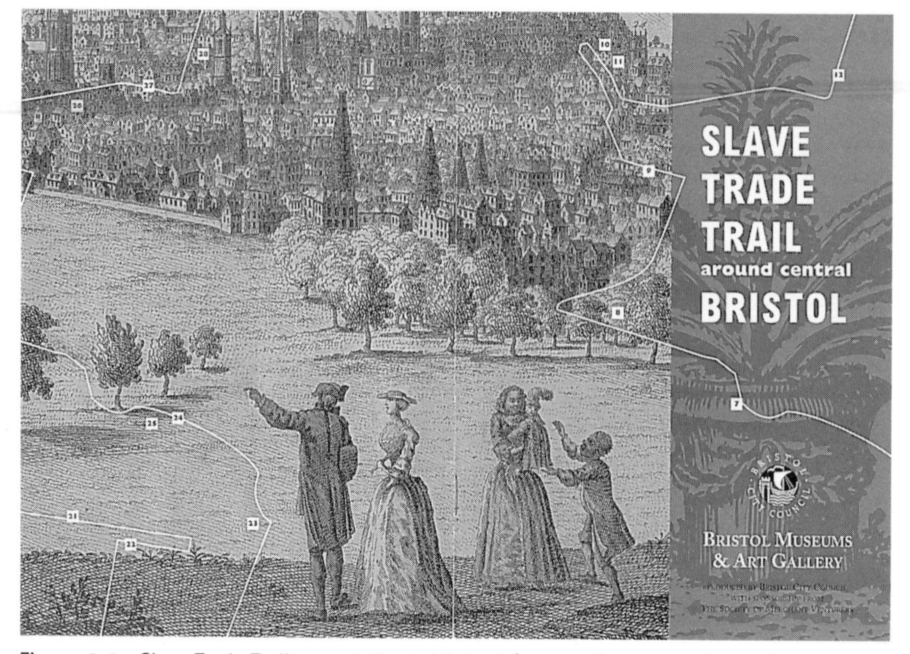

Figure 1.4 Slave Trade Trail around Central Bristol (*source*: Dresser *et al.*, 1998) (published with the permission of Bristol Museums and Art Gallery).

They spoke of the suffering of the enslaved, but also of the poor people of Bristol unwillingly caught up as well as complicit in a system of oppression – the sailors forced to work and die on slave ships and whose accounts of its terrors fuelled the efforts of those involved in Bristol's abolition movement. This kind of remembering recognises interconnection and interdependence, albeit in relationships structured through power, rather than a simple opposition between white and black histories. This kind of critical and constructive sense of a shared but also painful history was again evident in the Canadian celebrations of 1997 surrounding the arrival of the replica of Cabot's ship. Here the significance of the *Matthew* was tempered by the emphasis on the many journeys in which ordinary people from Scotland, Ireland and France travelled to establish early settlements there. Its programme of events celebrated a mixed folk culture more than a moment of heroic discovery and included plans for a new centre in Newfoundland Bay to commemorate the 'disappearance' of the aboriginal peoples of the area. In turn, other versions of the maritime are now being officially sponsored in Bristol. The City Council's Bristol Slave Trade Action Group have begun a series of initiatives including new museum displays on the slave trade and its heritage, and a Slavery Trade Trail which links the sites connected to Bristol's involvement in slavery and the anti-slavery movement (Figure 1.4). The historical geography of the city is being re-shaped as a constructive dialogue is opened up between the events of the past and the imperatives of the present.

Conclusion

Like modernity itself, this chapter, in its attempt to think through the historical geographies of modernising processes and the concept of the modern, seems both coherent and fragmented. Yet, the sometimes incongruous juxtaposition of examples such as the gendered geographies of enclosure and native Canadian museums, and its somewhat reckless combination of material from Bristol, India, Birmingham, Canada, London, the Caribbean, and New York, and its chronological irregularities, skipping about in time and space, is a product of a sustained and specific strategy. The apparent eclecticism of the sources used reflects the argument for differentiated and interconnected historical geographies of modernity that are informed by a critical sense of the contemporary political implications of how its historical geographies are written and represented. It also reflects the usefulness and limits of using the concept of modernity as a thematic umbrella to bring together diverse processes and patterns. When the term is apparently able to shelter every event and process (economic, political, social, cultural, environmental) from the micro-scale of everyday life to the macro-scale of global change, its usefulness seems redundant, generalising and totalising, if that means that the differences within and between those processes are not also recognised. Yet, better historical geographies are written as a result of rejecting, challenging or refining its meaning. The chapters to follow are a product of this process and through their concentration on specific themes provide a greater sense of the subtleties and complexities of modernity than the schematic framework of this introductory chapter has allowed.

References

Allen, J. (1995) Global Worlds, in Allen, J. and Massey, D. (eds) *Geographical Worlds*, Oxford University Press, Oxford, 105–42.

Atkinson, D. and Laurier, E. (1998) A Sanitised City? Social Exclusion at Bristol's 1996 *International Festival of the Sea. Geoforum*, **29**, 199–206.

Barrell, J. (1988) *Poetry, Language and Politics*, Manchester University Press, Manchester.

Barrell, J. (1990), The public prospect and the private view: the politics of taste in eighteenth-century Britain. In Pugh, S. (ed.) *Reading Landscape: Country – City – Capital*, Manchester University Press, Manchester, 19–40.

Bennett, T. (1996) The Exhibitionary Complex. In Greenberg, R., Ferguson, B. W. and Nairne, S. (eds) *Thinking about Exhibitions*, Routledge, London, 81–113.

Berman, M. (1982) *All That is Solid Melts Into Air: The Experience of Modernity*, Sage, London.

Bhabha, H. (1994) *The Location of Culture*, Routledge, London.

Breckenridge, C. A. (1989) The Aesthetics and Politics of Colonial Collecting: India at World Fairs. *Comparative Studies in Society and History*, **31**, 195–216.

Burt, R. and Archer, J. M. (eds) (1994) *Enclosure Acts: Sexuality, Property, and Culture in Early Modern England*, Cornell University Press, Ithaca.

Burton, A. (1992) The White Woman's Burden: British Feminists and 'The Indian Woman', 1865–1915. In Chaudhuri, N. and Strobel, M. (eds) *Western Women and Imperialism: Complicity and Resistance*, Indiana University Press, Bloomington, 137–57.

Burton, A. (1996) A 'Pilgrim Reformer' at the Heart of the Empire: Behramji Malabari in Late-Victorian London. *Gender and History*, **8**, 175–96.

Bush, J. (1994) The Right Sort of Women: female emigrators and emigration to the British Empire, 1890–1910. *Women's History Review*, **3**, 385–409.

Chakrabarty, D. (1992) Postcoloniality and the Artifice of History. *Representations*, **32**, 1–27.

Chapman, W. R. (1985) Arranging Ethnology: A. H. L. F. Pitt Rivers and the Typological Tradition. In Stocking, G. W. (ed.), *Objects and Others: Essays on Museums and Material Culture*, University of Wisconsin Press, Madison, Wisconsin, 15–48.

Chaudhuri, N. (1992) Shawls, Jewellery, Curry and Rice in Victorian Britain. In Chaudhuri, N. and Strobel, M. (eds) *Western Women and Imperialism: Complicity and Resistance*, Indiana University Press, Bloomington.

Chaudhuri, N. and Strobel, M. (eds) (1992) *Western Women and Imperialism: Complicity and Resistance*, Indiana University Press, Bloomington, 1–15.

Clifford, J. (1988) On collecting art and culture. In Clifford, J. *The Predicament of Culture: Twentieth-century Ethnography, Literature, and Art*, Harvard University Press, Cambridge, Mass., 215–51.

Clifford, J. (1991) Four Northwest Coast museums: travel reflections. In Karp, I. and Lavine, S. D. (eds) *Exhibiting Cultures: The Poetics and Politics of Museum Display*, Smithsonian Institution Press, Washington, 212–54.

Clifford, J. (1997) *Routes: Travel and Translation in the Late Twentieth Century*, Harvard University Press, Cambridge, Mass.

Coombes, A. E. (1991) Ethnography and the formation of national and cultural identities. In Hiller, S. (ed.) *The Myth of Primitivism, Perspectives on Art*, Routledge, London, 189–214.

Creighton, M. S. and Norling, L. (eds) (1996) *Iron Men, Wooden Women: Gender and Seafaring in the Atlantic World*, John Hopkins University Press, Baltimore.

Cronon, W. (1990) Modes of Prophecy and Production: Placing Nature in History. *Journal of American History*, **76**, 1122–31.

Daniels, S. (1992) Loutherbourg's Chemical Theatre: Coalbrookdale By Night. In Barrell, J. (ed.) *Painting and the Politics of Culture: New Essays on British Art 1700–1850*, Oxford University Press, Oxford, 195–230.

Dias, N. (1994) Looking at objects: memory, knowledge in nineteenth-century ethnographic displays. In Robertson, G., Mash, M., Tickner, L., Bird, B., Curtis, B. and Putman, T. (eds) *Travellers' Tales: Narratives of Home and Displacement*, Routledge, London, 164–76.

Dresser, M., Jordan, C. and Taylor, D. (1998) *Slave Trade Trail Around Central Bristol*, Bristol Museums and Art Gallery, Bristol.

Driver, F. and Samuel, R. (1995) Rethinking the idea of place. *History Workshop Journal*, **39**, vi–vii.

Duncan, C. (1995) *Civilizing Rituals: Inside Public Art Museums*, Routledge, London.

Felski, R. (1994) The Gender of Modernity. In Ledger, S., MacDonagh, J. and Spencer, J. (eds) *Political Gender: Texts and Contexts*, Harvester Wheatsheaf, New York, 144–55.

Foster, H. (1985) The 'Primitive' Unconscious of Modern Art, of White Skin Black Masks. In Foster, H. *Recodings; Art, Spectacle, Cultural Politics*, Bay Press, Port Townsend, Washington, 181–208.

Fuller, N. J. (1992) The Museum as a vehicle for community empowerment; the Akchin Indian Community Eco-museum project. In Karp, I., Kreamer, C. M. and Lavine, S. D. (eds), *Museums and Communities: The Politics of Public Culture*, Smithsonian Institution Press, Washington, 327–66.

Gilroy, P. (1993) *The Black Atlantic: Modernity and Double Consciousness*, Verso, London.

Greenhalgh, P. (1988) *The Expositions Universelles, Great Exhibitions and World's Fairs, 1851–1939*, Manchester University Press, Manchester.

Greenhalgh, P. (1993) Education, entertainment and politics: lessons from the Great International Exhibitions. In Vergo, P. (ed.) *The New Museology*, Reaktion Books, London, 74–98.

Hall, C. (1992) Missionary Stories: gender and ethnicity in England in the 1830s and 1840s. In Hall, C. (ed.) *White, Male and Middle-class: Explorations in Feminism and History*, Polity, Cambridge, 205–53.

Hall, C. (1993) 'From Greenland's Icy Mountains . . . to Afric's Golden Sand': Ethnicity, Race and Nation in Mid-Nineteenth Century England. *Gender and History*, **5**, 212–30.

Hall, C. (1994) Rethinking Imperial Histories: The Reform Act of 1867. *New Left Review*, **208**, 3–29.

Hall, C. (1996) Histories, empires and the post-colonial moment. In Chambers, I. and Curti, L. (eds) *The Post-Colonial Question: Common Skies, Divided Horizons*, Routledge, London, 65–77.

Hall, S. (1995) New cultures for old. In Massey, D. and Jess, P. (eds) *A Place in the World: Places, Cultures and Globalization*, Oxford University Press, Oxford, 175–213.

Hiller, S. (ed.) (1991) *The Myth of Primitivism; Perspectives on Art*, Routledge, London.

Howkins, A. (1986) The Discovery of Rural England. In Colls, R. and Dodd, P. (eds) *Englishness, Politics and Culture 1880–1920*, Croom Helm, London, 62–88.

Hoyles, M. (1991) *The Story of Gardening*, Journeyman, London.

Humphries, J. (1990) Enclosures, common rights and women: the proletarization of families in the late eighteenth and early nineteenth centuries. *Journal of Economic History*, **2**, 17–42.

Hurle, A. (1994) Museums and First Nations in Canada. *Journal of Museum Ethnography*, **6**, 39–64.

International Festival of the Sea (1996) *Official Programme*, Venue Publishing, Bristol.

Karp, I. and Lavine, S. D. (eds) (1991) *Exhibiting Cultures: The Poetics and Politics of Museum Display*, Smithsonian Institution Press, Washington.

Karp, I., Kreamer, C. M. and Lavine, S. D. (eds), (1992) *Museums and Communities: The Politics of Public Culture*, Smithsonian Institution Press, Washington.

Kern, S. (1983) *The Culture of Time and Space, 1880–1918*, Harvard University Press, Cambridge, Mass.

Lidchi, H. (1997) The Poetics and Politics of Exhibiting Other Cultures. In Hall, S. (ed.) *Representation: Cultural Representations and Signifying Practices*, Sage/Open University, London, 151–222.

Linebaugh, P. and Rediker, M. (1990) The Many-Headed Hydra: Sailors, Slaves and the Atlantic Working Class in the Eighteenth Century. *Journal of Historical Sociology*, **3**, 225–52.

Mahony, L. M. (1987) Doxies at Dockside: prostitution and American Maritime Society, 1800–1900. In Runyan, T. J. (ed.) *Ships, Seafaring and Society: Essays in Maritime History*, Wanye State University Press, Detroit.

Massey, D. (1994) *Place, Space, Gender*, Polity, Cambridge.

Massey, D. (1995) Places and their pasts. *History Workshop Journal*, **39**, 182–92.

Meyer, S. L. (1990) Colonialism and the Figurative Strategy of *Jane Eyre*. *Victorian Studies*, **33**, 247–68.

Mitchell, T. (1989) The World as Exhibition. *Comparative Studies in Society and History*, **31**, 217–36.

Ogborn, M. (1998) *Spaces of Modernity: London's Geographies 1680–1780*, Guilford Press, New York.

Orton, F. and Pollock, G. (1980) Les Données Bretonnantes: La Praire de Répresentation. *Art History*, **3**, 314–44.

Peirson Jones, J. (1992) The colonial legacy and the community; the Gallery 33 project. In Karp, I., Kreamer, C. M. and Lavine, S. D. (eds), *Museums and Communities: The Politics of Public Culture*, Smithsonian Institution Press, Washington, 221–41.

Pred, A. (1995) *Recognising European Modernities*, Routledge, London.

Perry, G. (1993) Primitivism and the 'modern'. In Harrison, C., Francina, F. and Perry, G. (eds) *Primitivism, Cubism and Abstraction; The Early Twentieth Century*, Yale and Open University Press, New Haven and London, 3–85.

Pointon, M. (1994) *Art Apart: Art Institutions and Ideology Across England and North America*, Manchester University Press, Manchester.

Pollock, G. (1988) Modernity and the spaces of femininity. In Pollock, G. *Vision and Difference: Femininity, Feminism and the Histories of Art*, London, Routledge, 50–90.

Pollock, G. (1992) *Avant-Garde Gambits 1888–1893, Gender and the Colour of Art History*, Thames and Hudson, London.

Pollock, G. (1995) The 'View from Elsewhere': Extracts from a semi-public correspondence about the politics of feminist spectatorship. In Florence, P. and Reynolds, D. (eds) *Feminist Subjects, Multi-media: Cultural Methodologies*, Manchester University Press, Manchester, 2–38.

Ramusack, B. N. (1992) Cultural Missionaries, Maternal Imperialists, Feminist Allies: British Women Activists in India, 1865–1945. In Chaudhuri, N. and Strobel, M. (eds) *Western Women and Imperialism: Complicity and Resistance*, Indiana University Press, Bloomington, 119–36.

Rediker, M. (1987) *Between the Devil and the Deep Blue Sea: Merchant Seamen, Pirates and the Anglo-American Maritime World, 1700–1750*, Cambridge University Press, Cambridge.

Rose, G. and Ogborn, M. (1988) Feminism and Historical Geography. *Journal of Historical Geography*, **14**, 405–9.

Samuel, R. (1994) *Theatres of Memory, Vol. 1: Past and Present in Contemporary Culture*, Verso, London.

Shohat, E. and Stam, R. (1994) *Unthinking Eurocentricism: Multi-culturalism and the Media.* Routledge, London.

Tchen, J. H. W. (1992) Creating a dialogic museum: the Chinatown History Museum Experiment. In Karp, I. *et al.*, (eds), *Museums and Communities: The Politics of Public Culture*, Smithsonian Institution Press, Washington, 285–326.

Varnedoe, K. (1990) *A Fine Disregard: What Makes Modern Art Modern*, Thames and Hudson, London.

Vergo, P. (ed.) (1993) *The New Museology*, Reaktion Books, London.

Walton, J. R. (1990) Agriculture and rural society, 1730–1914. In Dodgshon, R. A. and Butlin, R. A. (eds) *An Historical Geography of England and Wales*, 2nd edn, Academic Press, London, 323–50.

Ware, V. (1992) Britannia's Other Daughters: feminism and the Age of Imperialism. In Ware, V. (ed.) *Beyond the Pale: White Women, Racism and History*, Verso, London, 117–66.

Withers, C. W. J. (1995) Geography, Natural History and the Eighteenth-Century Enlightenment: Putting the World in Place. *History Workshop Journal*, **39**, 137–64.

Wynn, G. (1997) Remapping Tutira: contours in the environmental history of New Zealand. *Journal of Historical Geography*, **23**, 418–446.

Young, J. E. (1983) *The Texture of Memory: Holocaust Memorials and Meaning*, Yale University Press, Yale.

Yelling, J. (1990) Agriculture 1500–1730. In Dodgshon, R. A. and Butlin, R. A. (eds) *An Historical Geography of England and Wales*, 2nd edn, Academic Press, London, 181–98.

Zonana, J. (1993) The Sultan and the Slave: Feminist Orientalism and the Structure of *Jane Eyre. Signs: Journal of Women in Culture and Society*, **18**, 592–617.

PART II

MODERNITY AND ITS CONSEQUENCES

Chapter 2

Historical geographies of globalisation, c. 1500–1800

Miles Ogborn

Introduction

Tea, coffee, chocolate, tobacco, sugar, rice and potatoes: all of these once 'exotic' products are now mundane and routine parts of European everyday life without which the world would seem strange. They became regular features of European diets and forms of sociability and pleasure during the eighteenth century, and in doing so transformed the relationships between many parts of the world (Kowaleski-Wallace, 1997; Walvin, 1997). Gold and silver from Africa and South America were traded with Asia for tea and coffee. Tobacco and sugar were grown on newly created plantations in the Caribbean and North America by slaves from Africa owned by recent European settlers. Potatoes came to form the basis of the diets of workers who produced the manufactured goods traded for furs in North America, men, women and children in West Africa, and the produce of the New World plantations. By 1800 these different areas of the globe were tied together into a new set of relationships, which changed what was grown, made and consumed in each part of the world. These ties also involved the forced and unforced movement of millions of people, the introduction of plants and animals into new habitats, changing forms of imperial politics, and the flow of capital to new and profitable uses and places and away from others. The aim of this chapter is to understand these changes by considering the historical geographies of globalisation.

It has become commonplace to understand the late twentieth century as a time of globalisation, when people and places are bound into relationships – economic, political and cultural – with other, distant people and places (Harvey, 1989; Hall, 1991). Yet if this term can also be applied to the whole period since 1500 (and perhaps even earlier, see Blaut, 1993) then it is necessary to look carefully at the technologies, institutions and relationships which shape global historical geographies in different ways in different periods (Hugill, 1993). However, it is soon apparent that studying globalisation is not simply a matter of dividing historical time into segments. It also needs to be recognised that in any period there are a variety of forms of globalisation, involving different, intersecting global processes which produce a variegated historical geography. To demonstrate this, and to put some historical flesh onto these bare bones, this chapter begins by considering an approach – Immanuel

Wallerstein's world-systems theory – which argues that the world since 1500 can be understood in terms of the emergence and expansion of a single global capitalist system (Wallerstein, 1974, 1980). Presenting a critique of Wallerstein's theory opens up a consideration of the ways in which the forms of globalisation brought about by seventeenth- and eighteenth-century capitalist merchants and states can be conceptualised, not as a single system, but as a series of networks which had to be built, extended and sustained, and along which people, goods, ships, capital and information moved. In turn, this focus on what was moving globally, and how it moved, brings into view those movements which may have run in the grooves of dominant forms of globalisation, but which also ran counter to them in various ways. Studying globalisation's historical geographies reveals many different worlds.

The modern world-system: capitalism and global change

Wallerstein's theory of the modern world-system has been a very influential way of interpreting the historical geography of globalisation. He argues that as a result of the 'crisis of feudalism' in the period 1300–1450, there was an 'expansion of Europe' overseas (Wallerstein, 1974: 23 and 44). Driven by falling feudal revenues and funded by Genoese capital, it was the Portuguese who explored the west coast of Africa and colonised the Atlantic islands to grow sugar. The subsequent centuries of capitalist expansion (1450–*c.* 1650) saw the Spanish establishing an extensive empire in the Americas, from which they exported vast amounts of bullion (particularly silver) for trade with Asia, and the division of the globe between their empire and the Portuguese (Brotton, 1997). This global structure was challenged by the Dutch during the economic downturn of the seventeenth century. They established themselves as the strongest economic power in the period 1625–1675, trading everywhere from the Caribbean to the East Indies, and particularly between the Baltic and Northern Europe. Dutch colonial power replaced the Portuguese in Brazil and the spice islands. Between 1650 and 1750 they were, in turn, challenged by the English (British after the 1707 union of England and Scotland) and the French. These economic and military battles for hegemony across the globe were eventually won by the British, who established an empire with colonies in North America, the Caribbean and India, and trade between them, Africa, China and Europe in sugar, slaves, textiles and tea (Wallerstein, 1980).

For Wallerstein, this historical geography of globalisation – ever expanding to take in new areas of the globe and ever deepening the exploitative relationships between places – is to be explained in one way: by detailing the working of what he calls the 'modern world-system'. Fundamental to this are the economics of capitalism which structure the system. What he terms the 'capitalist world-economy' (and it should be noted that in the past there have been other forms of world-economy outside Europe) is devoted to the accumulation of capital. Social and spatial relations are ordered globally to facilitate this accumulation by certain people in certain places, and the conflicts between classes and between states are struggles over the control of these processes of accumulation and the benefits that they bring. Thus 'commodity chains' of primary producers, traders, manufacturers and consumers are stretched across the globe so that some gain from them and others lose (Wallerstein, 1983: 30). As a result the capitalist world-economy takes on a particular geographical structure. Wallerstein describes its sixteenth-century expansion as creating areas which can be designated as core, periphery, semiperiphery and external arena. It is the structured relationships between these geographical areas that underpins the workings of the world-system, and it is the shifting locations of these areas across the globe that defines the fortunes of states and people (Wallerstein, 1974).

In this scheme, the relationships between these areas are based on a single hierarchical and geographical division of labour. Core areas are those which control the system and benefit most from it. At the heart of the core is the hegemonic power: the Netherlands in the seventeenth century, Britain in the nineteenth century, and the United States of America in the twentieth century (Taylor, 1996). Here labour is most varied and specialised. It is based on wage labour and self-employment, and concentrated in manufacturing, efficient agriculture, and the commercial and financial functions necessary to control trade. The periphery, whose surplus is drawn to the core, is increasingly given over to monoculture (for example, sugar on the Caribbean islands or tobacco in the Chesapeake) and to capitalist forms of unfree labour such as slavery and coerced cash-cropping (whereby peasants, such as those in sixteenth-century Eastern Europe and Hispanic America, were legally required to work for part of their time on large estates producing for the world market). The semi-peripheral areas (Southern France and Northern Italy in the sixteenth century; Spain, Portugal, Sweden, British North America, Prussia and the spine of Europe from Flanders to Northern Italy in the seventeenth century) adopt intermediate forms such as sharecropping and domestic industry organised and financed by foreign merchants. They are also tied to the core, but in a less dependent relationship than the periphery. Each area is a different part of a single system, the benefits of which flow to the core. Only external areas – such as Russia and Asia for most of the period – are outside, linked by trade, but not bound into the system by relations of dependency (Wallerstein, 1974, 1980).

As a whole, the world-economy goes through phases of expansion and contraction. Expansion in the sixteenth century, contraction from 1650 to 1750, expansion after 1750, and so on. Wallerstein likens these phases to 'the breathing mechanism of the capitalist organism', 'inhaling the purifying oxygen' of a more efficient allocation of resources for the accumulation of capital and 'exhaling [the] poisonous waste' of economic inefficiencies, which had become established in earlier phases of expansion (Wallerstein, 1983: 34). In each phase the hierarchical and geographical relationships are reshuffled, with enormous implications for people and places around the globe. The core shifts – moving from Iberia to North-west Europe. New peripheries are brought into the system, particularly in times of contraction. For example, the seventeenth century saw an area from coastal Brazil to Maryland – the 'extended Caribbean' – brought into the world-economy as a new periphery (Wallerstein, 1980: 167). The economic downturn prompted the rising powers of the Netherlands, England

and France to attempt to plunder the profits of the declining Spanish empire in the Americas by drawing it into relations of dependency. Elsewhere, older peripheries were restructured as they responded to contraction. Initial increases in production and exploitation of land and labour gave way to partial withdrawal from world markets and social conflict in Poland and Hungary. In the semiperiphery, the situation is more complex and transitional. The contraction may attach new areas to the semiperiphery as former core areas such as Spain and Portugal go into rapid decline. Others are added as former peripheral areas seize the opportunities that the downturn presents. For example, eighteenth-century Prussia was able, against the odds, to become the semiperipheral region for Eastern Europe rather than Sweden (Wallerstein, 1980). As the economy 'breathes' the fortunes of places rise and fall. The global geography shifts, but the structural elements remain the same only ever more polarised.

This understanding of the fortunes of states within the world-economy depends upon another crucial feature of the modern world-system, the ways in which the economic arena and the political arena have different geographies. The former – where decisions about capital investment and accumulation are made – is a global arena. The latter, however, is less than global. Increasingly, the political arena is the level of the nation-state within a competitive system of states. These nation-states may go to war, colonise each other, or weaken each other in other ways but none have been able, despite attempts by Spain and France in the sixteenth century, to create a world empire which would establish a global political arena (Wallerstein, 1974). Wallerstein argues that this situation promotes the accumulation of capital in the core. Strong states were necessary to begin the creation of the world-system (as in the case of Portugal), and they have been necessary to those classes in the core seeking to create the best conditions for capital accumulation. Their weakening of states in peripheral and semiperipheral areas, through warfare and colonisation, and their establishment of the infrastructures of violence and communication necessary to enable exploitative exchange between areas of the world-system, gives the lie to the idea that capitalism thrives with minimal state intervention (Wallerstein, 1983). Wallerstein's aim (1980) is to show how state policies such as seventeenth-century mercantilism have, within the limits set by the structures of the world-economy, been crucial to the accumulation of capital in certain areas and the rise and fall of nation-states within the world system. This combination of a world-economy based on a global division of labour and a multiplicity of states of different strengths and capacities gives the world-system its peculiar dynamic and has been responsible for globalisation and uneven development. For Wallerstein, the historical geography of the globe is the history of the capitalist world-system.

Questioning the modern world-system

Wallerstein's work certainly offers a powerful way of thinking about globalisation, which emphasises the exploitative social relations that have connected places together for many hundreds of years. There are, however, problems with his interpretation, which are often inseparable from the useful lessons that it teaches. I want to explore three problems: Wallerstein's simplification of Europeans' relationships to other peoples; the way the world-systems approach reduces everything to a single set of economic relationships; and the dangers of a structural analysis. Discussing them begins to open up other ways of exploring global historical geographies – the focus of the rest of the chapter.

Europeans and other peoples

Wallerstein's delineation of the macro-structures of the world-system comes at the expense of an extended discussion of the different ways in which European expansion changed other areas of the globe. This, however, has been provided by Eric Wolf (1982). He seeks to understand the histories of those people in Africa, Asia and the Americas who were drawn into trade and production through circuits of mercantile wealth and the worldwide reorganisation of production from the fifteenth century onwards. This means setting out what the world was like before European expansion. It also means demonstrating the ways in which social structures adapted to and were changed by that expansion in different ways in different places.

In the fifteenth century there was already an extensive set of connections between certain parts of the world (Wolf, 1982). In Asia, large areas had been brought under the control of political and military empires which extracted tribute from their subject populations in the chain of cultivated regions from the Moroccan Atlas to China. These empires waxed and waned. The Ottoman Empire controlled the eastern Mediterranean from 1453 until 1914, the Mughals established dominance in India after their invasion from Turkmenestan in 1525, and the Ming dynasty was established in China by 1370 having pushed out the Mongols. As these political powers rose and fell over the centuries, trade routes were extended across the land mass to exchange valuable luxury goods as part of the establishment of élite power (Figure 2.1). The Silk Road ran from Antioch in Northern Syria to Kashgar, and then into China. Spices and gold had been carried between India and the West since the early Roman Empire. East Africans traded gold, ivory, copper and slaves for Indian beads and cloth and Chinese porcelain. Much of this trade was in Arab hands. By the fourth century there were colonies of Arab merchants in Canton, by the twelfth century most of the wealthy people there were said to have black slaves. Regional trading networks bound other areas into this circulation of goods across the landmass. In West Africa trade connected forest, savannah, desert and the coastal belts through exchanges of slaves, cloth, ivory, pepper, kola nuts and gold for horses, brass, copper, glassware, beads, leather and textiles (Figure 2.2). While the Americas were not linked by trade to Africa and Asia, they had their own trading networks and, by the sixteenth century, the Incas and Aztecs had established extensive empires. Until then, it was Europe which was on the periphery of the world-system (Abu-Lughod, 1989; Blaut, 1993).

It was into this that the armed European merchants sailed, their initial voyages no different from the fifteenth-century sea-borne explorations from

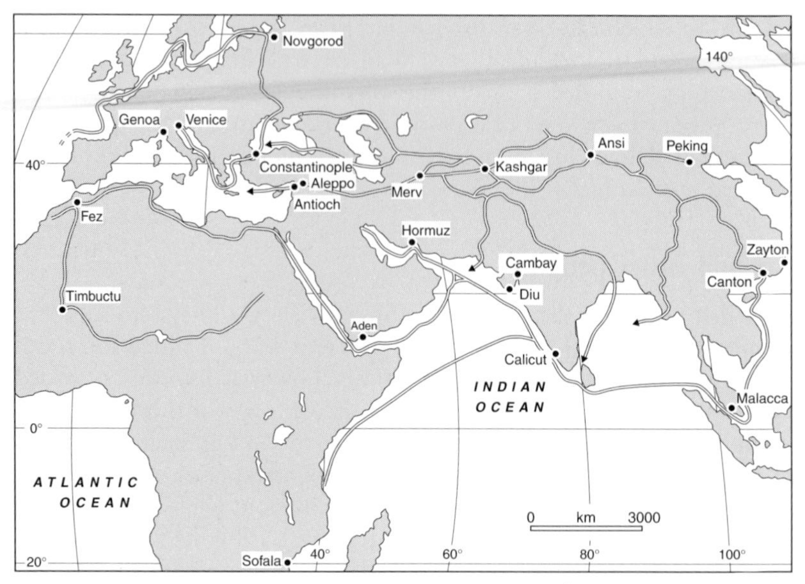

Figure 2.1 Major trade routes of the Old World, c. 1400 (after Wolf, 1982).

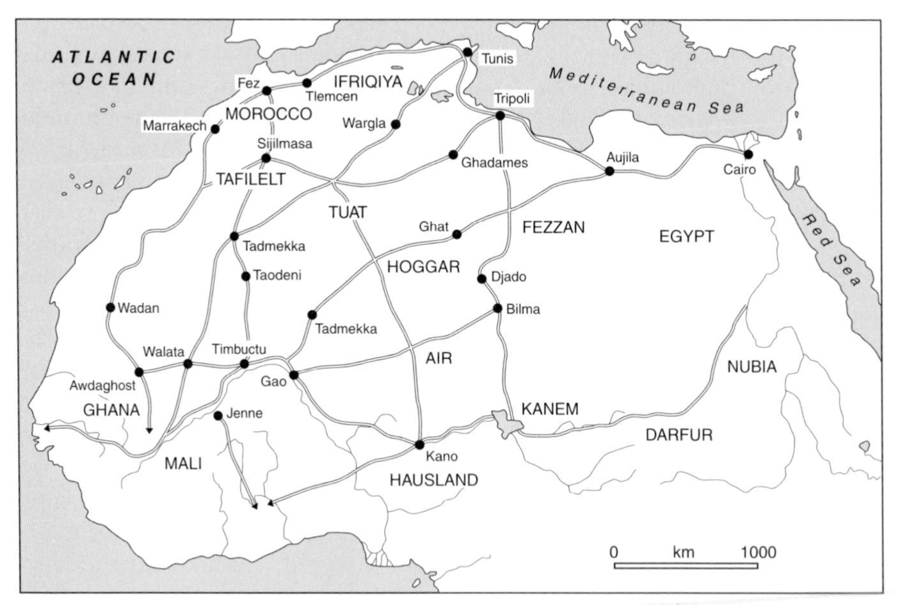

Figure 2.2 Trade routes of West Africa, c. 1400 (after Wolf, 1982).

China and India (Blaut, 1993). The Portuguese search for a spice route which would 'bypass the Turkish roadblock' (Wolf, 1982: 128) took them to Madeira (1420), the Bight of Benin (1482), the mouth of the Congo (1483), round the Cape of Good Hope (1487) and on to India (1497). The same search took Columbus, sailing for Castile-Aragon, to the Caribbean in 1492 and Cabral to

Brazil in 1500. By 1540, sugar and slaves were being moved across the Atlantic and Cortés and Pizarro had conquered Mexico and Peru. Although their empires remained, albeit weakened by plunder, the global trade routes established by the Iberians were later taken over and added to by the Dutch, and then the English and French.

This new geography of accumulation and power had dramatic impacts on the social systems and landscapes of Asia, Africa and the Americas. The societies of what became Hispanic America were devastated by new diseases, particularly smallpox and measles, brought by the Europeans (Newson 1996; Walvin, 1997). Diseases carried by trade and traders also swept through the indigenous populations of North America (Mancall, 1998). New sets of social relations emerged. In Hispanic America a new political economy shifted population away from the previous zones of cultivation to the Sierra Madre and the Bolivian *altiplano*. It organised silver mining, food production on extensive Spanish-owned *haciendas*, and native settlements into a hierarchical system of dependence and exploitation within which 'the emerging Indian communities came to occupy the lowest rung' (Wolf, 1982: 145). Elsewhere, new social formations emerged. The Miskito of Central America's Mosquito Coast were an amalgam of a kin-ordered native American group and large numbers of runaway slaves and buccaneers (Europeans who had lived by ranching wild cattle and then by piracy) who found a new role raiding and trading inland and then exchanging gold, tobacco, gum and tortoise shells for European manufactured goods at the coast. In each place – through the slave trade, the fur trade, and the trade in spices, cloth and tea with Asia – new geographies and new ways of life were created within these circuits of wealth and power. However, the degree of change and the nature of change – how much violence was used, whether new forms were grafted onto old ones or quickly undermined them, whether new political relations were created and who they benefited – varied in time and space allowing an examination of the complexities of this 'globalisation'.

In North America the fur trade led to a social and political reorganisation of relations within and between indigenous societies (Wolf, 1982). In the north east, beavers (wanted by hat makers in England and France) were hunted to exchange for manufactured goods. Archaeological evidence shows European trade goods on the Niagara frontier as early as 1570, and that by 1670 the Iroquois were dependent upon trade with Europe for arms, metal tools, kettles, clothing and jewellery. The demand for fur increased competition for hunting grounds and led to attempts at exclusive territorial control by small family groups. This increased the intensity and scope of warfare between native groupings armed with European guns, hardened a gendered division of labour between men who hunted and fought and women agriculturalists, and displaced whole populations. It also established new political entities. The Iroquois federation (made up of the Ganiengehaga – or Mohawk, the Oneida, the Onondaga, the Cayuga and the Seneca matrilineal clusters) had formed before the fur trade began but was given new impetus by it. Federation enabled dominance over other native groups and some room for manoeuvre in negotiations with the

English and the French. Further west, the trade and the guns it brought gave the Cree–Assiniboin alliance an advantage over the Dakota Sioux, the Gros Ventre and the Blackfoot. As they turned more to hunting they were fed by pastoralists (Apache, Comanche and Ute), whose new access to horses and guns altered their social ecology and meant that they could kill buffalo in much larger numbers. In turn, the trade in pemmican – preserved buffalo meat – transformed the relationships of power within these societies and led to internal challenges to traditional authority. Some groups did well. While the Europeans wanted to trade with native Americans rather than coveting their land, and while they were sought as allies in the regional versions of the global conflict between England and France, reorganised polities like the Iroquois were able to maintain some power and to integrate trade goods into 'new cultural configurations that combined native and European artefacts and patterns' (Wolf, 1982: 193). Others were not so fortunate.

In understanding the variation in forms of social and political change the differences between Africa and Asia are instructive. The slave trade transported over ten million people from Africa between 1660 and 1867, with some 8.9 million surviving the Atlantic crossing (Curtin, 1969; Postma, 1990; Eltis and Richardson, 1997; Richardson, 1998). Europeans financed and controlled the shipping and sale, primarily in the Caribbean. Africans captured the slaves and brought them to the coast in exchange for textiles and other manufactured goods. This trade may have 'dovetailed with pre-existing African circuits of exchange, not altering their basic structure but merely adding to the flow of goods through them' (Wolf, 1982: 206). It also allowed existing practices of slavery, warfare and pawnship to be turned to newly profitable uses. Moreover, trade prompted a process of state formation as African polities sought to control the trade. Small, predatory and militaristic states, their power based on European firearms which were being imported at a rate of 180,000 a year by 1730, emerged and waged war. For example, the Asante, whose origins as a political entity only date back to the late seventeenth century, used Dutch muskets to conquer Western Gonja (1722–3), Eastern Gonja (1732–3), Accra (1742) and Akyem Abuakwa (1744). Pre-existing polities were undermined. The Kongo kingdom (*c.* 60,000 square miles) fell apart as the Portuguese began to trade directly with local chiefs, bypassing the king (Wolf, 1982).

In Asia the situation was different. There the Europeans were faced with large, powerful tributary states which controlled the landmass. The Portuguese, Dutch and English were able to use their superior naval strength to control the seas and the long-distance trade routes, but they were able only to establish sites for their coastal 'factories' (sites of trade rather than production) by taking them from weaker, often island, states (such as at Malacca, Ternate and Amboina), or through negotiations with the Asian empires (such as at Goa, Surat and Canton) upon whose goodwill they were dependent (Marshall, 1998). As such, they tapped into the extensive and long-established sea-borne trade rather than controlling it. They were able to begin extending control over the production of cloth in India only in the late eighteenth century, and could only prevent the drain of bullion to China (exchanged for silks and then for vast amounts of tea) with the establishment of the opium trade in early 1800s. It was not until the nineteenth century that capitalism transformed Asia into a series of plantation economies (Wolf, 1982). The question of how to understand the history of capitalism raises the two other problems with Wallerstein's account, which will be dealt with much more briefly.

The singular story of capitalism

This attention to the impact of European capitalist globalisation, and to the differentiated history of destruction, transformation, accommodation and resistance, which Wolf sets out in detail, complicates and questions the simple historical geography of core and periphery. It does not, however, avoid the problems of understanding global historical geography as a single story: the story of capitalism. Although Wallerstein and Wolf differ on how to conceptualise the history of capitalism – Wolf insisting that the term should only designate production using wage labour and not mercantile exchange – they are both in danger of reducing everything to the demands of the accumulation of capital. It is a powerful and important point to make that the social relations of capitalism have transformed the globe, but that is not the same as presenting this as a single narrative of expansion and change driven by the logic of capital. For example, in his discussion of the state and politics, Wallerstein often presents the state as responding only to the demands of capital accumulation. Although he does suggest that the strong nation-state has some room for manoeuvre, and can be seen as both cause and consequence of the rise of the capitalist world-economy, he is only interested in the state to the extent that it can be used to explain the changing historical geographies of cores, peripheries and semiperipheries, which are economic constructions (Wallerstein, 1983). It is important to make room for global histories of politics, science, or culture which are not reducible to histories of capitalism, while recognising the very real imperatives and consequences of the economic processes with which they were bound up (see, for example, Miller and Reill, 1996).

The limits of structuralism

One reason why Wallerstein does not provide this lies in his theoretical approach. He presents a structural history of the capitalist world-system, which usefully sets out the relationships between the elements of that system – states and economies, cores and peripheries – and rereads the histories of global expansion and change through that lens. This often makes for complex accounts of events such as the rise of Prussia, the decline of Spain or the success of the Reformation, but although he is keen to stress that events might have turned out differently, this is at the level of which states succeeded or failed, which areas became the core, which the semiperiphery. Overall there is an inevitability about his history (or historical geography) which leaves little room for the actions of the people involved in making and resisting these global forces (Kearns, 1988).

These ways of studying globalisation shape how Wallerstein conceptualises the historical geography of the globe. Essentially this involves understanding the world, which is produced by capitalism's division of labour, as a single pattern of *areas* – core, periphery, semiperiphery – albeit with fluid and changing boundaries. Its geography is one of bounded, functional zones controlled either economically or politically. However, there are other ways of seeing the matter, which open up different possibilities. Wallerstein (1983; 30) is also interested in the 'commodity chains' that connect these areas – the flows of goods, capital, ships and people which make up the world-economy. Other movements across boundaries might also be added – to fight wars, to undertake scientific exploration, or to produce cultural artefacts (Brotton, 1997). These also depend upon travel and trade. Concentrating on the geographies of these movements offers another lens with which to view the historical geography of globalisation, one that can deal with some of the problems of encounter, reductionism and the seeming inevitability of global structures which world-systems theory faces.

Networking: putting the global together

Considering the historical geographies of globalisation in terms of movement means understanding how and why these movements occurred in the ways that they did. This means attending to the 'networks', which allowed, shaped and channelled movements of capital, people, ships, commodities, information and ideas around the globe. It means analysing the regularities of movement and the ways in which people worked to ensure that these movements were regular and profitable by putting stable networks together (Law, 1986). This raises certain questions: who makes these networks? For what purposes? What are their elements? How are they extended? How do they decline? The answers will be different in different circumstances and it is clear that there is not just one global capitalist network to account for. Instead, it is a matter of considering particular cases and delineating the nature and extent of the networks. In what follows, I examine two networks constructed at different periods in different parts of the world, albeit with the shared aim of accumulation through trade. The first example is the English East India Company in the early seventeenth century, the second the activities of a group of Atlantic merchants in the mid-eighteenth century.

The English East India Company, 1600–1640

England's sixteenth-century trade overwhelmingly involved exchanging wool in Antwerp for continental Europe's linen, wine, oil, fruits and naval stores, and for the spices and other luxuries carried from the East Indies by the Portuguese and Dutch. A series of political dislocations of this trade from the 1560s encouraged English traders to begin long-distance trading in a search for alternative markets and supplies of goods. This was done via the formation of merchant companies. The Russia Company had already been founded in 1553 and was followed by the Eastland Company (1578), the Levant Company (1592), the

East India Company (1600), the Virginia Company (1606) and the Royal Adventurers in Africa (1660). I want to use the example of the East India Company to show how trading in a sustained and profitable way with India and South-East Asia involved putting together new forms of organisation – a network – which could operate effectively across unprecedented distances (Chaudhuri, 1965; Lawson, 1993). This also brought new problems which had to be dealt with.

The experience of previous mercantile ventures to the East Indies suggests the difficulties. In 1591 London merchants had sent three ships east. Only one reached the Indonesian archipelago but then had to be abandoned in the West Indies. The remaining crew returned empty handed. While success could bring substantial profits, the risks of no return were great. Also, since such voyages took over two years, capital would be tied up for a long time whatever happened. One way of reducing these risks was a company formed under royal charter on a joint-stock basis. Investors subscribed a portion of the capital managed by the company and received a corresponding proportion of the profits from the sale of the incoming cargo. In this way the substantial capital – much of it 'fixed' in heavily armed ships and, eventually, in trading stations in Asia – needed for long-distance trade could be gathered. The royal charter provided the company with a monopoly over the trade (necessary to its profitability, given the limited demand of the home market); exemption from the restrictive laws on the exportation of bullion (essential since there was no demand in Asia for English manufactures); and support in hostilities with the Portuguese and Dutch (important given the late entry of the English into the trade, but which often conflicted with other objectives of Stuart foreign policy). A chartered joint-stock company made long-distance trade possible in the early seventeenth century (Chaudhuri, 1965; Brenner, 1993).

However, sending ships out into a part of the world comparatively unknown to English merchants was still a risky business. The first East India Company voyage (1601) was still 'little more than a hesitant, semi-speculative financial venture' (Chaudhuri, 1965: 3). It was not until 1608 that the Company was sufficiently confident its ships would return that it could send the next fleet out before the previous one had reached London. That confidence was based on what their voyages had achieved by that time. This was a matter of the return on the capital invested, but equally important was the knowledge that had been gained of the trade in and from the region, and the establishment of forms of organisation in London and the East Indies, which reduced the risks and decreased losses. Affairs in London were managed by a system of committees who took direct charge of financing, equipping and organising the voyages. The first voyage had also left 'factors' – agents of the Company employed to gather goods – at various points. The second voyage was directed to pick up cargo from them and establish others further afield. This began to form the basis of a network of principal and subordinate 'factories' throughout the Indies operating with the permission of local rulers and receiving instructions from London.

By 1620 'the Company's factory organisation in Asia had reached its definitive form' (Chaudhuri, 1965: 60). The Surat factory on the west coast of India

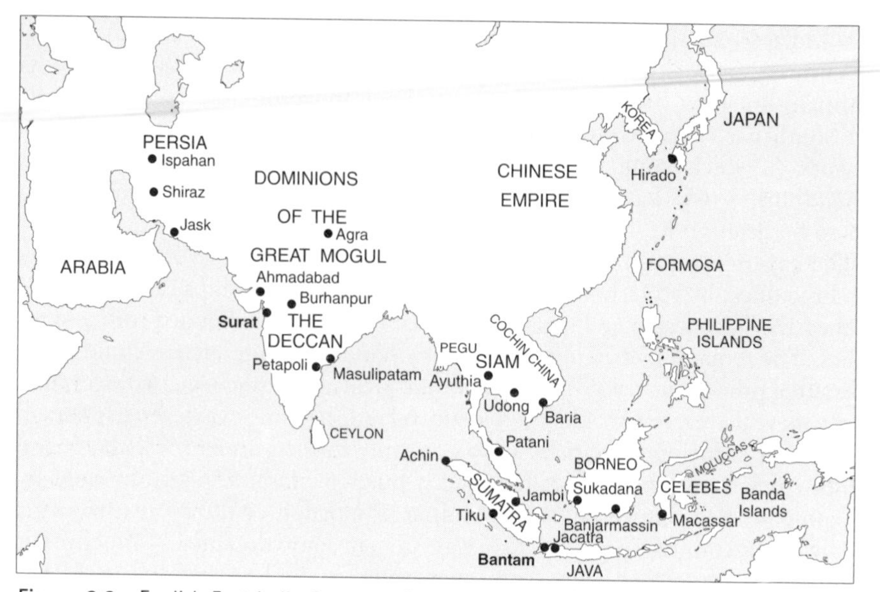

Figure 2.3 English East India Company factories in 1617 (after Chaudhuri, 1965).

controlled the Indian and Persian trade with subordinate factories at Ahmadabad, Agra, Burhanpur and on the Persian Gulf. The Bantam factory was responsible for outposts at Banda, the Celebes, Borneo, Java, Sumatra, Siam, Japan and on the Coromandel coast of India (Figure 2.3). These trading posts ensured that cargoes would be ready when the ships arrived. This reduced the time spent on the voyage overall, which lessened the capital outlay and the risk of attack by the Portuguese and Dutch. The factories therefore increased the return on the capital invested. Their geography was a response to what the Company had learned about how to make the trade profitable. While the East India Company had begun with the idea of simply taking part of the trade in pepper and other spices by trading directly with the spice islands of South-East Asia, it soon became apparent that the demand in those places was for cloth from India, a trade that was already conducted by Arab and Asian merchants. To make the spice trade pay, the East India Company had to involve themselves in what became called the 'country trade' between Surat and Bantam. Eventually, as the price of pepper dropped, they increasingly concentrated on shipping Indian cloth and other goods around Asia and to Europe (Chaudhuri, 1965).

Co-ordinating this trade and selling the imports (and gathering the bullion) in Europe required a high level of administrative efficiency in London and overseas. The East India Company became a large and powerful organisation – its shipbuilding yards at Deptford and Blackwall made it one of seventeenth-century London's largest employers – with a complex bureaucratic structure devoted to developing and implementing trading policy (Chaudhuri, 1965). All of this was necessary to operate across great distances, but it was not without problems. Having only limited communication with the factories thousands of miles away, the Company could never be sure that its instructions would be followed through. It tried to ensure this by establishing trusted men in positions of power in the East Indies and paying them well, but problems remained with what the Company saw as inefficiency and corruption. When, from 1618, its operations were hit by a combination of increased hostility from the much more highly capitalised Dutch East India Company, war in Europe, trade depression and plague in England, and famine in India, the Company found itself overextended. Several of the South-East Asian outposts were closed and the trade concentrated on Surat. The geography of its trading network was reorganised to cope with the decline in trade until conditions improved around 1640 (Chaudhuri, 1965).

Trading for pepper, spices and Indian cloth required the English East India Company to develop ways of organising capital, bullion, goods, ships, people, information and power which could make the trade both regular and profitable over the vast spans of space and time that they had to negotiate. It needed to be able to invest, circulate and reinvest all of these elements to make the trade pay and to produce a world in which the regular arrival of ships in London, Surat and Bantam became possible and predictable. That this network might be disrupted by internal problems of communication and trust, and external problems of economic decline and warfare, shows how febrile this network remained until the late seventeenth century (Chaudhuri, 1978). Looking west from London demonstrates how merchants responded to the same demands in the eighteenth-century Atlantic world.

The Associates: Atlantic trading, planting and slaving, 1735–1785

The middle of the eighteenth century was a time of great opportunity for merchants who entered into trade across the Atlantic. The growing colonies in North America and the Caribbean islands required supplies and labour. They also needed to transport food, lumber, sugar, tobacco, indigo and rice within the Americas and to Europe. The labour, at first supplied by indentured servants from the British Isles, was mainly performed by slaves from West Africa, most of whom were carried on British ships (Price, 1998). This cruel trade, along with the carriage of tropical staples produced by unfree labour and the movement of manufactures from Britain, Europe and India, which were bartered for slaves and fed American consumer demand, offered huge profits to those in a position to make them. David Hancock (1995), in an investigation of 23 London-based merchants, has reconstructed the biographies and business practices of some of the men who made those profits: Augustus Boyd, who began with a medium-sized plantation on St Christopher in the Caribbean and died in 1765 leaving an estate worth £50,000 (equivalent to over £3.3 million today); Alexander Grant, who began as a country doctor in Jamaica and died in 1772 worth £93,000 (over £5.4 million today); and Richard Oswald, the wealthiest of them all, who died in 1794 and left a legacy of £500,000 (around £28 million today). Hancock shows that these men made their money by constructing networks of trade which spanned the globe, tying people and places together within their business ventures and transforming lives and landscapes in the process. Reconstructing these networks shows the work of globalisation in the middle years of the eighteenth century.

Hancock (1995: 11) calls these merchants 'the associates'. They were a loose-knit group clustered around Boyd, Grant, Oswald, and John Sargent II who formed, usually in twos and threes, a series of short-term partnerships and ventures with each other. The associates began relatively small, often in Scotland and the Americas, moved to London in the 1730s and 1740s, and grew big by entering into the Atlantic trades which were not dominated by the established trading élites. In comparison with the distances to the East Indies, the Atlantic was relatively easy to cross. It took nine weeks to sail from Britain to the Chesapeake, with six weeks to get back, and ten weeks to travel to Jamaica, with fourteen weeks to get back (Figure 2.4). This meant that the associates were able to enter the trade on their own account after having worked for some years as London-based factors and agents for others, arranging the shipping of goods between Britain and the Americas. Between them they backed 456 trading voyages in the period 1745–1785. Of these 28 per cent were plantation supply voyages, particularly to the British West Indies, and 26 per cent were slave trading voyages on the triangular trade between Britain, Africa and the Americas (Hancock, 1995: 117).

Their arena of operations was the Atlantic. Their ships rarely visited Indian, Mediterranean or Levantine ports. The key to profitability lay in keeping their vessels full of cargo and keeping them moving. Unlike the East India Company, this system did not rely upon salaried employees and a fixed system of factories. The associates' ventures were much more flexible and opportunistic than that. Instead they sought to construct networks of clients, suppliers, correspondents, agents and employees, which stretched across the Atlantic and were made up of 'relatives, friends, friends of friends and fellow countrymen with an understanding of the larger commercial world' (Hancock, 1995: 140). These were people whom the associates could trust to fill their ships, buy their cargoes, follow their instructions and give them information on new markets and new opportunities. Indeed, it was information as much as cargo that flowed through these networks. In their London counting houses, these merchants and their clerks gathered, sorted and organised the commercial information on goods, prices, customs duties and sailing times, which allowed them to co-ordinate their trading ventures and make them pay. It was through these networks that a new Atlantic world was constructed.

As the associates grew richer – from trade and also from deploying their considerable organisational skills supplying armies with food during the Seven Years War (1756–1763) – they invested in other ventures which could be integrated profitably with their shipping operations. Gradually, plantations were acquired on the Caribbean islands and in South Carolina, Georgia and East Florida where they grew sugar, indigo and rice. The 9,000 acres the associates owned outside Britain in 1750 became 21,000 acres by 1763 and 130,000 acres by 1775 (Hancock, 1995: 144). Their investments in this land, the slaves needed to work it and the new buildings, tools and techniques which

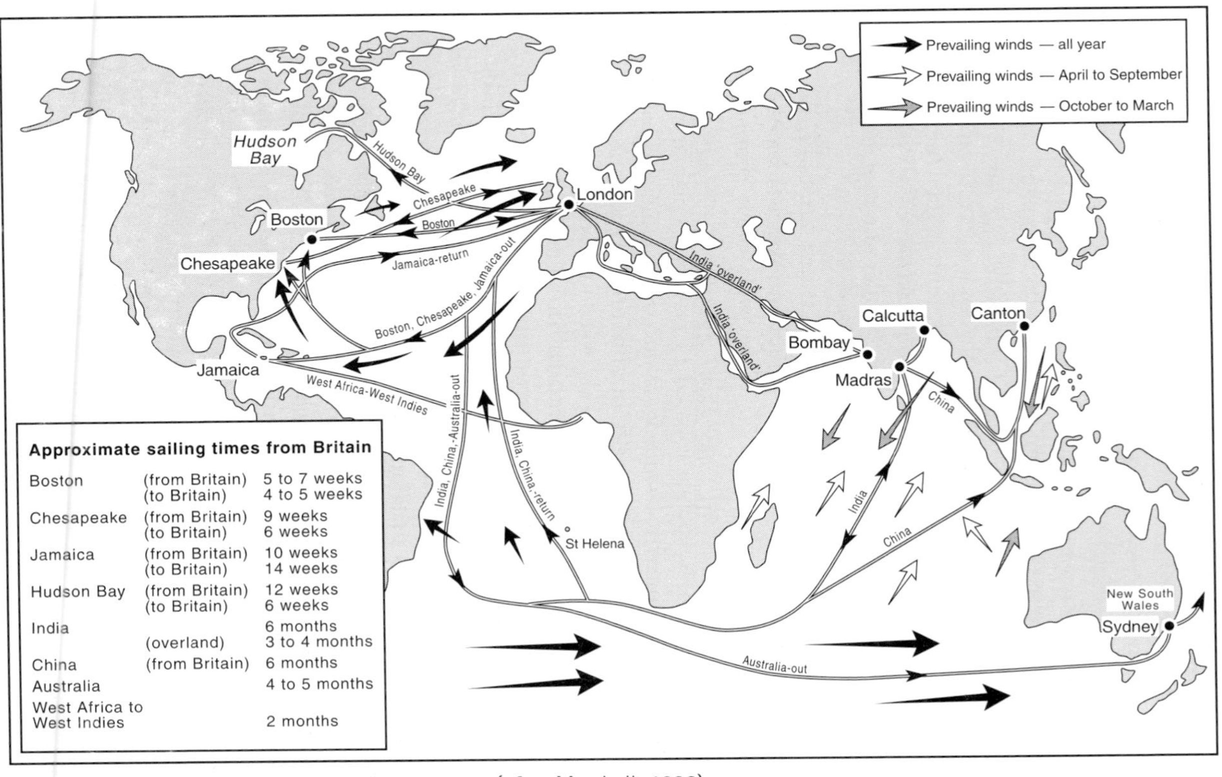

Approximate sailing times from Britain

Boston	(from Britain)	5 to 7 weeks
	(to Britain)	4 to 5 weeks
Chesapeake	(from Britain)	9 weeks
	(to Britain)	6 weeks
Jamaica	(from Britain)	10 weeks
	(to Britain)	14 weeks
Hudson Bay	(from Britain)	12 weeks
	(to Britain)	6 weeks
India		6 months
	(overland)	3 to 4 months
China	(from Britain)	6 months
Australia		4 to 5 months
West Africa to West Indies		2 months

Figure 2.4 London's global connections, c. 1770 (after Marshall, 1998).

would make it profitable, meant that they could now send cargoes to supply their own plantations and could load their own ships with their own produce, reducing costs and ensuring their ships were full. This 'backward integration' (Hancock, 1995: 19) also guided the associates' deeper involvement in the slave trade in 1748 when six of their number bought the Bance Island slave factory at the mouth of the Sierra Leone River. Substantial investment in buildings, personnel and armaments was required to establish the island as a lynch-pin in the Windward Coast slave trade. By the 1760s, slavery was booming and the island's proprietors were able to demand premium prices for the slaves for whom they had bartered cloth, arms, metalware (kettles, pans, tankards, knives, fire tongs, cutlery, nails and locks), sugar and tobacco with the local African rulers who controlled the trade. In return, the associates offered the slave ships – their own, as well as private coastal traders, and larger Dutch and French companies – a regular supply and shorter times on the coast, which translated into higher profits. They rarely paused to consider the morality of the trade (Hancock, 1995).

The Bance Island factory was the associates' greatest work of global network-building. The island, and the American plantations which it supplied, required the deployment of trusted personnel. These were often those relatives and friends who had proved effective in other ventures elsewhere in the world. Where there were failures, such as Oswald's East Florida plantations, they were often attributable to unreliable managers. Moreover, the work of the factory involved the co-ordination of capital, information, contacts and materials on an unprecedented scale. Augustus Boyd arranged for his brother Paul, a Waterford merchant, to supply the factory with Irish beef and butter, had slave sales managed by his South Carolina contacts, supplied his own St Christopher plantations, and arranged for favourable rates on the Indian cloth sold in London by the East India Company of which he was a director. John Sargent's connections brought in finance as well as cloth from India, Germany and the Levant (by 1751 they were trading 17 different textiles for slaves) and munitions from Northern Europe, Birmingham and Manchester. Richard Oswald brought his organisational skills and the American tobacco, Caribbean sugar and rum, and wine from Madeira which sealed the deals. The trade in human beings on Bance Island linked Africa to Britain, Europe, the Americas and Asia (Hancock, 1995).

A networked world?

Understanding the historical geographies of the East India Company and the associates as networks of people, capital, ships, goods and information that had to be constructed across space demonstrates different ways in which globalisation has taken place in different periods and places. It shows the globalisation of trade, not as the construction of a division between core and periphery, but as the extension of a network into new spaces already occupied by other networks such as African polities and Asian sea traders. It also demonstrates the work of 'networking' that had to be done for these global movements to occur, and that there could be contraction as well as expansion in what were the basic structures of the global economy. This offers another perspective on the relationships between the economics of capitalism and the politics of states and empires. In each case there is no easy separation between the economic and the political (O'Brien, 1998). The East India Company relied upon its royal charter and state support to operate in the East; when that failed it faltered too. Imperial power was part of the network. The associates made huge profits as a result of the Seven Years War, not only from supplying soldiers and sailors with bread and beef, but because it opened up land in the Americas for settlement, destroyed the French slave trade and established British control over the seas that they traversed. The British Empire was the environment in which the associates operated. It was inseparable from their actions and it was what they helped build as they integrated and improved their Atlantic world (Hancock, 1995). However, understanding the construction of these economic and political networks opens up the question of global movement more generally. Once the process of looking at what is moving around the world has begun, it is possible to ask more questions about what propelled these movements, what made them possible, what they might have meant for different people (the enslaved as well as the slavers), and how there were movements which ran counter to the power-laden networks of capital and empire even when they were organised by them.

Working and moving against the global grain

Workers of the world

For merchants to accumulate capital, it was not enough that they and their clerks worked in their counting houses, or that their correspondents, factors and agents worked on plantations, slave islands and docksides. They also had to put many other women and men to work around the world gathering furs, manufacturing cloth in villages in Yorkshire and Bengal, clearing and draining land, harvesting sugar cane and tobacco, and loading and sailing ships. In the eighteenth century, capital accumulated workers. It both brought them together in workplaces – plantations, merchant vessels and mines – and tied together their distant workplaces as never before (Rediker, 1987; Linebaugh, 1991). The global networks of production and trade were made of often back-breaking labour for workers of 'various degrees of freedom' (Linebaugh and Rediker, 1990: 226).

Nowhere was this new co-operative and international form of work more evident than in the workplaces which bound the sides of the Atlantic together – the merchant ships (Rediker, 1987). Sailors formed a maritime working class, the largest body of free wage labour in the eighteenth-century Atlantic world. The expansion of merchant capitalism increased the numbers of English sailors from around 5,000 in 1550 to over 60,000 by 1750. During that period they were also turned into workers who had nothing to sell but their labour power, working together on the complex and dangerous task of navigating

200 ton vessels across the oceans for an average of £1.46 per month in peacetime and £2.20 when at war. They were truly workers of the world. Ships' crews were made up of a multiplicity of nationalities and ethnicities – European, African, American, Indian – and their shared working lives took men like John Young from London to Barbados, Jamaica, Bristol, Africa, Virginia, Lisbon, Genoa, Leghorn and Cartagena (Rediker, 1987). It was their necessarily collective labour – loading and stowing the cargo safely; steering the ship, taking soundings and keeping lookout; and adjusting the rigging and sails, from on deck and aloft, with a knowledge of knots, wind and tides – which brought ships and their cargoes into harbour, decreased sailing times and tied tight the links in the international market chain.

Global trade depended on Jack Tar's labour. Thus those who sought to profit from trade tried to extend increasing control over that labour in order to make it more predictable, tractable and cheap. Captains were given 'near-dictatorial powers' over ordinary seamen and were quick to use violence to enforce their orders and to control the labour process (Rediker, 1987: 212). Together with their mates, stewards and pursers, captains also attempted to regulate the supplies of food and drink – in the face of sailors' notions of 'customary usage' – in order to ensure the profitability of voyages at the expense of the seamen's empty stomachs (Rediker, 1987: 126). Disputes over the embezzlement of small portions of the cargo, fishing and sailors' rights to carry their own merchandise for sale and the space needed for this, were all battles over the conditions of working life. The owners attempted to force sailors to rely increasingly on the money wage, which was in their hands. The sailors fought back – with violence, slander, rumours, threats, desertions and mutinies – to retain their autonomy and the protections to which they felt entitled by tradition. Their experience of collective labour gave them a sense of equality and justice, and a recognition of their shared opposition to those who would deny them these dignities for profit alone. Their collective rebellions by mutiny and Round Robins – lists of grievances presented to Captains with the crew's names in a circle so no leader could be identified – were forged alongside the global connections that capital made as it mobilised international wage labour (Rediker, 1987).

This internationalisation of conflict can be seen in two ways in which sailors and other workers made their complaints against merchants and imperial officials felt: rebellion and piracy. Peter Linebaugh and Marcus Rediker have begun to trace a history of riot and resistance among what they call the Atlantic working class. This account understands their rebellions as 'multiracial, multi-ethnic' events within 'a broad cycle of rebellion', which encompasses and connects slave revolts and conspiracies in the Caribbean, Irish agrarian movements, American revolutionary mobs and London's strikes and insurrections (Linebaugh and Rediker, 1990: 229 and 244). Thus Boston's King Street riot in 1770 brought together, in the words of John Adams, 'a motley rabble of saucy boys, negroes and molattoes, Irish teagues and outlandist jack tarrs' (quoted in Linebaugh, 1982: 112). London's Gordon riots in 1780 saw Newgate prison thrown open and burned by a mob led by two African-American ex-slaves, John Glover and Benjamin Bowsey, while another mob marched behind a black woman, Charlotte Gardiner (Linebaugh and Rediker, 1990; Linebaugh, 1991). These connections between slaves, sailors, free blacks and other working men and women, were forged in the working, drinking and singing cultures of port towns and cities which they shared, but which were also differentiated by gender and race. The linkages were extended as sailors carried ideas of liberty and autonomy as well as cargo between the Atlantic ports on board ships, which, as Linebaugh (1982: 11) argues, were 'an extra-ordinary forcing house of internationalism' where experiences were circulated among the many different people that international trade brought together (Bolster, 1997).

However, the most dramatic way for seamen to escape from the rigours of wage labour within international trade was to turn pirate and prey on it. Between 1716 and 1726, when piracy's successful attacks on merchant shipping were precipitating an international crisis, there were about 5,000 active pirates operating in the Atlantic. Most of them were going 'upon the account' by choice when their own ships were taken by freebooters. To sail under the Jolly Roger was to ply a trade of plunder, violence and revenge between Africa and the Caribbean within a social world whose 'hallmark was a rough, improvised, but effective egalitarianism that placed authority in the collective hands of the crew' (Rediker, 1987: 261; Thomson, 1994). The world of piracy turned the world of the merchant ship upside down. They insisted on equitable distributions of power, plunder and food; discipline was a matter of the often violent punishment of a collective sense of social transgression; and they were motivated by a desire for vengeance against exploitative captains, profiteering merchants and power-hungry imperial officials. As pirates, sailors made real their wish, albeit often only briefly, for autonomy and equality. This was a rough internationalism of shared purpose whose 'communitarian urge' was perhaps strongest in coastal strongholds such as the short-lived Pirate Republic of Libertalia on Madagascar (Rediker, 1987: 275; Rediker, 1997). Ironically, it was in another stronghold at the mouth of the Sierra Leone River in 1719 – thirty years before the associates bought Bance Island – that over two hundred pirates led by Thomas Cooklyn, Oliver LaBouche and Howell Davis were urged to 'remember their reasons for going pirating were to revenge themselves on base Merchants and cruel commanders of ships' (quoted in Rediker, 1987: 271–2).

Workers, including sailors, pirates and slaves, but also those who did not move so far, offer another global geography. This time it is one of hard labour and rebellion, which gives a view – from below – of the making of global networks and the ways in which people tried to challenge or step outside them. Globalisation is shown to be something that had to be constructed through many forms of work, and something that was, in its various manifestations, resisted by many of those being made to work in new ways. Not everything that moved along and constructed the networks of the early modern globe did so simply in order to further the accumulation of capital. Indeed, once these counter movements come into view there are a range of stories which begin to trace out global historical geographies with quite different contours. I want to end by focusing on these complex and often individual movements against the global grain by telling three brief stories – of a woman, a man, and a flag – and then showing how they offer other views of globalisation.

Moving in other worlds

The British flag and the politics of Tahiti

On 26 June 1767 Samuel Wallis and the sailors of the *Dolphin* 'took possession of Tahiti in the name of George III with a pennant and a pole, a turned sod, a toast to the King's good health and three British cheers' (Dening, 1994: 198). This ceremony on Matavai beach was attended by the islanders who had been there for many years and had their own rituals and stories of arrival and possession. They also conducted a ceremony with plantain branches symbolic of religious and political power, and they later took the British flag away with them to the other side of the island where a new *marae* (a temple) was being built at Mahaiatea by the Landward Teva chief Amo and his wife, Purea. When William Bligh returned to Tahiti in 1792 to finish the work disrupted by the dramatic events on the *Bounty*, he saw the pennant. By now it had been woven into a ceremonial wrap or girdle – a *maro ura* – of red and yellow feathers, along with a thatch of auburn hair belonging to Richard Skinner, one of the *Bounty* mutineers. The *maro ura* was a symbol of chieftaincy on the island. In feathers and stitches it recorded 'a history of sovereignty' (Dening, 1994: 205). It signified the momentous events of wars, sacrifices, peace treaties and, with its new incorporations, 'a history of the first native encounter with the [European] stranger' (Dening, 1994: 280). Indeed, this symbol of sovereignty on the island had, in the 25 years between Wallis and Bligh, been a crucial part of a political struggle for control over the island's peoples, which had eventually led to Purea's family being defeated by the Seaward Teva and their chief, Pomare. The British flag had travelled around the globe to be woven into Tahiti's politics. It had been possessed by the islanders as well as being part of the possession of the island by the British.

Maria Sibylla Merian's insect world

Maria Sibylla Merian was a seventeenth-century woman who painted and engraved pictures of insects and plants. Like most women artists of the time she was born into a family of artists where her talents were appreciated, despite her contemporaries' claims about what women could or could not do. By 1683 she had published a book of flower pictures and a two volume work on caterpillars. What was different about Merian was that – unlike her contemporaries – she was not interested in constructing exhaustive classifications of insects. Instead she studied and depicted their metamorphoses from egg, to caterpillar, pupa, and butterfly or moth. To show this, her pictures arranged each insect's metamorphosis around the plant on which it laid its eggs and fed. Hers was an 'ecological vision' (Davis, 1995: 155).

In 1699, after five years as part of a radical Protestant sect in Friesland – where she was able to continue but not publish her studies – and some years

Figure 2.5 Maria Sibylla Merian (1705): Caterpillar metamorphosis and moth on a palisade tree in Suriname (*source*: Plate 11 of Maria Sibylla Merian *Metamorphosis Insectorum Surinamensium*, Maria Sibylla Merian and Gerald Valck, Amsterdam, 1705) (published by permission of Houghton Library, Harvard University).

in Amsterdam among a circle of artists, collectors and naturalists, she and her eldest daughter boarded a ship for Suriname. In this Dutch sugar colony with its population of Amerindians (Arawaks, Waraos, Tairas and Waiyana), African slaves (about eight thousand), and Europeans (around six hundred Dutch Protestants and three hundred Portuguese and German Jews), she continued her work. In doing so, she drew upon the assistance of both European plantation owners and colonial administrators and African and Amerindian men and women – including her own slaves – who brought her specimens, led her through the forest, and described flora and fauna that she had not seen herself. The book she produced, *Metamorphosis of the Insects of Suriname* (1705), followed her earlier practice, but was now marked by these encounters (Figure 2.5). While Merian did not name them, she did recognise in her book

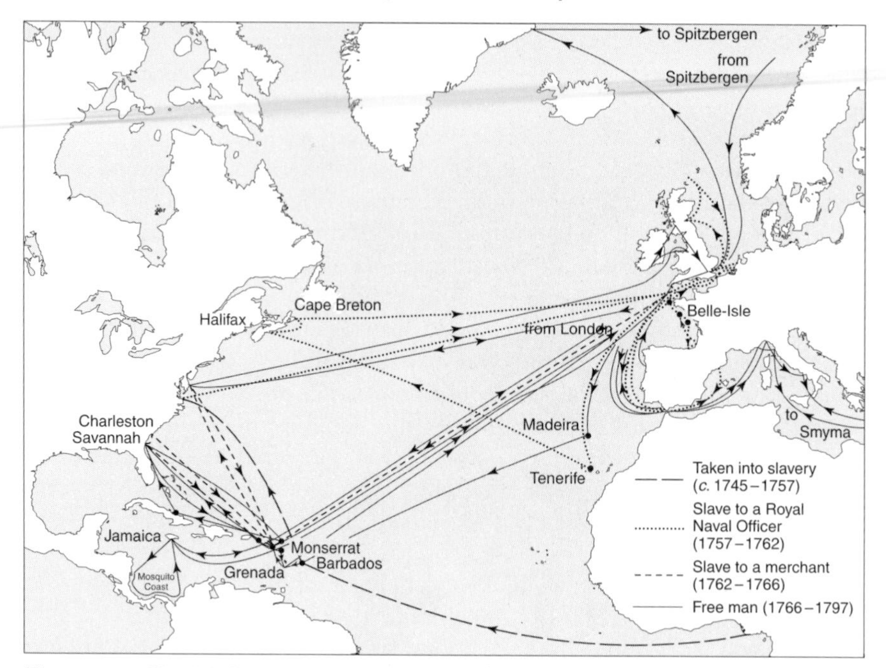

Figure 2.6 Olaudah Equiano's transatlantic life (c. 1745–97).

the work of the Africans and Amerindians, including the 'Indian woman' she brought back to Amsterdam, who had assisted her. She also incorporated their testimony on the uses of plants into the text, along with her own ethnographic observations, in ways which, contrary to contemporary practice, were 'indifferent to the savage/civilised boundary' (Davis, 1995: 190). Her 'ecological vision' allowed other relationships with nature to be included which could never find room in the neat classifications of her fellow naturalists. Maria Sibylla Merian's travels and the work produced through them were part of the construction of a world, which, although shaped by slavery and colonial domination, refused to conform to them.

Olaudah Equiano's transatlantic life

Figure 2.6 maps the description of his life that Olaudah Equiano (Figure 2.7) set out in his *Interesting Narrative* (1995; orig. 1789). Born in what is now Nigeria, he was kidnapped by African slavers and sold to Europeans at the coast. Forced to endure the horrors and cruelties of the 'middle passage' from Africa to the Caribbean, he was landed in Barbados and sold on to a plantation owner in Virginia. Having been renamed Michael on the ship and Jacob in America, he was sold to Captain Pascal of the Royal Navy who, despite resistance from Equiano, renamed him Gustavus Vassa after the Swedish liberator king. In five years with Pascal, Equiano sailed to Britain and Europe, and was present at the siege of Louisbourg (in what is now Canada) and naval

Figure 2.7 Olaudah Equiano (*source*: frontispiece of Olaudah Equiano, *The Interesting Life of O. Equiano Written by Himself*, London, 1789) (published by permission of the British Library).

engagements along the French coast. On board these ships he also learned to read and write and how to cut hair. Returning to London in 1762 he expected to be freed, but was instead sold to a Quaker merchant and, despite his fellow sailors' promises to rescue him, taken to Montserrat. For the next four years he worked on his master's ships as they traded – sometimes in slaves – between the Caribbean islands, Georgia and Philadelphia. Always with an eye for accumulation and self-improvement, and a memory of his master's throwaway promise that he could purchase his freedom for 40 pounds, he traded on his own account in a small way.

To his master's surprise, Equiano bought his freedom in 1766 and, after a shipwreck in the Bahamas, sailed for London. His years of freedom extended his travels. In search of money, and with a desire to see the world, he travelled to Italy and Turkey, took part in John Phipps' 1773 expedition to the Arctic (as far north as any such voyage had been), and worked as an overseer on a new slave plantation on the Mosquito Coast of Central America (Murphy, 1994). Returning to London in 1786 and guided by his personal convictions and deep Christian beliefs, Equiano campaigned for the abolition of the slave trade and was active in the Committee for the Relief of the Black Poor's plan for resettlement in Sierra Leone (Potkay, 1994; Gilroy, 1997). By the time his life story was published – as a vital part of his political activities – he was a well-known figure. He had also been shaped by his movements. Equiano and other black writers of the eighteenth century such as Ukawsaw Gronniosaw, Quobna Ottobah Cugoano and Phillis Wheatley 'led lives that were neither simply African nor American, West Indian nor British, but in succession all of these, and ultimately all of these at once' (Potkay, 1995: 2).

Moving stories

Each of these 'global stories' challenges conceptions of the nature of globalisation in the period before 1800. The British flag is not simply a symbol of possession which brings the Pacific islands into the European orbit and under imperial dominion. Instead it is part of a much more complicated history whereby each side endeavoured to take possession of the other for its own ends. Globalisation looks different from the perspective of the Tahitians on the beach (Dening, 1994). Similarly, Maria Sibylla Merian's *Metamorphosis* shows that European science was not only a matter of the classification and control of the exotic, but could be part of relationships which, if not equal, at least recognised that an encounter of different worlds had taken place (Davis, 1995; Martins, 1998). It also calls attention to the gendered histories of globalisation (see Chapter 1). Finally, Olaudah Equiano's remarkable story puts an African – and he was one among many (Bolster, 1997) – onto ships other than those of the middle passage, showing him at work in all corners of the Atlantic world and giving his view of it to his readers. It also highlights the ways in which opposition to the forms of globalisation driven by the slave trade were led by people like Equiano, who had become who they were by moving – and being moved – around that world. While these forms of globalisation did lead to imperial domination, racist ideologies of science and civilisation and plantation slavery in the Americas, it is also important to bear in mind tales such as these which offer other, parallel historical geographies of globalisation, which show that the world was not simply made one way.

Conclusion

This chapter has moved from a vision of the historical geography of globalisation in the period between 1500 and 1800, which attempts to order it into a single 'world-system', to a sense of globalisation which emphasises the construction of networks of movements through which the often unequal relationships between parts of the world could take place and, finally, on to a sense of the conflicts and resistances within those networks and the movements that ran across and against them. The aim here is not simply to argue for one way of viewing the historical geography of the globe. The structural approach to the history of capitalism has problems recognising some of the forms of globalisation which are not reducible to the logic of capital or organised in terms of the establishment of a spatial structure of core and periphery. However, a concentration on networks and counter-movements can miss the longer and larger histories of relationships between parts of the globe of which they are only the latest part. As well as recognising the differences between these ways of understanding globalisation it is important to consider how Wallerstein's structural histories can be more effectively 'peopled' with merchants and workers than I have allowed them to be here, and how the travels of Equiano, Merian and Wallis's flag were 'structured' by global economic and political relations. This also means considering the historical geographies of globalisation in terms of a range of different sorts of geography: Wallerstein's shifting but bounded and interdependent areas of advantage and disadvantage; Wolf's more nuanced insistence on peripheries within the core and core areas (both political and economic) within the periphery; the skein-like networks of ships, factories, sailors, information and goods which were extended and contracted across the oceans in both hemispheres; and the single tracks and complex entanglements of Wallis's flag, Maria Sibylla Merian's search for insects and Olaudah Equiano's bids for freedom. Keeping them all in view at the same time is an impossibility, but a range of historical geographies can be written by working between them.

Having started with European tastes, I will end with one place that had a role in supplying them. The Sierra Leone River and the land around it was, having been tied into the trans-Saharan trade routes, transformed into a peripheral area of the world-system. It became part of the triangular trade in slaves, sugar and manufactured goods between Europe, Africa and the Americas. In the process the political, economic and cultural relationships of the people who lived there were substantially altered, not least by a 'consumer revolution' in which many were shrewd and enthusiastic participants (Walvin, 1997). Indeed, the thousands of those who were sold into slavery, and the 17 varieties of cloth for which they were bartered, were both transported in the ships of rich London merchants. These men, through their work of organisation and network building, made a former pirate haven – where the business practices and profit-making of merchants like them had often been roundly rejected by escaped workers whose labour had once held these networks together – into a profitable slave factory with connections across the globe. It was also the place to which, 25 years later, a former slave and Atlantic traveller, merchant and worker sought to send black settlers as part of his campaign against the slave trade. The historical geography of Sierra Leone, and of all the other places which became part of these geographies of globalisation, needs to be understood in ways which are attentive to the variety of changing connections which map out its place in a shifting world.

References

Abu-Lughod, J. L. (1989) *Before European Hegemony: The World System A.D. 1250–1350*, Oxford University Press, New York.

Blaut, J. (1993) *The Colonizer's Model of the World: Geographical Diffusionism and Eurocentric History*, Guilford Press, New York.

Bolster, W. J. (1997) *Black Jacks: African American Seamen in the Age of Sail*, Harvard University Press, Cambridge, Mass.

Brenner, R. (1993) *Merchants and Revolution: Commercial Change, Political Conflict, and London's Overseas Traders, 1550–1653*, Cambridge University Press, Cambridge.

Brotton, J. (1997) *Trading Territories: Mapping the Early Modern World*, Reaktion Books, London.

Chaudhuri, K. N. (1965) *The English East India Company: The Study of an Early Joint-Stock Company 1600–1640*, Frank Cass, London.

Chaudhuri, K. N. (1978) *The Trading World of Asia and the English East India Company, 1660–1760*, Cambridge University Press, Cambridge.

Curtin, P. D. (1969) *The Atlantic Slave Trade: A Census*, The University of Wisconsin Press, Madison.

Davis, N. Z. (1995) *Women on the Margins: Three Seventeenth-Century Lives*, Harvard University Press, Cambridge, Mass.

Dening, G. (1994) *Mr Bligh's Bad Language: Passion, Power and Theatre on the Bounty*, Cambridge University Press, Cambridge.

Eltis, D. and Richardson, D. (eds) (1997) *Routes to Slavery: Direction, Ethnicity and Mortality in the Atlantic Slave Trade*, Frank Cass, London.

Equiano, O. (1995) *The Interesting Narrative and Other Writings* (orig. 1789), Penguin, Harmondsworth.

Gilroy, P. (1997) Diaspora and the detours of identity. In Woodward, K. (ed.) *Identity and Difference*, Sage, London, 299–343.

Hall, S. (1991) The local and the global: globalization and ethnicity. In King, A. (ed.) *Culture, Globalisation and the World System*, Macmillan, London, 19–39.

Hancock, D. (1995) *Citizens of the World: London Merchants and the Integration of the British Atlantic Community, 1735–1785*, Cambridge University Press, Cambridge.

Harvey, D. (1989) *The Condition of Postmodernity*, Blackwell, Oxford.

Hugill, P. J. (1993) *World Trade Since 1431: Geography, Technology, and Capitalism*, Johns Hopkins University Press, Baltimore.

Kearns, G. (1988) History, geography and world-systems theory. *Journal of Historical Geography*, **14**, 281–92.

Kowaleski-Wallace, E. (1997) *Consuming Subjects: Women, Shopping, and Business in the Eighteenth Century*, Columbia University Press, New York.

Law, J. (1986) On the methods of long-distance control: vessels, navigation and the Portuguese route to India. In Law, J. (ed.) *Power, Action and Belief: A New Sociology of Knowledge*, Routledge and Kegan Paul, London, 234–63.

Lawson, P. (1993) *The East India Company: A History*, Longman, London.

Linebaugh, P. (1982) All the Atlantic mountains shook. *Labour/Le Travailleur*, **10**, 87–121.

Linebaugh, P. (1991) *The London Hanged: Crime and Civil Society in the Eighteenth Century*, Penguin, Harmondsworth.

Linebaugh, P. and Rediker, M. (1990) The many-headed Hydra: sailors, slaves, and the Atlantic working class in the eighteenth century. *Journal of Historical Sociology*, **3**, 225–52.

Mancall, P. C. (1998) Native Americans and Europeans in English America. In Canny, N. (ed.) *The Oxford History of the British Empire. Volume I: The Origins of Empire*, Oxford University Press, Oxford, 328–50.

Marshall, P. J. (1998) The English in Asia to 1700. In Canny, N. (ed.) *The Oxford History of the British Empire. Volume I: The Origins of Empire*, Oxford University Press, Oxford, 264–85.

Martins, L. (1998) Navigating in tropical waters: British maritime views of Rio de Janeiro. *Imago Mundi*, **50**, 141–155.

Miller, D. P. and Reill, P. H. (eds) (1996) *Visions of Empire: Voyages, Botany, and Representations of Nature*, Cambridge University Press, Cambridge.

Murphy, G. (1994) Olaudah Equiano, accidental tourist. *Eighteenth-Century Studies*, **27**, 551–68.

Newson, L. A. (1996) The population of the Amazon basin in 1492: a view from the Ecuadorian headwaters. *Transactions of the Institute of British Geographers*, **21**, 5–26.

O'Brien, P. K. (1998) Inseparable connections: trade, economy, fiscal state, and the expansion of empire, 1688–1815. In Marshall, P. J. (ed.) *The Oxford History of the British Empire. Volume II: The Eighteenth Century*, Oxford University Press, Oxford, 53–77.

Postma, J. M. (1990) *The Dutch in the Atlantic Slave Trade, 1600–1815*, Cambridge University Press, Cambridge.

Potkay, A. (1994) Olaudah Equiano and the art of spiritual autobiography. *Eighteenth-Century Studies*, **27**, 677–92.

Potkay, A. (1995) Introduction. In Potkay, A. and Burr, S. (eds) *Black Atlantic Writers of the Eighteenth Century*, Macmillan, Basingstoke, 1–20.

Price, J. M. (1998) The imperial economy, 1700–1776. In Marshall, P. J. (ed.) *The Oxford History of the British Empire. Volume II: The Eighteenth Century*, Oxford University Press, Oxford, 78–104.

Rediker, M. (1987) *Between the Devil and the Deep Blue Sea: Merchant Seamen, Pirates, and the Anglo-American Maritime World, 1700–1750*, Cambridge University Press, Cambridge.

Rediker, M. (1997) Hydrarchy and Libertalia: the utopian dimensions of Atlantic piracy in the early eighteenth century. In Starkey, D. J., van Eyck van Heslinga, E. S. and de Moor, J. A. (eds) *Pirates and Privateers: New Perspectives on the War on Trade in the Eighteenth and Nineteenth Centuries*, University of Exeter Press, Exeter, 29–46.

Richardson, D. (1998) The British empire and the Atlantic slave trade. In Marshall, P. J. (ed.) *The Oxford History of the British Empire. Volume II: The Eighteenth Century*, Oxford University Press, Oxford, 440–64.

Taylor, P. J. (1996) *The Way the Modern World Works: World Hegemony to World Impasse*, John Wiley, Chichester.

Thomson, J. E. (1994) *Mercenaries, Pirates and Sovereigns: State-Building and Extraterritorial Violence in Early Modern Europe*, Princeton University Press, Princeton.

Wallerstein, I. (1974) *The Modern World-System I: Capitalist Agriculture and the Origins of the European World-Economy in the Sixteenth Century*, Academic Press, London.

Wallerstein, I. (1980) *The Modern World-System II: Mercantilism and the Consolidation of the European World-Economy, 1600–1750*, Academic Press, London.

Wallerstein, I. (1983) *Historical Capitalism*, Verso, London.

Walvin, J. (1997) *Fruits of Empire: Exotic Produce and British Taste, 1660–1800*, Macmillan, Basingstoke.

Wolf, E. R. (1982) *Europe and the People Without History*, University of California Press, Berkeley.

Chapter 3

The past in place: historical geographies of identity

Brian Graham

Introduction

According to Homi Bhabha (1994: 6), the currency of international connections 'is no longer the sovereignty of the national culture', so famously evoked in Benedict Anderson's concept of an 'imagined political community' rooted in an ethos of modernity and progress. Anderson (1991: 6–7) argues that any nation is imagined

> as both inherently limited and sovereign . . . [i]t is imagined because the members of even the smallest nation will never know most of their fellow-members . . . the nation is imagined as *limited* because even the largest . . . has finite, if elastic, boundaries, beyond which lie other nations . . . it is imagined as *sovereign* because the concept was born in [the] age [of] Enlightenment and Revolution . . . [and] [f]inally, it is imagined as a *community* because, regardless of the actual inequality and exploitation that might prevail . . . , the nation is always conceived as a deep horizontal comradeship.

This imagining of an internal national homogeneity – one which inevitably draws upon a mythology of the past for its coherence and legitimacy – is perhaps the most potent manifestation of the relationship between place and identity. As Handler (1994) argues, the Western world has been accustomed for more than two centuries to think of identity as an object bounded in time and space, with clear beginnings and endings, and its own territory. Nevertheless, identity remains a highly ambiguous concept. In general terms, it incorporates values, beliefs and aspirations, which are used to construct simplifying structures of sameness that identify the self with like-minded people. Collective identity is a multi-faceted phenomenon that embraces a range of human attributes, including language, religion, ethnicity, nationalism and shared interpretations of the past (Guibernau, 1996), and constructs these into discourses of inclusion and exclusion – who qualifies and who does not.

Central to the entire concept of identity, therefore, is the idea of the Other, manifested in groups with competing – and often conflicting – beliefs, values and aspirations (Said, 1978: see Chapter 4). 'Recognition of Otherness will help reinforce self-identity, but may also lead to distrust, avoidance, exclusion and distancing from groups so-defined' (Douglas, 1997: 151–2). National identity therefore shares with imperialism in invoking a pervasive and persistent ensemble of cultural attitudes towards the rest of the world (see Chapter 5). Identity is not a discrete social construction, defined in a coherent relationship with space; rather, it comprises a multiplicity of attributes, identities – and their defining criteria – overlapping in complex ways and at a variety of geographical scales. Identity, for example, possesses potentially conflicting supra-national, national, regional and local expressions, in turn fractured by other manifestations of belonging – religion, language, ethnicity, tribe, culture, gender – that are not necessarily defined in terms of those same spatial divisions. Inevitably, therefore, identity is socially and geographically diverse rather than neatly bundled, ethnicity and nationalism – historically, at least, its most potent expressions – existing to simplify such heterogeneity into simplifying representations, synecdoches, of sameness.

Weaving together the book's three themes – interconnected historical geographies, differential experiences of modernisation and, in particular, politics of the past in the present – in a succession of case studies, this chapter is located at the intersection of historiography and geopolitics. In pursuing its three objectives, it is concerned both with the construction and writing of identities and with their repercussions. Particular emphasis is given to historical geographies of nationalism. In the first instance, I discuss the relationships between nationalism, ethnicity, identity and place, which are mediated through contested narratives of the past used to construct particular geographies of belonging. The power of such narratives rests on their ability to evoke the accustomed, tropes that work by appealing to 'our desire to reduce the unfamiliar to the familiar' (Barnes and Duncan, 1992: 11–12). Paralleling the discussion of imperialism in Chapters 4, 5 and 6, historiography and identity politics can be conceived as interconnected cultural discourses. Secondly, the chapter demonstrates the situated nature of nationalism, which cannot be visualised as a fixed primordial identity rooted in the mists of history (Agnew, 1997), but as a contingent, constantly mutating process. Although its function is to impose a degree of uniformity upon the heterogeneity of cultural diversity, it can do so only for a while. Finally, the differential experiences of national identity are explored through the medium of the case studies. Like imperialism, nationalisms cannot be viewed as monolithic meta-narratives with immutable and predictable outcomes. They are, like modernity itself, 'differentiated geographies . . . made in the relationships *between* places and *across* spaces' (Ogborn, 1998: 19).

Interconnections: nationalism, ethnicity, identity and place

Identity and place

The results of what Agnew and Corbridge (1995) refer to as the 'territorial trap', which equates identity with state sovereignty and territorial space, are to be found in a geopolitics in which geography, history and other academic disciplines act as disseminators of identity discourses, most potently those of nationalism. Space is oversimplified into idealised constructs of tradition and modernity and a particular time-period is imbued with an idealised historical experience which stands as a metaphor of the state (Agnew, 1996). That allegory may well lack any congruence with empirical economic and social conditions. Space is further politicised through its treatment as the distinctive and historic territory of the nation-state, 'the receptacle of the past in the present, a unique region in which the nation has its homeland' (Anderson, 1988: 24). The rise of the nation-state in the eighteenth and nineteenth centuries was closely connected to Romantic notions of the mysticism of place, and of notions of belonging and not belonging. For Woolf (1996: 25–6):

> National identity is an abstract concept that sums up the collective expression of a subjective, individual sense of belonging to a socio-political unit: the nation state. Nationalist rhetoric assumes not only that individuals form part of a nation (through language, blood, choice, residence, or some other criterion), but that they identify with the territorial unit of the nation state.

Hobsbawm (1990) also defines nationalism as primarily a principle that holds that the political and national unit should be congruent, the essence of 'national' being defined within the cultural realm. Davies (1996: 813) concurs, citing the French historian, Ernest Renan, to the effect that 'the essential qualities' of the nation are spiritual. 'One is the common legacy of rich memories from the past. The other is the present consensus, the will to live together . . .'.

Four important qualifications must be made at the outset of the discussion. First, 'nationalism favours a distinctly homosocial form of male bonding' (Parker *et al.*, 1992: 6) as expressed by Anderson's argument that ultimately it is the 'fraternity' of the 'deep horizontal comradeship' of the imagined community 'that makes it possible, over the past two centuries, for so many millions of people, not so much to kill, as willingly to die for such limited imaginings (Anderson, 1991: 7). Pratt (1994: 30) argues that as a consequence,

> [w]omen inhabitants of nations were neither imagined as nor invited to imagine themselves as part of the horizontal brotherhood . . . rather, their value was specifically attached to . . . their reproductive roles [as] mothers of the nation. . . .

Thus it is the common case that the iconography of war and militarism has been to the fore in the invention of the imagined communality of the modern nation-state.

Secondly, both the legacy of nationalism and the particular terms of the communal will are situated temporally in a specific social and political ethos and are thus subject to continual modification through time as circumstances alter. Consequently, the 'sovereign territorial state is not a sacred unit beyond historical time' (Agnew and Corbridge, 1995: 89), even though its rhetoric may claim otherwise. Because the meanings of identity spaces are undergoing continual renegotiation, disjunctions often occur in which, for example, contemporary diasporic versions of national allegiance remain sited in past circumstances that pay little heed to the ongoing evolution of identity in the metropolitan state. For example, the visions held of Ireland in North America (and Australia) often remain those of a nineteenth-century nationalist discourse framed as a narrative of English oppression. Although this perspective has long been heavily contested, both within Ireland and North America itself, its malevolent and violent legacy is apparent in the substantial financial support from some of the descendants of the Irish diaspora for the Provisional IRA (PIRA) and other Republican paramilitary organisations, whose claim to wage war in Ireland has been formulated in terms of traditional nationalist rhetoric.

Thirdly, despite its importance to the modern era, particularly in the West, the hegemony of national identities has always been compromised by other – often contradictory – allegiances, including religion, tribe, class and gender. As Agnew (1998: 216) argues, 'the "sacralization" of the nation-state has never been total; even within totalitarian states, sites of religious and local celebration have had their place'. The ethnic mosaic, sometimes seen as characteristic of the multicultural postmodern world, has always been so. State boundaries were imposed upon this chiaroscuro, often conflicting with other identities. For example, state boundaries drawn by imperial powers in Central Africa are rarely congruent with tribal boundaries, one cause of the ethnic politics which led to the genocidal horrors in Rwanda and Burundi during 1994, and the continuing instability of Congo, which, in late 1998, was threatening to draw the entire region into another ethnic war.

Moreover, some forms of identity have always been extraterritorial in terms of the state, religious adherence being the best example. The Great Schism of the Christian Church in 1054 divided Europe between a Latin, Roman West and an Orthodox East, a fracturing which maintains its importance to the present day. More widely, Christianity has been locked in conflict with Islam since the seventh century, a rivalry that has now become transmuted into the struggle between the West and a resurgent, militant version of Islam and its transnational network of terrorist groups. In part, this struggle represents the resistance of Islam to secularisation, which means that in an Islamic state still defined by its religious beliefs, these latter can compete with nationalism because they perform most of its functions (Gellner, 1997). However, the monolithic nature of religions should not be overstated for they too are fractured into multiple, historically based allegiances, themselves often the cause of conflict. For example, the Sunni Taliban movement, which now controls Afghanistan, has been implicated in massacres of Shi'ite ethnic minority groups, a grim prolongation of a fissure that has existed in the Islamic world since the death of Mohammed in 632.

Finally, although this chapter concentrates on Europe, that is partly dictated by the overall structure of the book. The 'provincialising' of Europe discussed in Chapter 1, while decentring the centre, does not preclude seeking to understand it. But it is also the case that modernist nationalism evolved primarily as a European discourse, later mediated in much of the rest of the world through imperialism and migration. As Anderson (1991: 47) argues, for example, attempts to generate nationalism in the new American states of the late eighteenth and early nineteenth centuries were defined in relation to the struggles for national liberation from the European imperial metropoles. These states were led 'by people who shared a common language and common descent with those against whom they fought'. The idea that sovereignty over everything is bundled into territorial state parcels, reflects what Harvey (1989: 242) has called 'a radical reconstruction of views of space and time', which occurred only in the transition from pre-modern to modern societies in Europe (see, for example, Johnson, 1998). This process witnessed the collapse of well-established medieval systems of hierarchical subordination such as feudalism and the Roman Church (Ruggie, 1993). Instead, social membership became exclusive and the 'identification of citizenship with residence in a particular geographical space became the central fact of political identity' (Agnew and Corbridge, 1995: 85).

The European nation-state is thus circumscribed by criteria of inclusion, which support representations of communality, but is defined also by the Othering of those, both internal and external, who are excluded from entry. If this manifestation of identity is indeed withering as the currency of international connections, then it is also largely in Europe that those processes might be located. For Bhabha – and many other cultural theorists – nationalist place-oriented ideologies and modes of explanation are ostensibly rendered evermore irrelevant to a postcolonial world in which mass migration has created complex, multicultural societies, thereby disrupting the justification for allegories of identity constructed around place. In turn, Bhabha postulates a postnationalist process of 'DissemiNation', in which new communities of interest are evolving, testimony to the construction of changing identities and transnational communities of interest that undermine and eventually negate the 'out of many one' ideology of nationalism and the nation-state. He denies the horizontal conceptualisation of belonging fundamental to the 'imagined community', arguing instead for an alternative 'modernity' in which 'organic' ideologies are neither consistent or homogenous and subjects of ideology are not unitarily assigned to a singular social position. Again, Stuart Hall (1996: 4) sees the construction of diverse multicultural societies, and the concomitant fragmentation of belonging, as pointing to a world in which identities are 'never singular but [are] multiply constructed across different, often intersecting and antagonistic discourses, practices and positions'.

None the less, as Hobsbawm (1990: 187), for one, argues, the world remains characterised by a continuing strength of longing for group identity of which 'nationality is one expression [albeit, as we have already seen] . . . not the only one'. Anderson (1991: 3) goes even further in claiming that the

> 'end of the era of nationalism', so long prohesised, is not remotely in sight. Indeed, nation-ness is the most universally legitimate value in the political life of our time.

Even if this remains an overstatement, the placeless or fragmented collective identity posited by Bhabha and Hall arguably understates the enduring power of territorial states and their tropes of identity. Despite economic globalisation and the sanctification of free trade, states remain as powerful economic regulators, largely still providing the 'only existing organisational frameworks within which representative and participatory politics can be pursued' (Agnew, 1997: 319). Moreover, the national idea 'still represents the most relevant recourse for those who wish to restore order when they are faced with a serious disturbance, whether the threat comes from within or without' the state (Dijkink, 1996: 16). It is argued here that we are now encountering more nuanced allegories of identity and place, but that an interconnection of belonging and territorial space remains fundamental to such representations. Indeed, Grosby (1995: 143) goes so far as to claim that territoriality is *the* 'transcendental, primordial attachment' of modern societies. In turn, this dimension of belonging constitutes not one but several of Hall's 'intersecting and antagonistic discourses'. Necessarily powerful narratives of place, fixed within hegemonic representations of the past, remain fundamental to the modernistic ideas of legitimacy and authority underpinning the territorial state.

Nationalism, ethnicity, identity and place

A further salient reason for being sanguine about 'dissemination' is that 'out of many one' ideologies retain an enduring power to kill and mutilate in the name of nationalist imaginings, no matter how limited. Twenty-nine people were killed and over 200 injured (several of whom were maimed for life) when a car bomb, left by a renegade Irish Republican group, devastated the centre of Omagh, Northern Ireland, on 15 August 1998. The 'justification' was that the town's commercial core was a 'legitimate' target in the continuing war to free Ireland from the 'Brits'. The obvious question is: whose Ireland? In separate referenda, more than 90 per cent of the population of the Republic of Ireland, and in excess of 70 per cent of those resident in Northern Ireland, had earlier voted to support the 1998 Good Friday Belfast Peace Agreement, which seeks, in ways not yet clearly articulated, to create a diverse but integrated society in Ireland and end the violence, which has killed more than 3,500 people since 1969. The very fragility of this agreement encapsulates the quintessential difficulties of identity politics because the ultimate acceptance of peace will depend on the creation and acceptance of alternative Irish narratives of place, which must acknowledge and accept responsibility for past suffering but do not make a nationalism centred on oppression and religion into the nexus of Irish identity. Rewritten narratives of the past create the alternative tropes of present belonging that must underpin the negotiation of acceptable political structures (Graham, 1997a; Graham and Shirlow, 1998).

More widely, 'out of many one' ideologies are not restricted to nationalism alone, its very malevolence being exacerbated by the myriad ethnic and tribal conflicts in which contested nationalist mythologies are often implicated. An ethnic group can be defined as a socially distinct community of people who share a common history and culture and often language and religion as well (Sillitoe and White, 1992). Despite a mutual reliance on a shared history and values, nationalism is not necessarily an intrinsic feature of ethnicity (Anderson, 1991). Nevertheless, both are situated in time and place, both are complex and dynamic and both are fundamental to notions of identity (Nanton, 1992). They share also a mutual invocation of tropes of hatred, which, if often largely recent inventions, are based on simplifying linear historical narratives imposed on the complexities and ambiguities of the past and used to legitimate armed conflict. When it is recalled, for example, that the boundaries of European nation-states still reflect the geopolitics of exclusion and division of territorialised political identity generally achieved through or as a result of violence and war (Heffernan, 1998), it is easier to understand the persistence of such narratives. Again, according to Sadowski (1998), virtually all the wars which have occurred in the post-Cold War world have (or had) an ethnic or tribal basis. These range from small but still deadly internecine conflicts in Ireland and Spain, through the confused ambiguities of former USSR republics such as Azerbaijan, Georgia and Tajikistan, to the former Yugoslavia (as many as 200,000 people were killed in the Bosnian War of 1991–5) and Rwanda, where the 1994 genocide conceivably massacred one million people, mostly from the country's Tutsi minority.

Ethnic and national identity – both equally contingent – share a reliance on three distinct strands of definition: segregation of all sorts of human activities; the myth of a perceived common past; and the delimitation of the group by key social and cultural markers (Poole, 1997). Because of this congruence of characteristics, it has become customary to refer to ethnic nationalism. By definition, this is exclusive, often intolerant and frequently unstable. Based on narratives, which make membership a matter of blood and equate identity with a national territory that embodies the community of birth and native culture, the primary function of ethnic nationalism lies in its subsumption of diversity and denial of heterogeneity (Shaw, 1998). Because such narratives rely on their appeal to cultural symbols of identity and collective material goals in pursuit of political goals – primarily control over a state, that denial often takes the physical form of 'ethnic cleansing'.

Historically, the invented, imagined community depended on tropes of cultural exclusivity. It required a common spoken language, often sponsored by a cultural élite, and disseminated through state education systems. This was conceptualised in the work of the eighteenth-century philosopher, Johann Gottfried Herder, who insisted on the natural division of the earth into separate language groups; the stress on language as a primary ethnic marker became particularly apparent towards the end of the nineteenth century. The imagined community could require also a common religion but, above all, its defining characteristics had to be located firmly in readings of the past and place that legitimated the nation's claim to its territory. By its very nature, nationalism embodies a zero-sum conceptualisation of power in which possession of territory is absolute and non-negotiable. Only through war can territory be wrested from its rightful owners. That sense of belonging depended on forgetting as much as remembering, the past being reconstructed as a trajectory to the national present in order to guarantee a common future (Gillis, 1994). 'History was raked to furnish proof of the nation's age-long struggle for its rights and its lands' (Davies, 1996: 815), and constructed around the Othering of a common enemy (or enemies). As Linda Colley remarks (1992: 5–6), Great Britain, for example, can plausibly be regarded as an invented nation 'superimposed, if only for a while, onto much older alignments and loyalties':

> It was an invention forged above all by war. Time and time again, war with France brought Britons, whether they hailed from Wales or Scotland or England, into confrontation with an obviously hostile Other and encouraged them to define themselves collectively against it. They defined themselves as Protestants struggling for survival against the world's foremost Catholic power. They defined themselves against the French as they imagined them to be, superstitious, militarist, decadent and unfree. (Colley, 1992: 5)

Hobsbawm (1990: 65) argues that very few modern national movements are actually based on a strong ethnic consciousness, although 'they often invent one once they have got going, in the form of racism'. For example, Charles Maurras, the co-founder of *L'Action Français* in 1899, expounded a virulent form of racial nationalism, which equated Frenchness with being native-born and Catholic; as was also true then of Irish nationalism, a particular religious allegiance and national identity became inseparable. This more intolerant, dogmatic and ultimately deadly variant of nationalism was characteristic of late nineteenth-century Europe, succeeding to earlier more liberal variants, which often originated in the struggle for freedom from reactionary dynastic rule (Davies, 1996). When exported through imperialism's networks, the racial hues of such ideas were compounded by notions of European superiority that projected on colonised lands and their inhabitants, 'imperial codes expressing both an affinity with the colonizing country and an estrangement from it' (Daniels, 1993: 10). To cite Colley (1992: 5) again, 'increasingly . . . [Britons] defined themselves in contrast to the colonial peoples they conquered, peoples who were manifestly alien in terms of culture, religion and colour'. In Europe, the racial connotations of nationalism were later glossed by fascism and its inherent anti-Semitism, the apogee of racial nationalism in France occurring during the Second World War when the collaborationist Vichy government deported French citizens who were Jews to the Nazi death camps in Poland. (It has taken much of the intervening 50 years for France to remember these events, proof of the adage that nationalism is often as much about forgetting the past as commemorating it.) Racially shaped discourses of national identity ostensibly merge into notions such as Friedrich Ratzel's conceptualisation of the state as an organism, which consequently had to expand when population enlargement threatened resource exhaustion. His idea of *Lebensraum* (literally breathing space) led to the conviction that modern advanced states must aspire to *Grossraum* (large space), nation building blending into empire building. Although this has been interpreted as a source for later Nazi ideology, *Grossraum*, for Ratzel, 'actually fostered the racial mixing that produced new ethnic strains' (Livingstone, 1992: 202).

But ethnic nationalism demands more than an acceptably homogenising reading of the past. Intensely territorial by nature, its narratives have also to be set in allegories of place. Often constructed through particular readings of religion, art, literature and even music, the ideal national or representative landscape acts as a visual encapsulation of a group's occupation of a particular territory and the memory of the shared past that this conveys (Daniels, 1993; Graham, 1994; Agnew, 1998). The dilemma, however, was that a 'nation of even middling size' had to construct this manifestation of unity 'on the basis of evident disparity' (Hobsbawm, 1990: 91). The nineteenth-century historian, Jules Michelet, posed the central paradox; pointing to the example of a geographically diverse France in which unity was achieved only by force and civil war, he concluded: 'The material [of France] is essentially divisible and strains towards disunion and discord' (cited in Braudel, 1988: 120). Consequently, national landscapes, which subsume such diversity, 'give national identity a materiality it would otherwise lack' (Agnew, 1998: 216), and act as specific texts in nationalism's role of legitimising power relationships. Like the content of geography itself, an ideal landscape can act as a means of disseminating the particular terms of a nationalist discourse. Johnson (1993) uses the term 'hegemonic landscape' to convey this sense of an imagery of ideal place becoming the heartland of a collective cultural consciousness. As Agnew (1998) observes, however, such places must not be reified, for history does not end with the modernist story of a representative landscape created to secure national identity. Unlike England or Ireland, he points to the failed attempts to delineate an Italian landscape ideal, partly because Italy has lacked the dominant heroic event on which landscape ideals are based:

> In England the shock of industrialisation produced a romantic attachment to a rural/pastoral ideal that has outlasted the original historical context. For the United States, the myth of the frontier and the subjugation of 'wilderness' has likewise served to focus national identity around themes of survival, cornucopia and escape from the confines of city life. (Agnew, 1998: 232)

But the very survival of Italy intimates the existence of another dimension to national identity, one which can include other attributes of citizenship and territory that are not necessarily congruent with ethnic markers. In this civic manifestation, members of the nation self-identify with common institutions, a single code of rights and a well demarcated and bounded territory. Membership is by residence rather than blood and is legally inclusive (if less so in reality) – at least of those who live within the national territory. Thus the 'American people' is a multi-ethnic cultural mosaic, but one drawn together by the country's institutional framework, the basis of its national identity (Smith, 1991). Although notions of Western superiority should be resisted, it does seem that the contemporary circumstances of the former USSR and Balkans are best described by the ethnic model of nationalism. While still potent elsewhere, as the enduring agony of Northern Ireland testifies, in general terms, identity politics in the West are more effectively defined by the precepts of civic nationalism. As Shaw (1998: 136) warns, in reality, however, 'the two forms of nationalism are by no means mutually exclusive'.

Nationalism and the territorial state

Whether civic or ethnic (or simultaneously both), nationalisms are largely vested within particular constructions of the past, dependent on the invention of tradition and, as such, constructed within and not outside discourses

(Hall, 1996). They represent a particular conceptualisation of power, which embodies issues of legitimation and validation – themselves quintessential modernist constructs. Identity politics are thus central to the distinction between the exercise of power as authority or coercion. In the former circumstances, the source of power is accepted as legitimate by a population, which sees itself as bonded by common interests. Conversely, coercion defines the situation in which the exercise of power, widely regarded as being illegitimate, depends on armed force and violence. Not only is this expensive of resources but states that rely on coercion tend to be short-lived. The legitimacy accorded the state thus depends on the ways in which a nationalism creates:

- representations of homogeneous place in which tropes of similarity are sufficient to subsume the diversity of place and people;
- a spatial synonymy between culture, political allegiance and citizenship;
- a definition of identity as a collective memory, often evocative of patriotism and framed in terms of resistance and Othering.

As we have seen, most commentators accept the concept of a discourse of identity linking nation, territorial state and people as a distinctly modern post-Enlightenment phenomenon, one which informed the construction of Europe in the late eighteenth and nineteenth centuries:

> Civic nationalism is regarded by many scholars as having derived from the rationalism of the eighteenth-century Enlightenment, whereas ethnic nationalism emerged from organic notions of identity associated with the Romantic movement. (Shaw, 1998: 136)

Hobsbawm (1990), for example, regards nationalism as pre-eminently the product of triumphant bourgeois liberalism in the period *c.* 1830–1880, although other commentators place its origins rather earlier. Colley (1992), for example, locates the making of a Britishness vested in recurrent Protestant wars, commercial success and imperial conquest to the eighteenth century, while Hastings (1997) argues for a medieval origin of both nationalism and nations. It seems reasonable to generalise, however, that nationalism received its biggest boost from the French Revolution and was largely crystallised by the social and political changes of nineteenth- and early twentieth-century Europe. As Davies (1996: 821) argues, irrespective of the precise relationship between nationalism and modernisation, 'it is indisputable that the modernizing process expanded the role of nationalism beyond all previous limits'. Factors involved included: persistent warfare; the modernisation of economies; the slow fusion of local and urban loyalties into unified states in Germany and Italy; and the extension of European empires abroad. While the ethnic and civic dimensions of these nineteenth-century European nationalisms evolved many different trajectories, they shared in the mutual assumption of modernity that all people of 'a similar ethical rationality might agree on a system of norms to guide the operation of society' (Peet, 1998: 14). Hence, nationalist discourses are sited within that sense of limitless, self-directed progress, which, in turn, demanded the creation of modernistic, progressive, linear narratives and the assumptions of long-term continuities of culture, place and allegiance. These were constructed to lead directly to the contemporary nexus of power, providing the precedents and traditions, which underpin the legitimacy of that authority, but simultaneously underpinning what Gilroy (1992) sees as the fatal junction of concepts of nationality and culture in the intellectual history of the West since the Enlightenment.

Whatever the precise chronology, it is readily apparent that the evolution of the territorially defined state – which began in Europe in the Middle Ages (Bartlett, 1993; Davies, 1996; Graham, 1998a) – long predates nationalism and the nation-state. State formation in Europe was a long-term, volatile and contingent process. The earliest states began to evolve through the medieval accretion of peripheries to urban-based core regions. The first centralised monarchy – based in London – emerged in twelfth-century England and by the start of the fourteenth century, England was the most modern state in the Christian West. Conversely, the unification of France, for example, required centuries (de Planhol, 1994). The Hundred Years War, which began in 1337 and lasted until the 1430s, and finally ousted the English from their possessions in western France, was not sufficient to unite the country. The present territorial limits of the country – the 'hexagon' – were achieved only by a slow process of accretion and virtually endemic warfare, and were not finalised until the accession of Nice in 1860. Even then, Alsace-Lorraine was annexed by Germany in 1871, not to be returned until after the First World War.

If the formation of many European states was generally a long-term process, each involving the fusion of several disparate territories, it has also been an unstable one. Territorial boundaries have proved to be as contingent as the identities they contain. Germany, for example, finally unified only in 1871, has altered dramatically in extent several times during the intervening period (Figure 3.1). After the First World War, it lost territory to the newly established Polish state, only to reclaim this (and more) following Hitler's invasion of Czechoslovakia and Poland in 1938–9. Following the Second World War, a much shrunken Germany was divided by the victors – the United States, Britain, France and the USSR – between the capitalist Federal Republic (West) and the Communist German Democratic Republic (East). The country's present boundaries were established as recently as October 1990 when West and East were reunited. Each ideological – and spatial – repositioning of the German state has demanded an enduring and radical revision of national narratives; as Tunbridge (1998: 250) observes:

> A ninety-year-old Leipziger (in 1998) will remember: the authoritarian Second Empire of Kaiser Wilhelm II; the first democratic Weimar Republic; the totalitarian Nazism of Hitler's Third Reich; the state socialism of the GDR; and the contemporary democratic Bundesrepublik.

As the early modern state evolved through these complex and often transient processes of fusion, the efficacy of centralised government depended on increasing levels of interaction and contact, the function of the territorial state being to impose centripetal forces on diversity and heterogeneity. This involved mechanisms as diverse as the dissemination of printed literature through

Figure 3.1 Salient changes in the boundaries of Germany, 1871–1990 (after Jess and Massey, 1995).

which the state could seek to influence its population, the evolution of taxation systems, and the mapping and naming the state's territory. Withers (1995: 392), for example, sees the 'framing of outer boundaries – putting physical limits to the sense of the nation', as a form of knowing crucial to the emergence of Scottish identity in the eighteenth century. Again, as Brayshay *et al.* (1998: 284) demonstrate, communications were important in that the creation of a Royal post in early modern England, a system designed 'to ensure the hegemony and articulation of the sovereign state . . . also enhanced the creation of a unified national consciousness'.

More widely, during these prolonged and shifting processes of European state formation, ethnically defined states have been imposed on a cultural mosaic, the complexity of which has been further exacerbated by the forced migrations caused by warfare. After the victory of the Red Army in Eastern Europe in 1945, for example, as many as twelve million Germans were forced to migrate west, only to be succeeded by Poles, in turn ejected from their eastern territories lost to the Soviet Union (Tunbridge, 1998). Such human costs reflected the reality that because of earlier migrations – either voluntary or forced – and wars, the construction of ethnically defined states in Europe could be no more than an approximate process. Nevertheless, the Treaty of Versailles of 1919 created a number of these polities in Europe, which, in the 1930s, proved incapable of defending themselves and were easily swept aside by Nazi Germany (Shaw, 1998). In numerous cases, these countries had to be substantially recast following the Second World War (Figure 3.2). In the Balkans, the ethnic mosaic was so complex as to prevent the creation of stable national states (Figure 3.3). The malevolent legacies of ethnic territoriality have been manifest. Boundaries failed in defining homogenous ethnic homelands within the mosaic. Numerous minorities were left on the wrong sides of boundaries and inevitably became – and may still remain – targets for persecution and sources of actual or potential conflict between countries. Most horrifically, the major European peoples not nationally defined, the Jews and Romanies, became the targets of Nazi genocides. By 1945, Hobsbawm claims that 'the homogeneous territorial nation' had become 'a programme that could be realised only by barbarians, or at least by barbarian means' (1990: 134).

If that is so, the barbarity has persisted to the present day as the geopolitical map of Eastern Europe is transformed in the wake of the collapse of the Soviet Empire. Pilkington (1998) estimates that around 25 million ethnic Russians have been displaced since 1991, around three million returning 'home' either voluntarily or as a result of force. Vicious conflicts, fuelled by nationalist and ethnic rivalries, as well as religion, have occurred in a succession of former Soviet republics. The wars in the former federal republic of Yugoslavia between 1991–5, reflected the worst excesses of ethnic nationalism (Figure 3.3). Ó Tuathail (1996: 219) describes the Bosnian War in particular as 'an irreducibly modern war over space, territory and identity', characterised by 'brutal and criminal campaigns of "ethnic cleansing"'. Serbian forces 'pushing for lebensraum and an ethnically pure state', were operationalising 'a fascistic form of modernity' and exhibiting 'a deadly intolerance' of the supposedly ambivalent allegiances

Figure 3.2 The redrawing of the political geography of North-central and North-eastern Europe during the later twentieth century (after Tunbridge, 1998).

of the postmodern era. The process has continued in the province of Kosovo, when in 1998, the Kosovars (predominantly ethnic Albanians) launched their own war of liberation against the Serbs. In early 1999, the Kosovo Liberation Army was still fighting Serbian forces engaged in a scorched-earth policy, aimed at terrorising the Kosovars into seeking refuge across the border in Albania and Macedonia. Meanwhile, military action by Western powers was being compromised by their unwillingness to commit ground troops in addition to mounting a sustained aerial attack on Serbia and Serbian targets in Kosovo.

Differential experiences

The dreadful fate of Yugoslavia, and the continuing suffering of its inhabitants, provides one notable example of the enduring strength of ethno-nationalist narratives of identity. In depicting a Bosnia 'understandable within the general history and practice of genocide in the twentieth century', Ó Tuathail (1996:

Figure 3.3 The constituent states of the former Federal Republic of Yugoslavia.

221–2) argues that 'incomplete' and 'questionable' as this reading may be, the agencies and governments of the West became the 'accomplices to genocide' because they would not take 'an unconditional moral stance in the war'. To deny the victory of the policy of 'ethnic cleansing' – the ultimate uncondonable outcome of identity politics – is to demonstrate the poverty of our own indifferent and postmoral society.

As observed in the introduction to the chapter, the enduring realities of zero-sum geopolitics, which depend on the gun and the targeting of ethnic minorities, suggests that allegories of identity constructed around place retain a profound importance. But clearly identity politics do not necessarily lead to

Bosnia or Rwanda. Instead, we must envisage a more nuanced interpretation in which an array of outcomes is possible. It is in this sense, rather than prescribing exaggerated denials of the relevance of place-oriented identity politics, that Hall's conceptualisation (1996: 4) of 'multiply constructed' . . . often intersecting and antagonistic discourses, practices and positions' carries most weight.

In his comparative global analysis of national identity, Dijkink (1996) found nothing to suggest a willingness to do away with national values or aims; indeed, the converse held in countries as diverse as Russia, Germany and Australia. It may be, he argues, that a belief in alternative universal or placeless values may not automatically serve the national interest. Conversely, the United States claims the universality of certain values, most notably free trade and market-driven economies, precisely because these are held to be congruent with its own national interests. If these factors still dictate contemporary geopolitics and power relationships, it again seems sanguine to dismiss nationalist territorial tropes of identity. Moreover, such are the ambiguities of nationalist narratives that they can either be geographically integrating or disintegrating. Their rhetoric can be used to claim unity across separate states but, within any one state, can also be directed at claiming independence for that territory's constituent regions (Agnew, 1997).

Paying particular attention to the ways in which constructions of the past underpin these identities, I now turn to Ireland, France and Spain as three examples of these differential outcomes. Western Europe seems to be the key region in any argument that nationalism and other place-oriented tropes of belonging are withering before the processes of 'dissemination'. The place allegories that describe these three countries allows us to explore something of the multiplicity of outcomes that attends the interconnections between spatiality and identity. Ireland is chosen as a small and apparently homogeneous entity (Northern Ireland excepted), but one which was also a colony within Europe. France, by contrast, is a large and diverse country, which, nevertheless, claims to possess a distinct and coherent national identity. This is in marked contradistinction to Spain, which, although sharing France's size and heterogeneity, is contested by a variety of 'sub-' or regional nationalisms.

Ireland

As we have seen, national identity is created in particular social, historical and political contexts and, as such, is a situated, socially constructed narrative capable of being read in conflicting ways at any one time and of being transformed through time. Irish nationalism provides a particularly good example of the ways in which particular imagined places, firmly fixed in time, are invented as the repository of collective cultural consciousness (Graham, 1997a). As Johnson (1993) argues, the hegemonic image of the west of Ireland as the cultural heartland of the country was an essential component of the late nineteenth-century construction of a discourse which, in its dependence on a Gaelic iconography, was to prove exclusive rather than inclusive, particularly when its representations became fused with Catholicism. Strongly reinforced

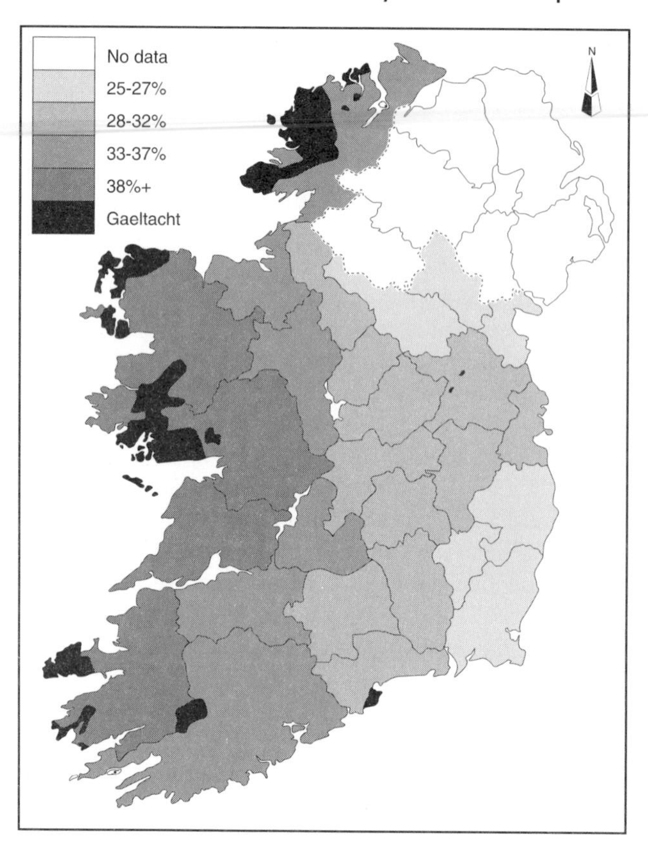

Figure 3.4 The percentage of Irish-speakers in 1981 and the location of the Gaeltachts (after Johnson, 1997).

by the intellectual élite of early twentieth-century Ireland, the 'west' became an idealised landscape, populated by an idealised people who invoked the representative, exclusive essence of the nation through their Otherness from Britain. The invented, manipulated geography of the West portrayed the unspoilt beauty of landscapes, where the influences of modernity were at their weakest and which evoked the mystic unity of Ireland prior to the chaos of conquest (Nash, 1993).

> The West was represented as containing the soul of Ireland – in [W. B.] Yeats's construction, a fairyland of mist, magic and legend, a repository of Celtic consciousness . . . Contrasting with the romantic idealisations . . . was the reality of emigration – as the myth of the West was being constructed, its population was leaving. (Duffy, 1997: 67 and 69)

Nevertheless, it became part of the ethos of the new Irish state that the Gaeltacht (Irish-speaking) regions of the West were 'the archive' of Irish identity (Johnson, 1997: 174) (Figure 3.4).

There is little exceptional in conceptual terms about the construction of Irish nationalism. It evokes politics of exclusion that in turn render the nationalist discourse incapable of assimilating change and resolving the conflicts engendered by exclusion. The creation of the symbolic premodern discourse of traditional Irish-Ireland, and its ultimate transformation into a construction of Irishness that was defined by Gaelicism and Catholicism, was 'a supreme imaginative achievement' that only began to dissolve in the 1960s (Lee, 1989: 653). By then, it bore only a 'tenuous relation to reality' in the Republic of Ireland, while it never had accommodated the Protestant, industrialised counties of north-east Ireland. On one hand, therefore, the invented geography of Irish-Ireland paralleled other dimensions of nationalism to create an Irishness that empowered and legitimised the new state. It was a powerful and exclusive ideology that – particularly through its Catholic ethos – imposed a startling degree of manipulated cultural homogeneity upon Ireland and became the acceptable diasporic imagery of Irishness. For the unionists of Northern Ireland, conversely, the whole structure of Irish-Ireland became such a powerful expression of Otherness that it was almost sufficient in itself to define 'Ulster' identity. The unionist administration never significantly addressed the cultural vacuum left by partition, preferring instead poorly articulated and even less clearly understood protestations of Britishness (Graham, 1998b). Ireland became divided less by the border than through the juxtaposition of an increasingly confident Irish identity and a confused and heavily qualified sense of Britishness in the north. It was the former which claimed the moral high-ground of legitimacy.

Lee (1989: 653) claims that no alternative self-portrait has yet emerged 'to command comparable conviction' with Irish-Ireland. A modernising, increasingly secular Irish state now looks beyond the Otherness of Britain toward inclusion within an ever more integrated Europe. The conflict in Northern Ireland, however, has been firmly fixed by the polarities of the historical nationalist discourse, ensuring that the successful outcome of peace politics demands further renegotiation of identities in both parts of the island. Attempts to transform the cultural closure of Irish-Ireland into something more congruent with these contemporary needs for non-exclusivist, outward-oriented, open-ended tropes of identity have engendered vitriolic debate, especially among historians and cultural theorists. Although a movement embracing many varieties, the revisionist perspective of Irish history embodies a common stress on the plurality, discontinuity and ambiguity of the Irish past, the antithesis to the narratives of manipulated homogeneity, which the monolithic Gaelic, rural and, later, Catholic representation of Irish-Ireland imposed on that diversity in the later nineteenth century. Revisionism warns against this retrojection of modern nationalism into the distant past and portrays instead a nation constructed from diverse and often contradictory elements (Boyce and O'Day, 1996). The need to take a less Anglocentric view of the past is also seen as being an important part of the revisionist agenda. Its critics, however, accuse revisionists of seeking to rehabilitate the British presence in Ireland, Deane (1996), for example, contending that revisionism's role is to legitimise partition, its proponents being no more than apologists for British misrule and oppression.

Gibbons (1996) seeks to reconcile these diverse claims through the concept of post-colonialism, arguing that Ireland – largely white, Anglophone and Westernised – was paradoxically also a colony within Europe. This past necessitates a present conditioned by post-colonial strategies of mixing and defined by notions of hybridity and syncretism rather than the 'obsolete ideas of nation, history or indigenous culture'. He writes:

> . . . there is no possibility of restoring a pristine, pre-colonial identity: the lack of historical closure [enduring partition] . . . is bound up with a similar incompleteness in the culture itself, so that instead of being based on narrow ideals of racial purity and exclusivism, [Irish] identity is open-ended and heterogeneous. But the important point in all of this is that the retention of the residues of conquest does not necessarily mean subscribing to the values which originally governed them. . . . (1996: 179)

Thus, above all, irrespective of the ongoing debate as to the future terms of Irishness, the historiography of Irish nationalism demonstrates the situated and contingent nature of the discourse, one that is framed in particular circumstances for specific purposes, but one that must be renegotiated and transformed as the nature of its society changes. It is the failure to accept the inevitability of this process, which produced Ireland's recent violence. Nationalism and ethnicity are not givens; particular constructions of both are situated in specific epochs and particular narratives of place.

France

As observed above, the impression that France is somehow a 'natural' state, defined by 'natural' boundaries on the Rhine, Alps and Pyrénées, is a fallacy, the country having grown only slowly, spasmodically, and largely through war, to attain those frontiers. The defining feature of that territory is its heterogeneity, so much so, indeed, that in *Tableau de la géographie de la France* (1903), Paul Vidal de la Blache argued that it was a diversity – rooted in the physical environment – which paradoxically provides France with its identity. The country's unity evolves from a beneficent force of commonality, derived from the ways in which civic life combines with – and transcends – this diversity.

> To be French was experienced at two levels: one was a member of a local or regional community and, because of that, was integrated into the national whole. Diversity appeared as a necessary component of . . . identity and was thus accepted and valued. (Claval, 1994: 50)

This relationship was encapsulated in the idea of the *pays* (literally an area with its own identity derived, not only from divisions of physical geography, but also from ethnic and linguistic divisions imposed on a region by its history) as the geographical mediation of synthesis and continuity, the product of human interaction with the environment over many centuries. In his sweeping investigation of the identity of France, Braudel (1988: 41) takes up the same

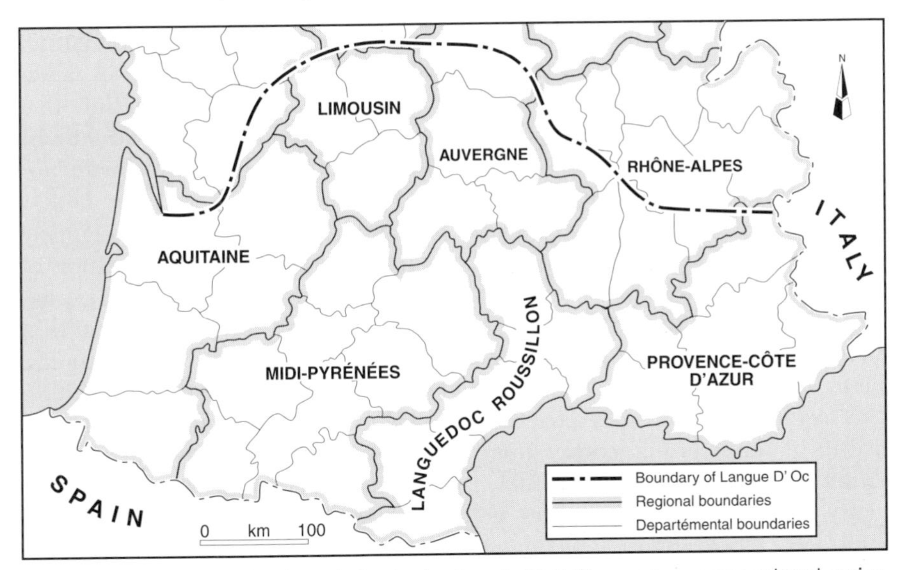

Figure 3.5 Administrative boundaries in the French Midi. The contemporary cultural region of Languedoc embraces not only Languedoc-Roussillon but also parts of Midi-Pyrénées.

theme. He sees a 'patchwork' country, a jigsaw of regions and *pays* in which 'the vital thing for every community is to avoid being confused with the next tiny "*patrie*", to remain *other*'.

How, then, was the unity of France invented to reconcile this geographical mosaic? This can be understood only if it is recalled that territorial identities are not absolute and never have been. Rather, they are layered one upon the other, frequently so in antagonistic discourses, each stratum defined by different sets of criteria. Hence the fierce localisation of the French *pays* and its administrative equivalent, the *commune*, is overlain by an intensely geographical affiliation with the *départements*, superimposed onto older alignments after the French Revolution but redolent of localised identities because they were generally allocated the names of rivers and mountains. Subsequently, the *départements* have been amalgamated into regions such as Languedoc-Roussillon or Midi-Pyrénées (Figure 3.5). That such scales of identity are defined largely by administrative and financial functions serves only to emphasise the enduring importance of place-oriented politics. Beyond that again is the national scale of identity, which in France essentially takes a civic form. 'French nationality was French citizenship: ethnicity, history, the language or patois spoken at home, were irrelevant to the definition of the "nation"' (Hobsbawm, 1990: 88). Democratisation, the 'turning of subjects into citizens', was crucial to this process and the engendering of a patriotism that shades into nationalism. This legacy of the Revolution has much in common with the similarly civic nationalism of the United States.

But this multiple layering of identity does not in itself summarise the full complexity of the situation. It is readily apparent that modernist national identity is being eroded in many countries by forces of globalisation and localisation, themselves not readily reconcilable. Global 'time–space compression' is radically reconstructing our views of space, and leading to an accelerated unbundling of territorial sovereignty, with the growth of common markets and various transnational functional regimes and political communities, which are not delimited primarily in territorial terms (Anderson, 1996).

Although the strategies of European integration were not created originally in opposition to the nation-state, but as part of the post-war rehabilitation of a number of those polities (Bideleux, 1996), the European Union (EU) has been characterised as the world's first truly postmodern international political entity, distinct from the national and federal state forms of the modern era, but in some respects reminiscent of pre-modern territorialities (Anderson, 1996). This hypothesis, sometimes referred to as 'new medievalism', speculates that the growth of transnational corporations and networks, combined with sub-state nationalist and regionalist pressures, is producing overlapping forms of sovereignty analogous to the complex political arrangements of medieval Europe (Bull, 1977). Sovereignty is again ceasing to be a state monopoly. This erosion of the 'national' is by no means restricted to Europe. Dijkink (1996) found that Indians, for example, increasingly lack a clear conception of the role of their country in the modern world. The rhetoric of a nationalism located in colonial history and the struggle for independence is now an old and inadequate story, the ensuing vacuum of cultural identity being occupied in this instance by more assertive religions.

The diffusion of power from the nation-state as embodied in the EU is seemingly compounded by the resurgence of sub-state expressions of identity, often forcibly expressed by narratives which recycle the familiar rhetoric of nationalism at a regional scale. As evidence of 'Europe of the Regions', Harvie (1994), for example, points to the emergence of aggressive, urban-based and affluent regions, including Lombardy, Baden-Württemberg and Rhône-Alps. He argues, however, that their distinctive consciousness largely lies in a shared affluence rather than being vested in any cultural realm. For example, Lombardy's material wealth was reflected in the emergence during the 1980s of a political movement – the Northern League (*Lega Lombarda*) – which portrays itself as a new clean start for Italy, based on northern productivity that contrasts markedly with the corruption of a southern Rome-based bureaucracy. Agnew (1995) argues, however, that while the League espouses a rhetoric of regionalism, it is still dragged into the axis of the Italian state as a whole because decision making remains concentrated at that level.

Returning to the example of France, it is readily apparent, however, that this dimension to belonging is itself a very ambiguous and fractured form of identity. The south of the country – the Midi – is partly defined by the linguistic divide between Teutonic northern France (*lange d'oil*) and the Latin, Mediterranean south (*langue d'oc*). But the Midi itself is divided into a number of cultural regions with their own distinct dialects. To the west of the Rhône, Occitan, for example, is one of the factors that distinguishes Languedoc-Roussillon from Provence to the east, two discrete regions with very different histories and contrasting trajectories of belonging (Figure 3.5). Languedoc-Roussillon, however, cannot itself claim to be a coherent cultural entity, its southernmost portion asserting a much closer trans-Pyrénéan affinity – based on a common historical narrative – with Catalonia than with any part of France at all. In turn, the identity imagery of contemporary Languedoc is heavily informed by depictions of the savage oppression suffered by the region in the early thirteenth century, when Capetian France, centred on the Île-de-France around Paris, allied with the Papacy to launch the brutal Albigensian Crusade against the Cathar Church. Ostensibly justified by the need to extirpate heresy, the underlying – and ultimately successful – political agenda sought to secure the submission of the Count of Toulouse to the French crown and more fully integrate his territories within its domain. Today, the Cathars, and their fortress eyries in the Pyrénées, have become symbols of a reborn regional consciousness that seeks to distance the Languedoc from Parisian ascendancy over the terms of Frenchness (Figure 3.6).

At the same time, however, despite these complex manifestations of a fragmentation of place-centred identity, it would be foolish to underestimate the enduring importance of that Frenchness as a central motif of the country's identity. France is divided still by the vitriolic debate on precisely who qualifies for membership. The contemporary legacy of Maurras's *L'Action Française* is Jean-Marie Le Pen's *Front Nationale* and its blatant recourse to anti-immigrant rhetoric. In particular, Le Pen exploits the presence in metropolitan France of peoples of North African descent (largely Muslim). On one hand, they can be seen as symbolic of a new multicultural France that negates the 'out of many one' conceptualisation of nationalism, but does not deny the identity of, and allegiance to, the French state (for which they died in large numbers, particularly in the First World War as the Islamic gravestones in French military cemeteries attest). It does well to remember that multiculturalism – itself a invented narrative – is commonly articulated within the national context. (This was precisely what occurred when France won the 1998 Soccer World Cup with a team containing many immigrants and a play-maker, Zinedine Zidane, who is of Algerian descent.) Conversely, for Le Pen, immigrants from Francophone North (and West) Africa stand as the Other within, an affront to a true Frenchness still essentially racist in its definition. In sum, the complexities of territorial identity in France are not amenable to explanation through the medium of nationalism alone. Place is still fundamental to belonging, but this occurs today – as indeed it always has – in a complex and nuanced fashion. While the terms of the interconnections between spatiality and identity have changed, that does not render allegories of place irrelevant, nor modernist nationalism redundant.

Spain

It may well be the case that both Ireland (bar Northern Ireland) and France achieved legitimacy because processes of democratisation and national identity were spatially congruent with the extent of the state. Numerous nationalisms,

Figure 3.6 The Cathar fortress of Montségur, Ariége. This has become the pivotal focus of the resurgence of interest in the Cathars and a symbol of regional identity. After a nine-month siege in 1243–4, the castle fell to the forces of France and the Church. More than 200 Cathars were burnt alive in the ensuing *auto-da-fé*.

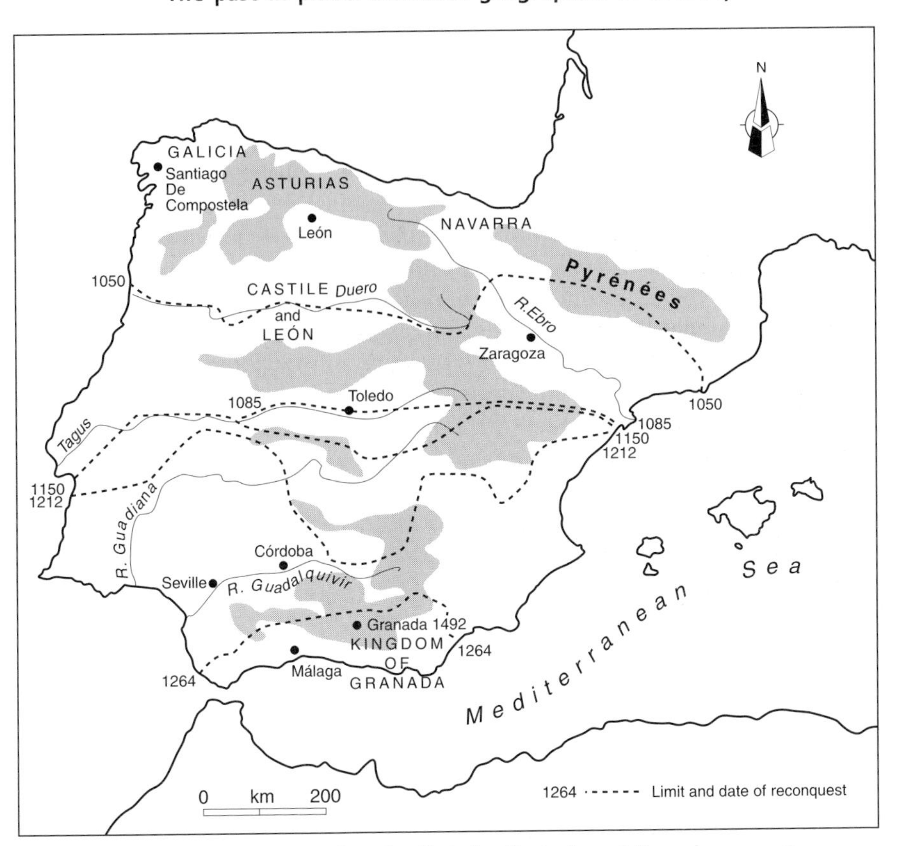

Figure 3.7 The traditional narrative of a Christian Spain forged through conquest: the progress of the *Reconquista*, c. 1000–1492 (after Holmes, 1988).

however, have been independent of states and thus inevitably a challenge to them. The Jewish nation in Europe had no state identity, which compounded its exposure, not only to Nazism, but also to the enduring anti-Semitism that continues to inform, for example, Polish nationalism. Less brutally but still violently, Spain provides an example of a state contested by several nationalisms and an array of regional identities. Further, its foundation myth also encapsulates a wider arena of conflict and one that is apparently gaining wider currency today – the age-old struggle between the West and Islam.

Traditional nationalist narratives of Spain depict a Christian country forged from Holy War against the Islamic Moors who had invaded Iberia from North Africa during the eighth century and soon conquered virtually all the peninsula (Graham, 1997b) (Figure 3.7). From their surviving footholds along Spain's northern coast, slowly, the forces of Christianity began the process of reconquest (the *Reconquista*) that was to occupy the ensuing seven centuries. By 1085, it had reached the Tagus where the fall of Toledo proved a pivotal defeat for the Muslims. The reconquest of the subsequent four centuries was a spasmodic process, completed only in 1492 when the Kingdom of Granada fell to the 'Catholic monarchs', Ferdinand (of Castile) and Isabella (of Aragon). In the same year, Christopher Columbus sailed west to begin Spain's imperial ventures in the Americas. That these linked symbolic events created a unified Spanish polity in the fifteenth century is, however, largely the perspective of a more recent early modern nationalism. Ferdinand and Isabella and their successors held enormous power but it did not lead to political integration. The constituent kingdoms of Spain preserved their particular institutional structures until the eighteenth century when a centralised monarchy, based in Madrid, was able finally to define the country as it now exists (Fernández Albaladejo, 1994). The imagery of a state, forged from the north by five centuries of holy war against the infidel was fundamental to underpinning Castilian hegemony (Graham and Murray, 1997). In the twentieth century, this orthodox construction of Spanish nationalism, accompanied by much imperial verbiage invoking the memory of Ferdinand and Isabella, was employed to reinforce the legitimacy of the Falangist (Fascist) leader, General Franco, in his guise as the heroic leader of a crusade to liberate Spain from the godless Communist hordes of Moscow (Preston, 1993).

Nevertheless, Spain has rarely been a convincing nation-state. Regional identities have always been strong, not least in the Basque Country (Euskadi) and Catalonia. Even under the 40 years of Fascist control that followed the defeat of the Spanish Republic in the Civil War of 1936–9, the devolutionist instinct was apparent. Since Spain's return to democracy, it has appeared to be among the most convincing illustrations of the concept of 'Europe of the Regions'. Ostensibly Catalonia and Euskadi offer more secure evidence than the examples cited above of the feasibility of 'region-states' – and also of the ubiquity of nationalist rhetoric, whatever the scale of the polity. In the former, a broad nationalist consensus was established around a few – largely cultural – central themes, most notably the Catalan language, while, in Euskadi, internal cultural, ideological and political fragmentation led to radicals resorting to confrontation with the Spanish state, the repression of which was required, in turn, for violence to begin and spread (Conversi, 1997). Euskadi has undergone a 'war of liberation' waged by ETA, which provides the closest parallel in Europe to PIRA. Catalan nationalism has remained moderate and broadly united around its cultural platform, while clearly Catalans also continue to profess a distinct empathy with Spanish tropes of identity. Above all, their nationalism is not separatist, nor is it violent, the principal objective being to renegotiate the region's relationship with the Castilian centre in Madrid. In marked contrast, Conversi argues that the fragmentation of the Basque nationalist movement from its earliest origins made violence inevitable. In the absence of: a coherent and unifying nationalist ideology, sufficiently powerful in cultural terms to transcend the effects of urbanisation; the Castilianisation of the Basque élite; and the effects of immigration; Euskadi's cultural space became plural and fragmented. Language is equally important to Basque nationalism but whereas Euskara – one of Europe's notably remote and enigmatic tongues – is in decline, Catalan is among the most dynamic regional languages in Europe, heady symbol of Catalonia's cultural regeneration. In that failure of the Basque nationalist movement to devise a similar programme of nationalist regeneration lies the propensity to violence expressed in ETA's action–repression–action theory, the Spanish state being provoked into the oppression and violence that helped cement a common identity out of chaos.

Although the armed struggle waged by ETA in Euskadi is clearly aimed at secession from Spain, in neither region does a majority support independence, even within the EU. Catalans and Basques alike apparently wish to retain their distinct cultural identities and regional economic prosperity, but within a highly decentralised Spain. The survival of even this most fissiparous nation-state illustrates the enduring relevance of this level of governance, but the emergence of region-states in Spain (and less convincingly elsewhere in Europe) also points to the continuing importance of constructions of place in identity politics.

Conclusions

The differentiated geographies of nationalism in Ireland, France and Spain demonstrate how such discourses are made in the relationships between places and across spaces. All three examples show too the contingent and situated nature of nationalism, vested in particular readings of the past and tropes of exclusivity, but inevitably being subjected to processes of continuous renegotiation to face the demands of changing circumstances. Any assessment of the continuing importance of nationalist tropes of identity is bound up with the parallel debate on the importance of the territorial state, drawing its legitimacy from linear narratives that connect representations of the past to the present nexus of power.

Ostensibly, the previous sentence is redolent of the assumptions of modernity that surround the entire construction of the nation-state. None the less, one theme of this chapter has been that the state, largely defined by historical narratives, continues to exert its influence, albeit heavily compromised, as a primary focus of political decision making on a global basis. A second has been that despite the apparent postmodern fragmentation of identity, discourses of belonging constructed around place remain important. However, instead of being organised primarily around geographies of nationalism, these have become much more complex.

Anderson (1995), for one, considers reports of the death of the nation-state to be exaggerated. First, even in the context of the EU, the unbundling of territoriality and sovereignty is limited and partial, affecting different state activities very unevenly. The politics of economic development is the sphere where state power has been most affected by globalisation and, clearly, territoriality is becoming less important in some fields (for example, financial markets). Nevertheless, the state remains as the principal spatial framework for many aspects of social, cultural and indeed political life, not least because of the enduring power of language differences which help construct and maintain identities. Secondly, Anderson argues, the EU, although a new political form, is itself territorial, and in many respects traditional conceptions of sovereignty remain dominant, whether exercised by Member States or by the EU as a whole. Very belatedly, the European Commission has recognised the cultural vacuum at the heart of a European project driven by economic integration. The 1997 Treaty of Amsterdam called upon the EU to deepen the solidarity between its peoples, while respecting their separate histories, cultures and traditions. Empty rhetoric this may be but the serious point is that the EU requires its own narrative of place, perhaps because the ostensibly modernist notion of legitimacy remains as valid as territorial politics (Graham and Hart, 1999).

Thus it may well be the case that the renegotiation of the political control of space in contemporary Europe is concerned with ongoing *re*territorialisation rather than the *de*territorialisation implied by Hall, Bhabha and other cultural theorists. Historical geographies of identity are implicated in these processes, reterritorialisation requiring new or adapted representations of place. The most challenging geographical manifestation of these concerns is to be found in Soja's adaptation (1996) of Bhaba's concept of Thirdspace. As Peet (1998: 224–5) argues, Soja 'wants to set aside either/or choices and contemplate the possibility of a "both/and also" logic which effectively combines postmodernist with modernist perspectives', the aim being to build on 'a Firstspace perspective focused on the "real" material world, and a Secondspace perspective that interprets this reality through imagining representations, to reach a Thirdspace of multiple real-and-imagined places'. The examples of Ireland, France and Spain addressed in this chapter support the contention that the representations of multiculturalism and dissemination will be expressed most coherently within a Thirdspace that remains defined by territorial, political boundaries. Even forces such as militant Islam, which depict themselves as being extraterritorial, are forced into this arena by the very organisation of international geopolitics.

I have sought throughout this chapter to demonstrate how the past is a resource which can be interrogated for different purposes and with varying results in the construction of identities. Representations of the past are fundamental to the situated cultural artefacts which are imagined communities. The argument has also demonstrated the enduring power of spatiality in these constructs. The contemporary stress on the multicultural construction of societies, and the fragmentation of identity along many diverse axes, is in part a reaction to the perceived hegemony of nationalist discourses in the recent past. But identity has never been framed entirely within those bounds. Modernist identity was concerned with more than nationalism but equally, postmodern identity remains interconnected with territorial allegories of belonging that inevitably draw upon allegories of the past. They are more diverse, and more numerous – at least in the West – but they are no less important. In that Thirdspace of multiple real-and-imagined places and their narratives of the past, people still die in the name of identity politics. Hall's claim (1996: 4) that identities are 'multiply constructed' has perhaps always been so, but a multiplicity of places and their meanings remain located amongst his 'intersecting and antagonistic discourses, practices and positions'.

References

Agnew, J. A. (1995) The rhetoric of regionalism: the Northern League in Italian politics, 1983–1994. *Transactions of the Institute of British Geographers*, **20**, 156–72.

Agnew, J. A. (1996) Time into space: the myth of 'backward' Italy in modern Europe. *Time and Society*, **5**, 27–45.

Agnew, J. A. (ed.) (1997) *Political Geography: A Reader*, Arnold, London.

Agnew, J. A. (1998) European landscape and identity. In Graham, B. (ed.) *Modern Europe: Place, Culture and Identity*, Arnold, London, 213–35.

Agnew, J. A. and Corbridge, S. (1995) *Mastering Space: Hegemony, Territory and International Political Economy*, Routledge, London.

Anderson, B. (1991) *Imagined Communities: Reflections on the Origins and Spread of Nationalism* (revised edition), Verso, London.

Anderson, J. (1988) Nationalist ideology and territory. In Johnston, R. J., Knight, D. B. and Kofman, E. (eds) *Nationalism, Self-determination and Political Geography*, Croom Helm, London, 18–39.

Anderson, J. (1995) The exaggerated death of the nation-state. In Anderson, J., Brook, C. and Cochrane, A. (eds) *A Global World? Reordering Political Space*, Open University/Oxford University Press, Oxford, 65–112.

Anderson, J. (1996) The shifting stage of politics: new medieval and postmodern territorialities. *Environment and Planning A: Society and Space*, **14**, 133–53.

Barnes, T. J. and Duncan, J. S. (1992) Introduction: writing worlds. In Barnes, T. J. and Duncan, J. S. (eds) *Writing Worlds: Discourse, Text and Metaphor in the Representation of Landscape*, Routledge, London, 1–17.

Bartlett, R. (1993) *The Making of Europe: Conquest, Colonization and Cultural Change, 950–1350*, Allen Lane, Harmondsworth.

Bhabha, H. K. (1994) *The Location of Culture*, Routledge, London.

Bideleux, R. (1996) Introduction: European integration and disintegration. In Bideleux, R. and Taylor, R. (eds), *European Integration and Disintegration: East and West*, Routledge, London, 1–21.

Boyce, D. G. and O'Day, A. (eds) (1996) *Modern Irish History: Revisionism and the Revisionist Controversy*, London, Routledge.

Braudel, F. (1988) *The Identity of France: Vol. 1: History and Environment*, Collins, London.

Brayshay, M., Harrison, P. and Chalkley, B. (1998) Knowledge, nationhood and governance; the speed of the Royal post in early-modern England. *Journal of Historical Geography*, **24**, 265–88.

Bull, H. (1977) *The Anarchical Society*, Macmillan, London.

Claval, P. (1994) From Michelet to Braudel: personality, identity and organization of France. In Hooson, D. (ed.), *Geography and National Identity*, Blackwell, Oxford, 39–57.

Colley, L. (1992) *Britons: Forging the Nation*, Yale University Press, London.

Conversi, D. (1997) *The Basques, the Catalans and Spain: Alternative Routes to Nationalist Mobilisation*, University of Nevada Press, Reno.

Daniels, S. (1993) *Fields of Vision: Landscape Imagery and National Identity in England and the United States*, Polity Press, Cambridge.

Davies, N. (1996) *Europe: A History*, Oxford University Press, Oxford.

Deane, S. (1996) *Strange Country: Modernity and Nationhood in Irish Writing Since 1790*, Clarendon Press, Oxford.

Dijkink, G. (1996) *National Identity and Geopolitical Visions: Maps of Pride and Pain*, Routledge, London.

Douglas, N. (1997) Political structures, social interaction and identity changes in Northern Ireland. In Graham, B. (ed.), *In Search of Ireland: A Cultural Geography*, Routledge, London, 151–73.

Duffy, P. J. (1997) Writing Ireland: literature and art in the representation of Irish place. In Graham, B. (ed.), *In Search of Ireland: A Cultural Geography*, Routledge, London, 64–83.

Fernández Albaladejo, P. (1994) Cities and the state in Spain. In Tilly, C. and Blockmans, W. P. (eds), *Cities and the Rise of States in Europe, A.D. 1000 to 1800*, Westview Press, Boulder, Colo., 168–83.

Gellner, E. (1997) *Nationalism*, Weidenfeld and Nicolson, London.

Gibbons, L. (1996) *Transformations in Irish Culture*, Cork University Press/Field Day, Cork.

Gillis, J. R. (1994) Memory and identity: the history of a relationship. In Gillis, J. R. (ed.), *Commemorations: The Politics of National Identity*, Princeton University Press, Princeton, N.J., 3–24.

Gilroy, P. (1993) *The Black Atlantic: Modernity and Double Consciousness*, Verso, London.

Graham, B. (1994) No place of the mind: contested Protestant representations of Ulster. *Ecumene*, **1**, 257–281.

Graham, B. (ed.) (1997a) *In Search of Ireland: A Cultural Geography*, Routledge, London.

Graham, B. (1997b) The Mediterranean in the medieval and renaissance world. In King, R., Proudfoot, L. J. and Smith, B. (eds), *The Mediterranean: Economy and Society*, Arnold, London, 75–93.

Graham, B. (1998a) The past in Europe's present: diversity, identity and the construction of place. In Graham, B. (ed.) *Modern Europe: Place, Culture and Identity*, Arnold, London, 19–49.

Graham, B. (1998b) Contested images of place among Protestants in Northern Ireland. *Political Geography*, **17**, 129–44.

Graham, B. and Murray, M. (1997) The spiritual and the profane: the pilgrimage to Santiago de Compostela. *Ecumene*, **4**, 389–409.

Graham, B. and Shirlow, P. (1998) An elusive agenda: the devlopment of a middle ground in Northern Ireland. *Area*, **30**, 245–54.

Graham, B. and Hart, M. (1999) Cohesion and diversity in the European Union: irreconcilable forces. *Regional Studies*, **33**, in press.

Grosby, S. (1995) Territoriality: the transcendental, primordial feature of modern societies. *Nations and Nationalism*, **1**, 143–62.

Guibernau, M. (1996) *Nationalisms: The Nation State and Nationalism in the Twentieth Century*, Polity Press, Oxford.

Hall, S. (1996) Introduction: who needs 'identity'? In Hall, S. and Du Gay, P. (eds) *Questions of Cultural Identity*, Sage, London, 1–17.

Handler, R. (1994) Is 'identity' a useful cross-cultural concept? In Gillis, J. R. (ed.) *Commemorations: The Politics of National Identity*, Princeton University Press, Princeton, N.J., 27–40.

Harvey, D. (1989) *The Condition of Postmodernity*, Blackwell, Oxford.

Harvie, C. (1994) *The Rise of Regional Europe*, Routledge, London.

Hastings, A. (1997) *The Construction of Nationhood: Ethnicity, Religion and Nationalism*, Cambridge University Press, Cambridge.

Heffernan, M. (1998) War and the shaping of Europe. In Graham, B. (ed.) *Modern Europe: Place, Culture and Identity*, Arnold, London, 89–120.

Hobsbawm, E. J. (1990) *Nations and Nationalism Since 1780: Programme, Myth, Reality*, Cambridge University Press, Cambridge.

Holmes, G. (ed.) (1988) *The Oxford Illustrated History of Modern Europe*, Oxford University Press, Oxford.

Jess, P. and Massey, D. (1995) The contestation of place. In Massey, D. and Jess, P. (eds) *A Place in the World? Place, Cultures and Globalization*, Open University/Oxford University Press, Oxford, 133–74.

Johnson, N. C. (1993) Building a nation: an examination of the Irish Gaeltacht Commission Report of 1926. *Journal of Historical Geography*, **19**, 157–68.

Johnson, N. C. (1997) Making space: Gaeltacht policy and the politics of identity. In Graham, B. (ed.), *In Search of Ireland: A Cultural Geography*, Routledge, London, 174–91.

Johnson, N. C. (1998) Nations and peoples. In Unwin, T. (ed.) *A European Geography*, Longman, Harlow, 85–99.

Lee, J. J. (1989) *Ireland, 1912–1985: Politics and Society*, Cambridge University Press, Cambridge.

Livingstone, D. N. (1992) *The Geographical Tradition: Episodes in the History of a Contested Enterprise*, Blackwell, Oxford.

Nash, C. (1993) 'Embodying the nation': the west of Ireland landscape and Irish identity. In O'Connor, B. and Cronin, M. (eds) *Tourism in Ireland: A Critical Analysis*. Cork University Press, Cork, 86–112.

Nanton, P. (1992) Official statistics and problems of inappropriate ethnic categorisation. *Policy and Politics*, **20**, 277–85.

Ogborn, M. (1998) *Spaces of Modernity: London's Geographies*, 1680–1780, Guilford Press, New York.

Ó Tuathail, G. (1996) *Critical Geopolitics*, Routledge, London.

Parker, A., Russo, M., Sommer, D. and Yaeger, P. (1992) Introduction. In Parker, A., Russo, M., Sommer, D. and Yaeger, P. (eds), *Nationalism and Sexualities*, Routledge, London, 1–18.

Peet, R. (1998) *Modern Geographical Thought*, Blackwell, Oxford.

Pilkington, H. (1998) *Migration, Displacement and Identity in Post-Soviet Russia*, Routledge, London.

Planhol, X. de (1994) *An Historical Geography of France*, Cambridge University Press, Cambridge.

Poole, M. A. (1997) In search of ethnicity in Ireland. In Graham, B. (ed.), *In Search of Ireland: A Cultural Geography*, Routledge, London, 151–73.

Pratt, M. L. (1994) Women, literature and national brotherhood. *Nineteenth-Century Contexts: An Interdisciplinary Journal*, **18**, 29–45.

Preston, P. (1993) *Franco: A Biography*, HarperCollins, London.

Ruggie, J. (1993) Territoriality and beyond: problematicizing modernity in international relationships. *International Organization*, **47**, 139–74.

Sadowski, Y. (1998) Think again: ethnic conflict. *Foreign Policy*, **111**, 24–8.

Said, E. (1978) *Orientalism*, Columbia University Press, New York.

Shaw, D. J. B. (1998) 'The chickens of Versailles': the new Central and Eastern Europe. In Graham, B. (ed.) *Modern Europe: Place, Culture and Identity*, Arnold, London, 121–42.

Sillitoe, K. and White, P. H. (1992) Ethnic group and the British census: the search for a question. *Journal of the Royal Statistical Society A*, **155**, 141–63.

Smith, A. D. (1991) *National Identity*, Penguin, Harmondsworth.

Soja, E. (1996) *Thirdspace: Journeys to Los Angeles and Other Real-and-Imagined Places*, Blackwell, Oxford.

Tunbridge, J. E. (1998) The question of heritage in European cultural conflict. In Graham, B. (ed.) *Modern Europe: Place, Culture and Identity*, Arnold, London, 236–60.

Vidal de la Blache, P. (1903) *Le Tableau de la Géographie de la France*, Hachette, Paris.

Withers, C. W. J. (1995) How Scotland came to know itself: geography, national identity and the making of a nation, 1680–1790. *Journal of Historical Geography*, **21**, 371–397.

Woolf, S. (1996) Introduction. In Woolf, S. (ed.) *Nationalism in Europe, 1815 to the Present: a Reader*, Routledge, London, 1–39.

Chapter 4

Historical geographies of imperialism

Alan Lester

Introduction: geographers and imperialism

Despite multiple interpretations of what precisely imperialism is and where it has been practised, it has conventionally been viewed as the means by which Europeans acquired privilege over non-Europeans since the fifteenth century. Over the last two decades, although the temporal and spatial focus has largely been retained, the emphasis in analysis has shifted. A profusion of post-colonial studies has focused on the forms of knowledge and the cultural encounters and transactions involved in the extension of the European empires, rather than simply in their military conflicts, economics and politics.

Post-colonialism is perhaps most associated with literary and cultural studies. These examine the ways that other peoples and places (often capitalised in post-colonial writing as the Other) were and are represented in the European arts, and especially in Europe's canonical literature. However, European representations of Others are also seen as being fundamental to the shaping of the modern world order by many historians, sociologists and geographers, since they informed Europeans' 'knowledge', about both the wider world and themselves. By defining Others in certain ways, Europe was also defining its own cultural identity in contrast to them. For instance, while non-Europeans were represented as suffering from varying degrees of barbarism or savagery, Europeans considered themselves civilised. While the Others were defined as heathens of various kinds, Europeans were Christians, and while non-Europeans supposedly had various degrading kinds of gender relations, those between European men and women were mutually beneficial. As we will see, within post-colonial studies, most would argue that in creating such distinctions, Europe was prefiguring the power relations which characterised imperialism and shaped modernity.

Post-colonial scholars then, have identified the cultural transactions between Europeans and those whom they conquered, dispossessed, co-opted, incorporated or destroyed in their colonies, as being just as important as the more obvious military, mercantile, evangelical and diplomatic techniques by which those colonies were acquired and administered. The European ways of representing and understanding the colonised have come to be known as colonial or imperial discourse and an entire academic industry has arisen in recent years centring on its analysis.

Despite this burgeoning interest in imperialism and its discourse, however, the focus of primarily English-speaking historical geographers has remained limited. While some, such as Crush (1987) and Christopher (1988), have dealt with aspects of imperialism in the colonies themselves (see Chapter 6 for a critical post-colonial approach), most have concentrated on imperialism as it was understood in the European metropoles. The edited collection, *Geography and Empire* (Godlewska and Smith, 1994), is indicative of this. While such studies are subtle and sophisticated, often engaging deliberately with imperial discourse more generally, they nevertheless tend to take its existence largely for granted. Assuming its prior historical construction, they concentrate on interrogating that discourse's effects on individual subjectivity and on specifically geographical understandings among Europeans.

Recent geographical analyses, for instance, have highlighted the nuanced ways in which gender, race and class intersected to shape European individuals', and especially travellers' and explorers' conceptions of themselves and of others within an imperial context that is taken as given (Mills, 1991; Blunt, 1994; McEwan, 1994; Gregory, 1995; Kearns, 1997). Other accounts have focused on the kinds of knowledge about racial Others, tropical climates and colonial landscapes produced within this context by European geographers (Livingstone, 1991, 1992; Driver, 1992; Driver and Rose, 1992; Bell *et al.*, 1995; Barnett, 1998). These latter studies stem, at least partially, from an imperative to highlight the ways in which geographers and their discipline were implicated in the extension and legitimation of imperial power during the period when it was at its most extensive.

Rather than contributing to this body of literature I intend in this chapter to pursue a different agenda; one similarly focused on British imperial discourse, but set primarily by historical anthropologists. Anne Stoler (1989, 1995) and Jean and John Comaroff (1991, 1997), like geographers, have exposed the imperial complicity of their own discipline in the late nineteenth and early twentieth centuries. But they have also gone further, revising our understanding of the wider context in which their predecessors operated, and to which they contributed. They have done so by uncovering the critical circuits of information and knowledge which connected various interests, including officials, settlers and missionaries in the colonial peripheries to those in the imperial metropoles. Through this emphasis, historical anthropologists have indicated the ways in which the historical geographies of Europe and its colonial peripheries were interconnected (see also Stoler and Cooper, 1997, and for two historians' perspectives, Marks, 1990; Colley, 1992a). I wish to argue that it was primarily these global connections and the flows of information traversing them, as well as the movements of people and goods between Europe, the Caribbean, Africa, the Far East and the Antipodes, which shaped the practices and discourse of imperialism, both on the margins and at the centre of empire. They conditioned the milieu in which late nineteenth-century travellers, explorers and geographers formulated their understandings.

Colonial/imperial discourse

The recent academic concern with colonial discourse stems from Edward Said's influential book *Orientalism* (1978), itself inspired by Michel Foucault's writings on the nature of power and knowledge (Foucault, 1972, 1977). *Orientalism* analysed a large corpus of representations of the East produced by academics, novelists and others situated in the West, especially during the last two centuries. Collectively, Said argued, these representations comprised a *discourse.* He defines this as 'a tradition . . . whose material presence or weight, not the originality of a given author, is really responsible for the texts produced out of it' (Said, 1978: 94). Through this 'tradition', the 'Orient' was established in the European imagination as a 'contrasting image, idea, personality, experience' (1978: 2) – as a place that was made the Other of Europe.

The implications of this Othering are far more than merely conceptual. Said claimed that the discourse of the Orient produced a 'Western style for dominating, restructuring, and having authority over the Orient' (1978: 3). The more overt European military, economic and political ventures in the East, with all their implications for conflict, exploitation and domination, were guided by and entrapped within this 'style', even as they contributed to its reformulation. For Said, European ways of seeing and understanding the Orient were not simply conscious justifications for its exploitation. Rather, the discourse which they comprised in itself shaped 'knowledge' of the Orient. That 'knowledge' was both incentive and legitimation for exercising power there.

Said's work was soon applied to imperialism in other contexts. Timothy Mitchell (1988), analysing the colonisation of Egypt for instance, associated the European discourse of the Orient with late nineteenth-century Parisian Exhibitions. These exhibitions, which displayed places and peoples, especially from the French empire, for the education and entertainment of the metropolitan public, were 'a vast theatre or exhibition of the real' which helped shape further metropolitan knowledge. As such, they too helped to marshal European power over the objects, peoples and places which they represented. For Mitchell, as for Said, the European representation of 'external reality' in the non-European world, whether through texts or exhibitions, was thus 'itself a mechanism of power' (Mitchell, 1988: 168 and 19).

However, while I appreciate the importance of discourse, representation and knowledge for the exercise of European power in the colonies, I wish to take issue with Said and Mitchell on certain points. First, the image of an overarching metropolitan representation of other places and peoples, or of a uniform European agenda, needs to be disaggregated. There was no one European representation of 'external reality', and no uncontested 'mechanism of power' in the nineteenth century. Such a homogeneous interpretation of discourse was never intended by Foucault. As Stoler points out, in his treatment of discourse concerning racially defined others, 'Foucault is concerned with a more general racial grammar', one that allowed various groups to 'infuse a shared vocabulary with different political meaning' (Stoler, 1995: 72). Recent critiques of Said's *Orientalism* and some of the work inspired by it, have accordingly emphasised the need for differentiation. Lisa Lowe (1991), for example, explores the heterogeneity within and between French and British representations of the 'Orient', holding that Orientalist discourse was flexible and capable of sustaining diverse political projects. Both Orientalist discourse and that pertaining to other colonies could contain 'competing agendas for using power, competing strategies for maintaining control and doubts about the legitimacy of the venture' (Stoler and Cooper, 1997: 6).

Secondly, discourse analysts like Said and some of his followers tend to conceive of colonial and imperial discourse as emanating out from the European metropolises. Discourses of other places and peoples are seen as being created by scholars, writers, artists and travellers who may have visited the places which they represented, but were based in Europe itself. I view their activities, however, as only part of the picture. In the case study of British colonial discourse during the nineteenth century that I set out below, I wish to recover those networks of knowledge and power which spanned the empire as a whole, connecting the colonies to the metropolis. I wish to conceive of discourse being constructed not simply in the metropoles, but mutually and contingently by groups located both there and in the colonies. These groups included colonial settlers, officials and missionaries. They also included the colonised themselves, since their resistance and co-operation informed colonial culture, although this dimension is not explored thoroughly here (see Chapter 6; Comaroff and Comaroff, 1991, 1997; Lester, 1998a).

My conception of British colonial discourse does involve a general vocabulary of 'civilisation' which was shared between imperial interests, and which encompassed the British empire during the nineteenth century (De Kock, 1996). Africans for instance, were positioned, by British officials, missionaries, settlers, merchants and metropolitan interests alike, below Europeans on a consensual scale of 'civilisation'. Their 'heathenism' was universally regretted or condemned and the necessity for some form of British intervention if their 'improvement' was to be brought about, was universally identified. However, I seek to show that such a vocabulary of difference could be used in flexible and differentiated ways, and towards competing economic and political ends. The struggles which I trace between imperial and colonial officials, liberal humanitarians and capitalist settlers over the meaning and purpose of colonising Africans on the fringes of the Cape Colony illustrate a plurality of imperial projects or, in John Comaroff's phrase (1997: 16), 'models of colonialism'.

By analysing these 'models' and their interaction, I intend to indicate how British interests, exchanging representations across the spaces of the empire, understood and created imperialism in different ways. Their respective variants of imperial discourse were influenced, I will suggest, by their differentiated material positions and programmes. These meant that, even within the colonising population, experiences of and relations to imperialism were heterogeneous. I also want to argue, however, that the eventual ascendancy of settler interests over those of the early nineteenth-century humanitarians helped ultimately to establish the dominant late nineteenth-century imperial discourse to which metropolitan travellers, explorers and geographers contributed.

Figure 4.1 The Cape Colony in the nineteenth century.

The Cape Colony and Britain: officials, humanitarians and settlers

Officials

The Dutch East India Company's Cape Colony at the tip of southern Africa was seized by British forces in 1806, during the Napoleonic Wars, due to its strategic value as a naval base (Figure 4.1). Its capture, undertaken in order to ensure that Napoleon's ships could not base themselves there, was an example of the way that European politics could have lasting repercussions far from Europe itself. Dutch colonial rule had been built in the Cape since the mid-seventeenth century using slaves imported from the Company's South-East Asian bases (Elphick and Giliomee, 1988; Keegan, 1996). During the late seventeenth and eighteenth centuries, the Company allowed European farmers to appropriate territory spreading east and north of its harbour at Cape Town and the relatively small groups of indigenous Khoisan pastoralists and hunter-gatherers were killed, expelled or absorbed as clients and farm labourers.

While colonial control was fairly secure by 1806 in the western districts around Cape Town, on the colony's eastern frontier, some four hundred miles inland, colonial farms were intermingled with African Xhosa homesteads (Figure 4.2). By the end of the eighteenth century, competition over grazing land had already resulted in three frontier wars (Mostert, 1992). After their takeover,

Figure 4.2 The eastern Cape frontier, 1778–1865.

British officials in the War and Colonial Office were interested only the critical harbours of the western Cape and their immediate hinterland, and they instructed the military governors posted to the Cape to maintain order on the turbulent frontier at minimal expense to the British exchequer (Galbraith, 1963).

The concept of order shared by the autocratic British governors however, was unlike those of the frontier colonists and Xhosa themselves. While the latter groups engaged on a daily basis in labour, trade and occasional sexual transactions, simultaneously contesting possession of grazing land by raiding each others' cattle, colonial governors desired a neat and discrete boundary, separating colonial subjects from the members of independent chiefdoms. As Governor Sir John Cradock put it, 'it should be our invariable object to establish the separation from them [the Xhosa], as intercourse can never subsist to the advantage of one party, or the other' (Cape Archives, CO 5807 Government Proclamation, 21 August 1810). From 1809 the British military carried out a series of expulsions in order to secure the desired boundary, the most dramatic occurring in 1812, when the Ndlambe and other minor Xhosa chiefdoms were forced east across the Fish River (Lester, 1997). Their homesteads and fields were burnt and those who resisted were shot, bringing unprecedented levels of violence to the frontier (Maclennan, 1986). Following these expulsions, and in the face of retaliatory cattle raids, Governor Lord Charles Somerset intended to seal the border more effectively through denser colonial settlement along the frontier. This was the strategy which lay behind the importation of some 4,000 settlers from the British Isles in 1820.

Humanitarians

While the aristocratic and military governors tended to see the frontier as the frustratingly troubled front line of a colonial order, for many largely middle class British humanitarians it came to be conceived in a very different light, as an opportunity to extend Christianity and 'civilisation'. In order to understand humanitarian approaches to the Cape frontier, we need briefly to examine developments in Britain and its empire as a whole.

British society was undergoing a dramatic transition as industrialisation gathered pace and a more powerful middle class, or bourgeoisie was formed during the early nineteenth century. Associated with this transition was the rise of middle class evangelical influence. During the late eighteenth century, nonconformist evangelicalism had emerged among the bourgeoisie as an alternative cultural focus to the High Church Anglicanism of the aristocracy. At first, it was eyed warily by the élite, since its stress on the universal qualities of humankind threatened an erosion of upper class privilege. However, as the Napoleonic War continued, evangelicalism was appropriated by the ruling classes themselves. It served as a useful rallying point, securing the loyalty of the 'lower orders' in what was now portrayed as a pious Protestant struggle against despotic French Catholicism (Colley, 1992b; Thorne, 1997). In the aftermath of the war, the upper classes accepted the need for some kind of continuing alliance with the bourgeoisie, so that the propertied classes as a whole might secure a prosperous and stable hierarchy in the face of working-class radicalism (Thompson, 1980). Middle-class evangelicals and other reformers were thus given greater access to the levers of power during the 1820s (Elbourne, 1991; Stoler and Cooper, 1997).

In this context, colonial missionaries were able to forge a powerful humanitarian alliance with the bourgeois reform movement in Britain. While colonial humanitarians concentrated on 'saving the souls' of the heathen within the empire and on getting the Atlantic slave trade and then slavery itself abolished, the metropolitan reformers focused on a range of domestic issues, including securing the vote for the middle classes, removing restraints on free trade, regulating factory conditions and providing a form of poor relief. The colonial and metropolitan humanitarian programmes seemed complementary since each involved political reform to allow for broadened access to citizenship, and each envisaged the replacement of market constraints (of which forced labour was one) by free trade and free labour (Lee, 1994; Evans, 1996; Keegan, 1996).

During the 1820s and 1830s, with the campaign for the abolition of slavery at its peak, colonial and metropolitan reformers were able to maintain a certain political and discursive coherence (Davis, 1975). They formed a powerful network spanning the British empire. Its hub was in London, where influential members of parliament such as Thomas Fowell Buxton co-ordinated political strategy, but lines of communication, information and comment connected that hub to reformers based in the British provinces and in the distant colonies (Lester, 1998a and b).

At the peak of their campaign, abolitionist missionaries and their allies formulated a 'racial imagery' of African slaves 'that permeated the metropolitan social imagination' (Stoler and Cooper, 1997: 28). Although it challenged the notions of slaves' savagery which had previously been generated by slave traders and travellers, this imagery was nevertheless qualified by the assumption of a hierarchy of civilisation with Europeans, and particularly propertied Europeans, positioned at the top.

While slaves were portrayed as child-like and innocent victims of European brutality, it was still taken for granted that only a better European example would enable them to improve upon their current cultural backwardness. Once freed, the slaves, in many respects like Britain's own working classes, would have to be converted to what the bourgeoisie considered 'virtuous' behaviour. They would have to respect authority, adopt a methodical routine, construct a well-regulated landscape of dwellings and fields and develop a confining domesticity for their women (Davidoff and Hall, 1987; Comaroff, 1997). Emancipated slaves, like British workers, would 'be free to pursue their own self-interest but not free to reject the cultural conditioning that defined what that self-interest should be. They would have opportunities for social mobility, but only after they learned their proper place' (Holt, 1992: 53). Thus, when the British government finally freed all its colonial slaves in 1834, they remained with their former masters as 'apprentices' for a further four years, so that they could be prepared for the responsibilities of freedom.

Humanitarians based in the Cape, notably John Philip, Director of the London Missionary Society (LMS) and his son-in-law and newspaper editor, John Fairbairn, were key figures within the imperial humanitarian network. Riding the back of the campaign against slavery during the late 1820s, they were able to draw attention to the continuing plight of the Khoisan within the colony (Botha, 1984; Ross, 1986). Philip and his allies in the British parliament, especially Buxton, succeeded in 1828 in getting Ordinance 50 ratified by the Cape and British governments. The Ordinance abolished the pass laws which had restricted the Khoisan's mobility under the Dutch and early British colonial governments, released them from the legal requirements which bound them to serve the colonists and explicitly recognised their right to own land. For humanitarians, Ordinance 50 represented the Khoisan's charter of freedom, but for the colonists who relied on their labour, it posed an immediate threat to their social privilege.

The 'liberation' of the Khoisan apparently behind them, during the early 1830s Philip and Fairbairn extended their critical gaze to the affairs of the eastern Cape frontier and the Xhosa. They represented the Xhosa in the metropolis in the same terms that the slaves were being portrayed by abolitionists. For Beverly MacKenzie, writing as 'Justus', for instance, it was important for Britons to appreciate the Xhosa's 'traits of generosity and kind feeling . . . their good-natured attentions to strangers and visitors, their quick and grateful perception of friendly feeling; and especially their placable dispositions' ('Justus', 1837: 59). But just as the similarly innocent slaves had, up until now, been mistreated by Europeans, so had the Xhosa. 'Justus' argued:

> If it could be proved, that after a thirty years' dominion we had done nothing at all . . . to bring the barbarian tribes under the mild and transforming power of the Gospel – that the government had in no way advanced the civilization, or improved the condition of the Aborigines of its colonies – then we are weighed in the balance and found wanting; but when we advance beyond this, and prove that our Christian sway has been applied to purposes of spoliation, wrong and cruelty, how shall we find words to express our guilt and degradation? (xii–xiii)

It was just such colonial 'guilt and degradation' that Philip and Fairbairn communicated to their metropolitan audiences. They attacked directly the strategies employed by a succession of governors. Using Philip's, Fairbairn's and other colonial humanitarians' testimony, their colleagues in the metropole condemned the series of Xhosa expulsions carried out since 1809, insisting that 'justice is a stronger wall for a frontier than . . . valleys gained by rapine, and that national integrity is worth a thousand cannons, and a hundred thousand soldiers' ('Justus', 1837: 129–30).

Settlers

In 1820, the 4,000 British settlers requested by Somerset to act as a buffer arrived on the frontier (Figure 4.3). As a group, they were fairly representative of the societies of the British Isles as a whole. They were drawn from England, Scotland, Wales and Ireland, and they came from almost the full range of social classes including the landowning gentry, the professional middle classes, the skilled artisanal classes, the labouring poor and the pauper population. They also had a more balanced gender ratio than was usual in early colonial settlements, being 36 per cent men, 20 per cent women and 44 per cent children (Lester, 1998c).

Regardless of their own class background, all of the settlers soon faced severe labour shortages in the new settlement. They thus came rapidly to appreciate the racial stratification which had long been established in the colony, using cheap Khoisan labour to build and protect their settlements. The humanitarian agitation to free the Khoisan of the pass and other regulations which guaranteed such dependency represented an immediate threat to the settlers' fragile prosperity. Thomas Stubbs spoke for most when he described the humanitarians' intervention as 'that abominable false philanthropy which made [the Khoisan] free and ruined them . . . They were a people that required to be

Figure 4.3 Cruickshank's view of the choices facing British emigrants.

under control, both for their own benefit and the public; the same as the slaves in this country' (Maxwell and McGeogh, 1978: 71).

A settler consensus on the Xhosa took longer to develop since some settlers who benefited from trans-frontier trading relations represented them positively, while the educated settler gentry often portrayed them in the noble light of the ancient Greeks and Romans, seeing them as being similarly uncorrupted by the vices of capitalism and industrialism (Lester, 1998c). However, the form that a broad settler consensus would eventually assume was apparent among those farmers on the most exposed lands of the frontier even from the early 1820s. The settlers were located in the district around Graham's Town which had been occupied until their expulsion by the Ndlambe Xhosa (Figure 4.4). Not surprisingly, these farmers found themselves the victims of retaliatory raiding. Within a few months, almost all of their livestock had been taken. A stereotype of the Xhosa similar to that held by the military officials with longer frontier experience was soon in circulation among these settlers. As one of them put it, the Xhosa were nothing but a 'cunning . . . and dangerous enemy' (Cape Archives, A602/2 Journal of S. H. Hudson, n.d., 1821). During the early 1830s, a broader group of settlers came to share the exposed farmers' constructions of the Xhosa. A mutual realisation of the eastern Cape's potential for sheep breeding and a massive increase in the demand for raw wool in Britain persuaded settlers with various class origins that wool production needed to be extended into Xhosa territory. Their vision of 'progress', articulated by

Figure 4.4 The settler capital, Graham's Town, in the early nineteenth century.

Robert Godlonton's frontier newspaper, *The Graham's Town Journal* (*GTJ*), became inextricably linked to capitalist expansion and the corresponding dispossession of the Xhosa. Contributors to the *GTJ* accordingly emphasised the Xhosa's 'depredations' against the colony in order to legitimate their expulsion from adjacent land (Keegan, 1996; Lester, 1998a).

In December 1834, frontier Xhosa chiefs (Figure 4.5) whose people had been forced from their lands on successive occasions, launched a co-ordinated counter-attack on the colony, killing twenty-four settlers within a few hours. The attack initiated a war which was won by colonial troops during the following year. A remarkably unified British settler identity now emerged, transcending the settlers' class, regional, and ethnic divisions and uniting all who were threatened by the Xhosa's outbreak of military resistance. This identity was defined above all in opposition to two mutual challenges. The first was the Xhosa themselves and the second, the humanitarians who sympathised with them.

Phrenological lectures, given in Graham's Town after the 1834–5 war and using Xhosa skulls taken during the fighting as models, proved unexpectedly popular among the settlers. As the South African historian Andrew Bank points out, such racial 'science', fixing and elaborating upon the supposedly inherent inferiorities of Africans through a comparative study of skull dimensions, had great 'appeal for those with experience of frontier conflict and an associated antipathy towards a "savage" enemy' (Bank, 1996: 402–3). In early nineteenth-century Australia and New Zealand, where settlers similarly sought to force capitalist agrarianism into new lands and suffered the consequences of resistance,

Figure 4.5 Maqoma, leader of the Xhosa attack on the Colony in 1834.

phrenology also became popular. Its adoption in the settler colonies fuelled interest in such racial science in the European metropolises and it even allowed early and mid-nineteenth-century European scientists to be supplied with the raw materials for their enquiries, in the form of a steady stream of skulls 'acquired' in various frontier conflicts (Bank, 1995).

After the war had swung in favour of the colony, British troops were instructed to secure the newly conquered Xhosa territory, now named Queen Adelaide Province. The Cape governor, Sir Benjamin D'Urban, anticipated that with the Xhosa chiefs disempowered and 'military posts of occupation . . . within, around and among their locations, the means will be ever at hand to subdue any serious resistance' (Cape Archives, A519 D'Urban to Smith, 17 September 1835). Most of the settlers were delighted, as official and settler agendas seemed to have converged. Since his plans for the colonisation of the Xhosa raised the prospect of considerably more land grants on which the local woollen industry could thrive, D'Urban was fêted throughout the British settlement. The Xhosa's total expulsion from the province had proved militarily impossible, but settlers reasoned that their colonisation and spatial confinement would be the next best thing (Lester, 1998d). Ultimately, the coincidence between settler and official imperatives was to become consolidated as settlers won greater autonomy from the metropolitan government, but not before a political and discursive struggle with humanitarianism, connecting the Cape and Britain, had been won.

Constructing colonial discourse

Philip, Fairbairn and the other colonial humanitarians disagreed vehemently with D'Urban over the causes of the war and the policy of colonisation. In his correspondence with London-based humanitarians, Philip made his view clear: the war had been caused by the series of Xhosa expulsions, and by the provocations of settlers intent on acquiring farms in Xhosa territory. To then envisage the outright domination of the Xhosa as punishment for their reaction was outrageous (Ross, 1986).

While settlers were battling to establish and aggrandise themselves on the imperial margins during the 1820s and 1830s, as we have seen, the bourgeois 'religious public' of Britain was being exposed to influential humanitarian representations of colonialism (Thorne, 1997: 239). In 1836, the Secretary of State for the Colonies was Lord Glenelg, a humanitarian raised within the 'Clapham Sect' of evangelicals led by the abolitionist William Wilberforce. Glenelg was determined not only to follow the government programme of cutting expenditure in the colonies, but also to inject popular humanitarianism more effectively into colonial policy. Encouraged by humanitarians in both the Cape and Britain, he wrote to D'Urban that the Xhosa, having suffered so long under the yoke of colonial oppression, had had 'a perfect right to hazard the experiment' of launching their attack (Cape Archives, GH 1/107 Glenelg to D'Urban, 26 December 1835). To the enormous chagrin of D'Urban and most of the British settlers on the Cape frontier, the scheme for the colonisation of Queen Adelaide Province was abandoned in December 1836 and the governor himself was recalled to London (Lester, 1998b).

Even while the dust of war was still settling, however, the self-appointed settler spokesman, Robert Godlonton was writing an account of the frontier which he hoped would carry more weight in Britain than *The Times*, which supported the settlers' cause, and the few copies of the *Graham's Town Journal*, which currently circulated among metropolitan officials and merchants. His *Narrative of the Irruption of the Kafir Hordes* (1836), was a history of persistent and united settler endeavour. The settlers' initial class and regional divisions, and their continuing gender inequalities were obscured in Godlonton's account. Instead, they were portrayed as being engaged in a mutual attempt to build up a civilised society and a productive landscape on a wild and inhospitable frontier – only for it to be destroyed by the unprovoked terror of the 'irreclaimable' Xhosa. Godlonton continued, not only had the fruits of their diligent labour been destroyed; in addition they now found themselves condemned by their fellow-Britons, the humanitarians (Godlonton, 1836).

While the Xhosa were being constructed by the settlers as biologically predisposed to savagery, their humanitarian advocates were described as 'unprincipled scoundrels' (Maxwell and McGeogh, 1978: 112). Holden Bowker expressed the settlers' grievances succinctly: 'with the assiduity of purpose that Satan himself might envy', he wrote, the humanitarians had 'gained their object in persuading our countrymen, to whom we looked for sympathy and succour, that we are monsters' (Bowker, 1864: 2). Settler representations proved powerless against humanitarian influence in the metropole during the 1830s, but propaganda such as Godlonton's and Bowker's was beginning to pay dividends by the end of the next decade. Godlonton's next attempt to publicise the fate of the settlers in Britain, his 1844 *Memorials of the British Settlers of South Africa*, was purchased tellingly by Queen Victoria herself.

Through exactly the kinds of political and discursive struggles in which the Cape settlers continued to engage, a wider imperial consensus on racial difference was being established by the mid-nineteenth century. It was one which, in practice, overshadowed early nineteenth-century humanitarian 'sentiment', even though it retained legitimating aspects of its rhetoric. In order to understand the creation of this more coherent late nineteenth-century imperial discourse, we have to broaden our focus again, situating struggles in the Cape alongside those in other colonies and in the metropolis, and indicating the flows of 'knowledge' between groups in these diverse locales.

Widespread British disillusionment with humanitarian visions for the colonised Others of empire began with the experience of abolition itself. When emancipated slaves, particularly in the West Indies, asserted their autonomy, when they refused to conform to the humanitarians' expectations of 'virtue', involving continued, diligent plantation work and sober Christian conversion, their 'inherent' racial characteristics were blamed by planters and their powerful metropolitan supporters. (For a study of a metropolitan-based West Indies planter, see Seymour *et al.*, 1998.) Planters' propaganda, filtered through parliament and the London press, succeeded in associating the 'greatest good' for the colonies, and for the empire as a whole, with the prosperity of the plantations. It suggested that the humanitarian notion of freed labourers rapidly learning the responsibilities of citizenship had been naïve.

By 1846 (when Xhosa chiefdoms were fighting once more against dispossession in the Cape) the decline of sugar production and the bankruptcy of West Indies estates had persuaded even Henry Taylor, one of the metropolitan architects of abolition, that 'negroes, like children, require a discipline which shall enforce upon them steadiness in their own conduct and consideration for the interests of others' (Holt, 1992: 285). As Holt (1992: 280) demonstrates, the influential writer Thomas Carlyle was already channelling the Jamaican planters' imagery of former slaves 'sitting . . . with their beautiful muzzles up to their ears in pumpkins . . . while the sugar crops rot around them uncut' effectively to metropolitan audiences by the late 1840s.

However, peripheral interpretations of far more widespread and violent indigenous resistance interacted with shifts in the basis of metropolitan power and proved decisive in the 1850s and 1860s. Prime Minister Palmerston's war in China indicated a new willingness on the part of Britain's increasingly secure bourgeois electorate to support peripheral capitalists with a more aggressive imperialism, and the 'Indian Mutiny' of the following year, 1857, was critical. Given India's prominence within the empire, its administrators and its colonial population had always enjoyed well-developed communications with the metropole and India had loomed large in British debates about cultural difference and colonial government (Stokes, 1959; Mehta, 1997). The shock of the rebellion generated a metropolitan 'fever of race hatred', giving rise to a more general assertion that 'the Indian could never be improved' (Porter, 1996: 37 and 44).

During the early 1860s, metropolitan notions of indigenous irreclaimability were consolidated by the Maori Wars. As James Belich points out, since the well-publicised humanitarian cause of 'salvageability' in New Zealand had come to be

> based partly on [Maori] readiness selectively to adopt European ways in commerce, agriculture, literacy and religion . . . resistance was seen as a reversal of this trend; evidence that the civilising mission had failed, or even that it had always been doomed to failure. (Belich, 1986: 328)

The 1865 Morant Bay rebellion in Jamaica and Governor Edward Eyre's brutal reaction to it, gave metropolitan humanitarians including John Stuart Mill an opportunity to challenge increasing British racism, to make a determined stand for the assimilationist principles inherited from their abolitionist forefathers (Hall, 1996). But, defended by influential figures such as Carlyle, Charles Dickens and Alfred Tennyson, Eyre discovered upon his recall to Britain 'that racist fears [now] found a receptive public'. As Holt puts it, 'In the aftermath of Morant Bay and on the eve of its greatest imperialist adventures, British public opinion accorded more closely with Eyre and Carlyle than with Mill' (1992: 305 and 307).

American slave-owners' propaganda during the American Civil War converged with the stream of representations of racial Others which had been pouring into the metropole from colonial capitalists and settlers like those in the Cape since at least the 1830s (Young, 1995). It was a stream which flowed via potent books and articles such as Carlyle's, parliamentary debates like Palmerston's, published settler memorials like Godlonton's and Bowker's and organs of the popular press such as *The Times*. In ambivalent ways, and through a variety of media including advertising (Pieterse, 1992; McClintock, 1995), music hall entertainment (Crowhurst, 1997) and, as Ploszaska indicates in Chapter 5, schooling, constructions of race forged on the peripheries of the empire were channelled during the second half of the nineteenth century beyond the ruling classes to become an integral component of popular culture (MacKenzie, 1984).

By the late nineteenth century, despite the persistence of certain debates, a notion of inherent racial inequality akin to that generated by the colonial settlers was firmly entrenched in the full range of metropolitan disciplines, including, as we have noted, that of geography (Livingstone, 1992). Cumulatively, the flow of racial representation from settlers and planters on the margins of empire undoubtedly enhanced the prestige and stimulated the enquiries of scientific racism, helping to establish it as a more pervasive model of social interaction in both colony and metropole (Stepan, 1982; Gilman, 1985; Goldberg, 1990; Dubow, 1995; Stoler, 1995; Young, 1995).

Images derived from settlers on the peripheries, and even from outside the empire, and from travellers who sympathised with them, then, had assaulted humanitarian arguments and revealed their contradictions more effectively by the 1860s than they had done in the 1830s. Colonists had managed to persuade those in the metropole more convincingly of the dangers involved in living among 'irreclaimable savages', and of a deeper, biological meaning behind indigenous peoples' resistance to the European 'civilising mission'. The persistence and the sheer volume of their representations, their utility to capitalist projects, and their more rapid communication through technological improvements did much to ensure their ultimate discursive dominance.

However, the triumph of settler imagery in the metropole was also profoundly assisted by its interaction with domestic, bourgeois concerns. By the 1860s, notions akin to the settlers' 'irreclaimable savage' were being used by a now economically and politically entrenched bourgeoisie, including utilitarian liberals, to link the racial Others of empire more coherently with those who threatened subversion 'at home'. We have seen how, even during the reformist period of the early nineteenth century, African 'backwardness' was represented as being analogous to that of the British labouring poor. Now that the British bourgeoisie had attained social, economic and political hegemony, it could afford to jettison the discourse of universal humanity, reform and incorporation which had assisted in its ascendancy and concentrate instead on consolidating its position against various threats, including that of potentially socialist workers.

Settler constructions of innate human difference, generated for decades on the violent frontiers of the empire, could assist. After all, as Stoler points out, they 'captured in one sustained image internal threats to the health and well-being of a social body where those deemed a threat lacked an ethics of "how to live" and thus the ability to govern themselves' (Stoler, 1995: 127). Similar constructions could serve to legitimate British male, bourgeois control over women, the Irish, the poor, the insane and criminals, among others (Hall, 1992). Together, colonial settlers and the British bourgeoisie had developed by the late nineteenth century a discourse which connected centre and peripheries, and that 'designated eligibility for citizenship, class membership, and gendered assignments', linking each of these social dimensions in direct or indirect ways to inherent difference (Stoler, 1995: 128; see also McClintock, 1995).

Imperial discourse and the Xhosa

On the Cape frontier, by 1844 Bowker already felt more confident of metropolitan support for the dispossession of the Xhosa. Asking, 'Is it just that a few thousands of ruthless, worthless savages are to sit like a nightmare upon a land that would support millions of civilised men happily?', he felt assured of the reply that he would receive from his metropolitan compatriots: 'Nay; Heaven forbids it' (Bowker, 1864: 125). It is worth noting that in his appeal to an audience based in Britain, Bowker frequently quoted Carlyle on the characteristics of 'Quashee', the stereotypical freed West Indian slave, whom Carlyle in turn had fashioned from the Jamaican planters' propaganda (Holt, 1992: 280–1). In his rationalisation of colonial expansion in the eastern Cape, Bowker was deploying a discursive construction of Africans which spanned the empire, from the West Indies to London, on to the Cape frontier and back again. It was such circuits of 'knowledge' which ultimately allowed colonial settlers and the metropolitan bourgeoisie mutually to fashion a powerful, imperial discourse of exclusion.

In the new economic, intellectual and moral climate of the late nineteenth-century empire, the Cape's settlers, like those elsewhere, were given a degree of autonomy under Representative Government. Supported by refashioned discourses of Xhosa Otherness, British and colonial troops had already crushed the renewed Xhosa resistance of 1846–7 and put down a combined Xhosa and Khoisan labour rebellion in 1850–2. Xhosa territory was now carved up into settler farms and 'locations', in which tenacious Xhosa occupation was reluctantly conceded. Despite their grip on remnants of the land, many more Xhosa were induced by taxation and other means to labour on the settlers' farms. Through the imposition of land loss, hunger, dispersal and poorly remunerated toil, the British empire was able to further, although never quite accomplish, the settlers' vision of 'civilisation' among the frontier Xhosa (Peires, 1982; Lester, 1996).

Conclusion

During the early nineteenth century, colonial officials, humanitarians and missionaries had agreed on the relative 'backwardness' of the Xhosa, but they had held out the prospect of markedly different strategies towards them. These strategies ranged from the military exclusion proffered by officials, through the professedly benign Christian and economic embrace of the humanitarians to the dispossession and subjugation of settler capitalists. In their diverse representations to each other and to the metropole, each of these colonial groups was informing an ambivalent discourse of imperialism. The colonial settler narrative of capitalist progress frustrated by the resistance of the 'irreclaimable savage' was ultimately imbued with greater power than the humanitarian narrative of mutual redemption and civilisation. This colonial outcome was mirrored by shifts in the power structure of Britain as the consolidation of bourgeois hegemony saw the dilution of reformist projects and the 'scientific' exclusion of the racially, gendered and class-defined Others who threatened that hegemony.

With the disillusionment and marginalisation of humanitarian visions, the convergence of official and settler aspirations and the political dominance of exclusive strategies in the colonies and in Britain, the terms of debate within imperial discourse were shifting by mid-century. The shift did not mean 'the end of one discourse and the emergence of another'. Rather, as Foucault, points out, discourse consists of vacillations. 'It operates at different levels and moves not only between different political projects but seizes upon *different* elements . . . reworked for new political ends' (cited in Stoler, 1995: 72). These 'political projects', with their different and often contradictory 'elements' of racial construction were grounded in localised conditions of material struggle. But their interaction, not only between metropole and colony, but also between the various peripheral colonies themselves, provided an imperial terrain across which knowledge was produced, contested and communicated.

Lowe concludes that contemporary literary discourse is one in which 'articulations and rearticulations emerge from a variety of positions and sites' (1991: 15). The same can be said of imperialist discourses of difference. While the formal mechanisms of imperialism have now been superseded, there is still a differentiated but spatially interconnected network of 'positions' and 'sites', in which racial and other cultural constructions are formulated, refashioned and conveyed to metropolitan populations. Formal imperialism has been outlived by the global circuits in which the West's Others, perhaps the most threatening of which is currently identified as the Muslim fundamentalist, come to be 'known'.

References

Bank, A. (1995) *Liberals and Their Enemies: Racial Ideology at the Cape of Good Hope, 1820–1850*, unpublished PhD thesis, Cambridge University.

Bank, A. (1996) Of 'native skulls' and 'noble Caucasians': phrenology in colonial South Africa. *Journal of Southern African Studies*, **22**, 387–404.

Barnett, C. (1998) Impure and worldly geography: the Africanist discourse of the Royal Geographical Society, 1831–73. *Transactions of the Institute of British Geographers*, **23**, 239–51.

Belich, J. (1986) *The Victorian Interpretation of Racial Conflict: The Maori, The British, and the New Zealand Wars*. McGill-Queen's University Press, Montreal and London.

Bell, M., Butlin, R. A. and Heffernan, M. (eds) (1995) *Geography and Imperialism 1820–1940*, Manchester University Press, Manchester.

Blunt, A. (1994) *Travel, Gender and Imperialism: Mary Kingsley and West Africa*, Guilford Press, London and New York.

Botha, H. (1984) *John Fairbairn in South Africa*, Historical Publication Society, Cape Town.

Bowker, J. M. (1864) *Speeches, Letters and Selections From Important Papers*, Struik, Graham's Town.

Bryer, L. and Hunt, K. S. (1984) *The 1820 Settlers*, Don Nelson, Cape Town.

Cape Archives, CO 5807 Government Proclamation, 21 August, 1810.

Cape Archives, A602/2 Journal of S. H. Hudson, n.d. (1821).

Cape Archives, A519 D'Urban to Smith, 17 September, 1835.

Cape Archives, GH 1/107 Glenelg to D'Urban, 26 December, 1835.

Christopher, A. J. (1988) *The British Empire at its Zenith*, Croom Helm, Beckenham.

Colley, L. (1992a) Britishness and otherness: an argument. *Journal of British Studies*, **31**, 309–29.

Colley, L. (1992b) *Britons: Forging the Nation, 1707–1837*, Yale University Press, London.

Comaroff, J. (1997) Images of empire, contests of conscience: models of colonial domination in South Africa. In Cooper, F. and Stoler, A. (eds) *Tensions of Empire: Colonial Cultures in a Bourgeois World*, University of California Press, Berkeley and London, 163–97.

Comaroff, J. and Comaroff, J. (1991) *Of Revelation and Revolution: Christianity, Colonialism and Consciousness in South Africa*, University of Chicago Press, Chicago and London.

Comaroff, J. and Comaroff, J. (1997) *Of Revelation and Revolution: the Dialectics of Modernity on a South African Frontier*, University of Chicago Press, Chicago and London.

Crowhurst, A. (1997) Empire theatres and the empire: the popular geographical imagination in the age of empire. *Environment and Planning D: Society and Space*, **15**, 155–73.

Crush, J. (1987) *The Struggle for Swazi Labour, 1890–1920*, McGill-Queen's University Press, Kingston and Montreal.

Davidoff, L. and Hall, C. (1987) *Family Fortunes: Men and Women of the English Middle Class, 1780–1850*, Routledge, London.

Davis, D. B. (1975) *The Problem of Slavery in the Age of Revolution, 1770–1823*, Cornell University Press, Ithaca.

De Kock, L. (1996) *Civilising Barbarians: Missionary Narrative and Textual Response in Nineteenth-Century South Africa*, Witwatersrand University Press and Lovedale Press, Johannesburg.

Driver, F. (1992) Geography's empire: histories of geographical knowledge. *Environment and Planning D: Society and Space*, **10**, 23–40.

Driver, F. and Rose, G. (eds) (1992) *Nature and Science: Essays in the History of Geographical Knowledge*, IBG, London (Historical Geography Research Series, No. 28).

Dubow, S. (1995) *Scientific Racism in Modern South Africa*, Cambridge University Press, Cambridge.

Elbourne, E. (1991) *'To Colonize the Mind': Evangelical Missionaries in Britain and the Eastern Cape, 1790–1837*, unpublished D. Phil thesis, Oxford University.

Elphick, R. and Giliomee, H. (1988) (eds) *The Shaping of South African Society, 1652–1840*, Wesleyan University Press, Middletown.

Evans, E. J. (1996) *The Forging of the Modern State: Early Industrial Britain*, Longman, London.

Foucault, M. (1972) *The Archaeology of Knowledge*, Harper and Row, New York.

Foucault, M. (1977) *Discipline and Punish: the Birth of the Prison*, Allen Lane, London.

Galbraith, J. S. (1963) *Reluctant Empire: British Policy on the South African Frontier, 1834–1854*, University of California Press, Berkeley.

Gilman, S. (1985) *Difference and Pathology: Stereotypes of Sexuality, Race and Madness*, Cornell University Press, Ithaca.

Godlewska, A. and Smith, N. (eds) (1994) *Geography and Empire*, Blackwell, Oxford.

Godlonton, R. (1836) *A Narrative of the Irruption of the Kafir Hordes into the Eastern Province of the Cape of Good Hope, 1834–35*, Meurant and Godlonton, Graham's Town.

Godlonton, R. (1844) *Memorials of the British Settlers of South Africa*, Godlonton, Graham's Town.

Goldberg, D. (1990) *Anatomy of Racism*, University of Minnesota Press, Minneapolis.

Gregory, D. (1995) Between the book and the lamp: imaginative geographies of Egypt, 1849–50. *Transactions of the Institute of British Geographers*, **20**, 29–57.

Hall, C. (1992) *White, Male and Middle Class: Explorations in Feminism and History*, Verso, Cambridge.

Hall, C. (1996) Imperial man: Edward Eyre in Australasia and the West Indies, 1833–1866. In Schwarz, B. (ed.) *The Expansion of England: Race, Ethnicity and Cultural History*, Routledge, London, 130–70.

Holt, T. C. (1992) *The Problem of Freedom: Race, Labour and Politics in Jamaica and Britain, 1832–1938*, Johns Hopkins University Press, Baltimore and London.

'Justus' (1837) *The Wrongs of the Caffre Nation*, James Duncan, London.

Kearns, G. (1997) The imperial subject: geography and travel in the work of Mary Kingsley and Halford Mackinder. *Transactions of the Institute of British Geographers*, **22**, 450–72.

Keegan, T. (1996) *Colonial South Africa and the Origins of the Racial Order*, Leicester University Press, London.

Lee, S. J. (1994) *Aspects of British Political History, 1815–1914*, Routledge, London and New York.

Lester, A. (1996) *From Colonization to Democracy: A New Historical Geography of South Africa*, Tauris Academic Studies, London.

Lester, A. (1997) The margins of order: strategies of segregation on the Eastern Cape frontier, 1806–*c*. 1850. *Journal of Southern African Studies*, **23**, 635–53.

Lester, A. (1998a) *Colonial Discourse and the Colonisation of Queen Adelaide Province, South Africa*, RGS-IBG, London (Historical Geography Research Series, No. 28).

Lester, A. (1998b) 'Otherness' and the frontiers of empire: the eastern Cape Colony, 1806–*c*. 1850. *Journal of Historical Geography*, **24**, 2–19.

Lester, A. (1998c) Reformulating identities: British settlers in early nineteenth-century South Africa. *Transactions of the Institute of British Geographers*, **23**, 515–31.

Lester, A. (1998d) Settlers, the state and colonial power: the colonization of Queen Adelaide Province, 1834–37. *Journal of African History*, **39**, 221–46.

Livingstone, D. (1991) The moral discourse of climate: historical considerations on race, place and virtue. *Journal of Historical Geography*, **17**, 413–34.

Livingstone, D. (1992) *The Geographical Tradition: Episodes in the History of a Contested Enterprise*, Blackwell, Oxford.

Lowe, L. (1991) *Critical Terrains: French and British Orientalisms*, Cornell University Press, Ithaca and London.

Maclennan, B. (1986) *A Proper Degree of Terror: John Graham and the Cape's Eastern Frontier*, Ravan Press, Johannesburg.

MacKenzie, J. (1984) *Propaganda and Empire: The Manipulation of British Public Opinion, 1880–1960*, Manchester University Press, Manchester.

Marks, S. (1990) History, the nation and empire: sniping from the periphery. *History Workshop Journal*, **29**, 111–19.

Maxwell, W. A. and McGeogh, R. T. (eds) (1978) *The Reminiscences of Thomas Stubbs*, Balkema, Cape Town.

McClintock, A. (1995) *Imperial Leather: Race, Gender and Sexuality in the Colonial Contest*, Routledge, New York and London.

McEwan, C. (1994) Encounters with West African women: textual representations of difference by white women abroad. In Blunt, A. and Rose, G. (eds) *Writing Women and Space: Colonial and Postcolonial Geographies*, Guilford Press, New York and London, 73–100.

Mehta, U. (1997) Liberal strategies of exclusion. In Cooper, F. and Stoler, A. (eds) *Tensions of Empire: Colonial Cultures in a Bourgeois World*, University of California Press, Berkeley and London, 59–86.

Mills, S. (1991) *Discourses of Difference: An Analysis of Women's Travel Writing and Colonialism*, Routledge, London.

Mitchell, T. (1988) *Colonising Egypt*, Cambridge University Press, Cambridge.

Mostert, N. (1992) *Frontiers: The Epic of South Africa's Creation and the Tragedy of the Xhosa People*, Jonathan Cape, London.

Peires, J. (1982) *The House of Phalo: A History of the Xhosa People in the Days of Their Independence*, University of California Press, Berkeley.

Pieterse, J. N. (1992) *White on Black: Images of Africa and Blacks in Western Popular Culture*, Yale University Press, New Haven and London.

Porter, B. (1996) *The Lion's Share: A Short History of British Imperialism, 1850–1995*, Longman, London.

Ross, A. (1986) *John Philip (1775–1851): Missions, Race and Politics in South Africa*, Aberdeen University Press, Aberdeen.

Said, E. (1978) *Orientalism: Western Conceptions of the Orient*, Penguin, London.

Seymour, S., Daniels, S. and Watkins, C. (1998) Estate and empire: Sir George Cornewall's management of Mocca, Herefordshire and La Taste, Grenada, 1771–1819. *Journal of Historical Geography*, **24**, 313–51.

Stapletone, T. J. (1994) *Maqoma: Xhosa Resistance to Colonial Advance*, Jonathan Ball, Johannesburg.

Stepan, N. (1982) *The Idea of Race in Science: Great Britain, 1800–1960*, Macmillan, London.

Stokes, E. (1959) *The English Utilitarians and India*, Clarendon Press, Oxford.

Stoler, A. L. (1989) Rethinking colonial categories: European communities and the boundaries of rule. *Comparative Studies in Society and History*, **13**, 134–61.

Stoler, A. L. (1995) *Race and the Education of Desire: Foucault's History of Sexuality and the Colonial Order of Things*, Duke University Press, Durham and London.

Stoler, A. L. and Cooper, F. (1997) Between metropole and colony: rethinking a research agenda. In Cooper, F. and Stoler, A. (eds) *Tensions of Empire: Colonial Cultures in a Bourgeois World*, University of California Press, Berkeley and London, 1–56.

Thompson, E. P. (1980) *The Making of the English Working Class*, Penguin, London.

Thorne, S. (1997) 'The conversion of Englishmen and the conversion of the world inseparable': missionary imperialism and the language of class in early industrial Britain. In Cooper, F. and Stoler, A. (eds) *Tensions of Empire: Colonial Cultures in a Bourgeois World*, University of California Press, Berkeley and London, 238–62.

Young, R. (1995) *Colonial Desire: Hybridity in Theory, Culture and Race*, Routledge, London and New York.

Chapter 5

Historiographies of geography and empire

Teresa Ploszajska

Introduction

This chapter offers a nuanced and contextual critical reappraisal of relations between popular geographical and imperial discourses. In so doing, it emphasises that experiences and perspectives were never all-pervasive or singular. Only recently have we begun to recognise that the interconnected histories of geography and imperialism were complex, variable and multiple, as are their legacies. In exploring these issues through a case study of school geography's role in popularising particular ideas about, and attitudes towards, the colonised world and its peoples, the chapter has three distinct sections. First, I offer an historiographical overview of the recent convergence of multidisciplinary interest in the intimate relations between geography and imperialism, pursuing the broad consensus that imperialism and geography may perhaps most fruitfully be thought of as interconnected cultural discourses. Secondly, I focus on the medium of textbooks, emphasising the subjective and contingent nature of such (indeed all) research and knowledge production. Descriptions of Australia and Australians are employed as an example of late nineteenth- and early twentieth-century textbook representations of imperial landscapes and peoples, images that are often remarkably persistent. Finally, I address the use of visual imagery as a means of furthering geographical 'knowledge' and of fostering particular imaginative geographies of empire. Following a broad discursive overview of the kinds of illustrations that appeared most frequently in textbooks, specific examples, which exemplified commonplace verbal and visual textbook narratives, are considered. This dimension to the discussion also encompasses the increasing range of additional visual sources with which teachers illustrated geography lessons, and finally addresses the central role that perceived powers of 'accurate' visualisation often occupied in discussions surrounding both geographical education and education for citizenship.

Geography and imperialism

The context

Traditionally, histories of British geography have focused upon the emergence of its national institutions and the establishment of the subject as a university discipline. Standard accounts include anniversary publications celebrating the foundation of the Royal Geographical Society (RGS) in 1830 (Cameron, 1980), the Geographical Association (GA) in 1893 (Balchin, 1993), and the Institute of British Geographers (IBG) in 1933 (Steel, 1984). Conventionally, such studies present uncritical narratives, which chronicle the institutions' progressive evolution. Disciplinary histories have tended to mirror this institutional focus. For example, discussions of the establishment of geography at Oxford and Cambridge Universities (taken by many to signify the birth of modern geography in this country) assign a pre-eminent role to the RGS in what is portrayed as an epic struggle to gain recognition and respect for the subject (Scargill, 1976; Stoddart, 1989). As Livingstone (1992: 4) points out, such accounts are 'in-house reviews of disciplinary developments for the geographical community'. Stories of the great, heroic figures and epic moments in the history of British geography have thus been reproduced uncritically by successive generations of scholars. In short, a Whiggish consensus of received wisdom regarding the discipline's history became embedded within the corpus of geographical knowledge. Until recently, this was particularly apparent in geographers' overwhelming failure to acknowledge, still less interrogate, the relationship between their discipline and imperialism.

It was Hudson (1977) who first correlated the chronological and materialistic relationships between the emergence of 'new geography' in late nineteenth- and early twentieth-century Europe and the impulses of 'new imperialism'. This study acted as a catalyst for geographers' critical and contextual reflection on their disciplinary genealogy. For example, Peet (1985) provided a detailed development of its suggestion that geography lent scientific legitimacy to imperialist ideologies such as environmental determinism. Adopting a less materialistic framework, Livingstone (1991) later explored the moral dimensions of geography's climatic discourses in the past. Other studies retained an economic focus, however, exploring the material impacts of European territorial expansion into the pre-capitalist world. Thus Britton (1980) argued that the growth of urban centres and the development of transport routes in Fiji was almost entirely a product of colonial export needs. Although primarily concerned with the effects of colonialism on the physical and economic landscapes of the 'Third World', Britton identified and began to explore the ways in which economic processes impacted upon, and were played out in, social relations. A growing number of historical geographers (for example, Godlewska and Smith, 1994; Bell *et al.*, 1995) have since begun to reflect upon the complex and multiple interactions between rhetoric and practices – the cultural, intellectual and ideological manifestations – of Victorian and Edwardian geography and imperialism. The symbolic and experiential significance of the spatial development of colonial cities has received particularly extensive attention. The focus of analysis has ranged from major administrative and commercial centres (both domestic and overseas) through to Indian hill stations. In all these contexts, residential segregation between colonisers and colonised is now recognised as a crucial means of imposing and maintaining power relationships and notions of 'difference', which were fundamental to imperialism (Mitchell, 1988; Duncan,

1989; Christopher, 1997). More recently, explorations of the historical geographies of imperial margins have challenged the assumption that ideological rhetoric and practice were invariably fabricated within, and diffused from, the metropolitan heartlands of empire. Investigation of the construction and maintenance of racial Otherness in diverse colonial peripheries reveals that localised discourses and experiences played an equally important role in shaping metropolitan thoughts and actions. As Lester (1998; see also Chapter 4) concludes in the South African context, metropolitan-centred and local discourses of difference competed with, and impacted upon, each other and were negotiated and experienced in variable ways over time and space. His conclusions reflect an increasing recognition that imperialism was not a monolithic meta-narrative with a single, inevitable and predictable historical geography (Driver, 1992). In turn, this echoes a more general broadening of purely historical analyses of imperialism, which – until relatively recently – were almost exclusively concerned with its economic and political dimensions.

Over the past two decades or so, historians have come to view imperialism as involving far more than economic interests, territorial acquisition and political ambition. Cross-disciplinary studies by historians, anthropologists, literary critics, sociologists and others have extended imperialism's defining parameters. Broad agreement has been reached that it may most accurately and fruitfully be interpreted as, above all, a *cultural* process. Attention has thus shifted from the official mind to popular attitudes, beliefs and practices that are now seen as crucial to imperial expansion. The acquisition and maintenance of overseas territories, it is argued, necessitated the harnessing of mass emotion and popular support. In Britain, historian John MacKenzie (1984) suggests, imperialist sentiments were conveyed to, and sustained among, every strata of society through vehicles as various as entertainment, advertising and marketing, youth movements and children's fiction. MacKenzie conceives of imperialism as a pervasive and persistent ensemble of cultural attitudes towards the rest of the world, one that played a central ideological role within British culture and society between 1880 and 1960. Since the mid-1980s, he and other scholars, representing a diversity of disciplinary perspectives, have been exploring the expression and promotion of emotive imperialist sentiments through a range of cultural forms (MacKenzie, 1986; Bristow, 1991). Rather than viewing popular culture as a simple set of material artefacts, it has proved more illuminating to conceive of it as a framework of shared practices through which individual and collective understandings are constructed and negotiated. The impact of this interpretation upon the historiography (and, as I shall argue below, historical geographies) of imperialism has been considerable. Not only has the quantity and diversity of specialist studies multiplied, but previously economically oriented chronological narratives have been fully revised to reflect cultural perspectives (e.g. Hyam, 1993).

What might be termed the 'cultural' reorientation that characterises recent historical studies of imperialism has had an equal impact upon contemporary historical geographies of empire. But before addressing this point, it is worth reflecting briefly upon the fact that much 'imperial propaganda' was consciously targeted at children (MacKenzie, 1984). Certainly it was widely supposed that boys and girls enjoyed: reading the imperialist fictions of Kipling, Henty and Charlotte Yonge; being members of the Boy Scouts, Lads' Drill Association or Girl Guides; or parading as St George, Britannia or the May Queen on Empire Day. Ultimately, however, it was hoped that these activities would prove educational. The boundary between what might be considered formal and informal education has always been blurred and permeable. But a growing number of historians suggest that the regular programmes of study and activity in schools of all types in Victorian and Edwardian Britain were suffused with gender- and class-specific imperialist purpose (MacKenzie, 1984; Mangan, 1988; Davin, 1989). In addition, analyses of ideological indoctrination through schooling in Britain's former dominions and colonies are becoming more common (Mangan, 1993). Today, the task of conveying an understanding of the world and the pupils' place within it is usually assigned to geography. Somewhat paradoxically, however, historical investigations of the connections between education and imperialism have tended to focus upon the history curriculum. The claim that it was history lessons, which instilled children with a vision of the wonders and glories of Britain's rise to territorial dominance, and the conviction that global dominance was the birthright of their 'race', remained largely unchallenged for over a quarter of a century (Chancellor, 1970). The treatment of imperialism by historians of geography was largely confined to the role of ideas (about environment and race, for example) or institutions other than schools (most notably the RGS and the universities).

Only very recently has there been any detailed consideration of the extent to which school geography was embroiled within, or lay outside, the broader ideological discourses of imperialism. This perspective reflects the unprecedented and long overdue broadening of the conceptual criteria defining 'geographical knowledge' and a burgeoning of revisionist approaches to its historical interpretation (Driver and Rose, 1992). No longer is 'geography' conceived of as being embodied in scholarly texts alone. Rather, a complex multiplicity of geographical knowledges, and their popular effects, are beginning to be explored through the widening range of formal and informal social practices and material culture, now acknowledged as being implicated within geographical discourses (Driver, 1995; Matless, 1995a). Thus the cultural re-orientation in the historiography of imperialism has impacted similarly upon histories of geography. Revisionist histories of both geography and imperialism are informed by an increasing range of social, cultural and literary theories that accommodate a diversity of nuanced perspectives. For example, informed by feminist theory, Domosh (1990) and Rose (1993) draw critical attention to the degree to which discussions of geography's past have tended to focus exclusively on white males, thus subsuming and perpetuating the gendered nature of geographical practices. By acknowledging the contingent nature of geographical 'knowledge' and broadening our conceptions accordingly, they argue, marginalised groups and individuals may achieve the recognition that they have long deserved. Likewise, institutional studies are focusing on long-marginalised aspects of geography's history. By exploring the evolution and activities of provincial geographical societies, for example, MacKenzie (1992) has re-oriented perspectives away from their conventional metropolitan focus. A similar adjustment of attention away from élite academic institutions challenges the traditional conviction that developments here eventually disseminated, with unproblematic inevitability, to teacher training colleges and ultimately reached schools in 'suitably simplified' form. Meanwhile, cultural practices such as photography (Ryan, 1997), and popular reading material such as juvenile fiction (Phillips, 1996) and *National Geographic* (Lutz and Collins, 1993), have achieved recognition as important components of geographical discourses.

One of the most prominent themes to have emerged within these critical and contextual reappraisals of geography's past is the complex and intimate relationship between histories of geographical knowledge and imperialism. Since Hudson's pioneering account (1977) of their chronological coincidence and material association, new perspectives and ever more nuanced, theoretically informed, accounts have added considerably to our understandings. It has become, in short, difficult for either geographers or historians to refute Smith and Godlewska's assertion (1994: 2) that imperialism was 'a quintessentially geographical project', serving imperialist impulses in both pragmatic and conceptual terms:

> Geographers – both the amateur explorers and the university-based scholars – were the essential midwives of European imperialism. They provided both the practical information necessary for overseas conquest and colonisation and the intellectual justification for expansion through their increasingly elaborate 'theoretical' writings on geo-politics and the impact of climatic and environmental factors on the evolution of different races. (Bell *et al.*, 1995: 6)

In a practical sense, European territorial expansion depended upon the exploration and, ultimately, the colonisation of the distant reaches of the globe. The geographer's most specialised tool, the map, was thus a potent imperialist device. By surveying hitherto 'unknown' physical and human environments, and representing them in a single, ordered, codified image, geographers made distant places 'known' to Europeans. Thus geography provided the techniques and assumed scientific objectivity through which the burgeoning of information about the world could be rationalised, ordered and classified. This empowered European nations to locate valuable exploitable resources and evaluate the feasibility of settlement in particular regions.

None the less, as we have already seen, imperialism entailed political, intellectual and cultural, as well as material, domination. Empires were built and maintained as much by attitudes and ideas as by territorial acquisition. There is increasing recognition that geography's role in these less tangible elements of the imperial agenda warrants greater critical attention. Thus there is a convergence of interest in the interconnected cultures of geography and empire among historical geographers and across the social sciences. It is broadly agreed that to leave the connections and articulations between geography and imperialism unexamined implicates us all in the continual recirculation of powerful and divisive cultural myths. As Edward Said (1993: 18) stresses in his argument that imperial ideologies remain at the heart of modern western culture, '[m]ore important than the past itself . . . is its bearing upon cultural attitudes in the present' (see also Chapter 4). Viewed from Said's perspective, the cultures of geography and imperialism become inextricably linked. Imagined geographical 'knowledge' was central to the maintenance (both at 'home' and abroad) of imperial domination, and crucial to the formation of cultural identities, both in and beyond Europe. Through what Hall (1992) calls a discourse of 'the West and the Rest', notions of difference were reinforced by terminology which emphasised geographical separation. 'Knowledge' of Britain and notions of Britishness were constructed in direct opposition to the assumed characteristics of 'other' places and peoples.

Historical geographers have been directing increasing attention to the manifold means by which these imagined geographies were created and disseminated. Studies of leisure activities as diverse as photography (Ryan, 1997), childhood reading (Phillips, 1996), and visiting zoos (Anderson, 1995), museums (Haraway, 1989) and exhibitions (Driver, 1994) are leading to more nuanced understandings. But with a few notable exceptions, the geographical 'knowledge' imbibed by children through formal schooling has remained marginal to the project of critical reappraisal. Thus, our understandings of geography's histories and its relationship with imperialism have remained partial and élitist. By focusing on school geography, the remainder of this chapter explores the geographies imagined by the masses during their formative years – surely among the most empirically significant and ideologically powerful dimensions of both geographical and imperial discourses.

Popular geographical education and imaginative geographies of empire

As we have seen, the parameters defining 'geographical knowledge' have been extended considerably in recent years. Geography is no longer conceived of as a monolithic body of unproblematic 'facts' and theories created exclusively within the academy. It is now considered as comprising diverse sets of contingent and subjective knowledges and understandings of the world. Furthermore, these are recognised as being constituted and transmitted through popular social and cultural practices, as well as through dedicated printed texts. The relationship between different geographical knowledges, for example, the extent to which they may be mutually supporting or in conflict, is as yet little understood. Recent studies of popular geographical education, however, are beginning to shed considerable light upon the construction and perpetuation of imaginative geographies of empire. Maddrell (1996), for instance, examines British elementary school textbooks published between 1880 and 1925 in the context of government emigration policies during the period. She argues that these were part of a broader imperial discourse valorising emigration to the colonies as an expression of patriotic and responsible citizenship. By contrast, Nash's study (1996) of the educational writings and career of one individual indicates that geographical education had the potential to convey distinctly anti-imperialist sentiments. In Ireland and India, she explains, James Cousins advocated the teaching of school geography as a source of ordered knowledge, mystical insight and resistance to imperialism. This, he contended, would instil notions of citizenship based upon non-hierarchical difference and global spiritual unity.

Maddrell and Nash are among a growing number of historical geographers to subject particular aspects of popular geographical education in Britain and the former colonies to critical and contextual analysis. For example, applying a broad and informal definition of 'geographical education', Matless (1995b) and Gruffudd (1994, 1996) explore the ways in which the British countryside was used to provide practical lessons in patriotism and citizenship during the interwar period. Others have examined the role played by school geography curricula in the consolidation of notions of national or imperial identities in Argentina, Australia, Canada, China, Ecuador, France, Japan, New Zealand, South Africa and Sweden (Lilly, 1993; Hooson, 1994; Radcliffe, 1996). Nevertheless, in contrast to school history, the formal study of geography in British schools in the past has attracted little sustained contextual historical analysis.

Textbook studies – making all the difference in the world?

Overview and critique

The earliest studies of school geography sought to identify and excise ideological biases in contemporary textbooks and teaching materials (for example, Hicks, 1980, 1981). More recently, a recognition of the need for improved understandings of the interconnected cultural discourses of imperialism and geography has inspired several historical studies of textbooks (Marsden, 1990; Butlin, 1995; Marsden, 1996). Initially, there was some danger of this laudable remit leading to a somewhat uncritical acceptance of a straightforward relationship between popular geographical knowledge and imperialism. Other studies, however, are adopting a more discriminating perspective, with geography textbooks being appraised contextually in relation to broader educational, disciplinary and political discourses (Ploszajska, 1995, 1996a; Maddrell, 1996). This is essential in achieving an enhanced understanding of the articulations between popular geographies and the culture of imperialism. Awareness of issues surrounding text selection, analysis and interpretation is equally vital. Quantitative methods of content analysis, for example, allow no latitude to 'read between the lines'. This, however, is essential to the study of ideological biases, which are as frequently manifested by omission of information as by its inclusion. Said (1978), Pratt (1992) and others have amply demonstrated the effects upon cultural identities and geographical imaginations of the exclusion of the personal statements or perspectives of the Other. Any attempt to evaluate the possible and probable cultural effects of textbooks therefore demands qualitative analytical reading rather than quantitative measurement. But reading is itself a contested field of research activity and the focus of considerable contemporary theoretical debate (see, for example, Scholes, 1989). Accordingly, there is a range of formally defined modes of reading, many different ways of interpreting even a seemingly unequivocal text. The analysis upon which the following discussion draws entailed a close and self-reflexive hermeneutic reading of both verbal and visual imagery in geography textbooks used in London schools between 1870 and 1944. Developing an appreciation of their prevailing tone and ideological messages demanded an assessment of content balance (in terms both of inclusions and omissions) and format of entire books and series of books. Those cited here are (in my necessarily subjective opinion) faithfully indicative of broader generalities.

Textbook representations of imperial landscapes and peoples – the case of Australia

Representations of Australia and Australians provide a particularly interesting case study for this necessarily brief discussion, since they were frequently described as being the very antithesis of Britain and Britons. Such rhetoric of difference is seen by many contemporary scholars as an essential mechanism through which nations counter-identify. Here, I discuss textbook representations of Australia's physical environment, flora and fauna, indigenous population, and colonial settlers. I conclude by reflecting upon the persistence of such imagery and its attendant attitudes today.

Physical environment

Pupils in Victorian London learned from their geography textbooks that Australia was 'an immense barren wilderness (which) belongs entirely to the British' (Mackay, n.d.: 42). Its dangerous and hostile environment was a common theme throughout the period under discussion. Many authors recounted tales of European exploration of the continent's interior, popularising an image of an unknown and unconquerable land that seemed to conspire against its own measurement, survey and classification as legitimated European 'knowledge'. Others emphasised the value of Australia's supposedly inexhaustible mineral resources and vast pastures. Illustrative of 'the moral discourse of climate', which Livingstone (1991) suggests is threaded through the corpus of modern geographical knowledge, climate assumed centrality in many descriptions of Australia. Text after text expounded its benefits (excepting in the tropical north) to Britons' health and vitality. As we will see, however, the climate was not perceived to have extended its extraordinarily enervating effects to the country's indigenous population. This was just one of the many paradoxes that children encountered while learning about Australia. Indeed, some authors appear to have emphasised the country's perversity and opposite-ness exclusively in order to excite children's imaginations. Green (*c.* 1924: 14), for example, began: 'Because Australia is on the opposite side of the world, the things that happen there are just the opposite from what happens in our country.' This startling statement was not followed by any explanation of seasonal and diurnal variations between the hemispheres (although some texts, even for the very young, achieved perfectly comprehensible accounts of these fundamental geographical phenomena). Numerous other authors resorted to fairy-tale rhetoric, describing Australia as 'the Never-Never Land' (Hardingham, 1934: 73), 'Topsy-Turvy Land' (Cameron, 1912: 86), 'Upside Down Country' and 'Contrary-Land' (Yates, 1893: 160). Handbooks instructed teachers to emphasise the peculiarities of the antipodes in their oral lessons too: 'point out, on the globe that (the inhabitants) must walk with their feet against ours, and that their heads must point in opposite directions' (Murché, 1902: 324). The extraordinary weirdness of Australia, 'a little world of its own, full of oddities' (Lyde, 1924: 67), was a recurrent theme in both introductory and advanced texts. This was most explicit of all in their descriptions of its native life forms.

Indigenous plant and animal life

Geography textbook authors were unanimous about the uniqueness of Australia's animals and plants. Texts for young children invited them to imagine a fantastic land where the trees gave no shade, dogs could not bark, animals laid eggs like hens, flowers had no scent, birds had no song, and 'most of the animals carry their young one about in a bag' (Cameron, 1912: 89). Textbooks intended for older pupils explicitly attached value judgements to these peculiarities, creating a composite image of monotonous, ugly and primitive flora and fauna. Authors variously described these as 'more curious than attractive' (Fairgrieve and Young, 1923: 104), 'living fossils' (Muir, 1924: 51), and 'the grotesque, the weird, the scribblings of Nature learning how to write' (Herbertson and Herbertson, 1903: 54). Many remarked upon Australia's failure to contribute any useful plant or animal to global and, more importantly, imperial food supplies or commerce. British children were taught that its value lay, rather, in the vast territories available there for the cultivation of British crops and as pasture for British livestock. And as the animal and plant life of the 'mother country' colonised this overseas 'possession', its peculiar, primitive, indigenous species were disappearing. The texts suggested, unequivocally, that this was all part of the natural world order.

Indigenous peoples

Reflecting discourses of environmental determinism, which were central to late-nineteenth and early-twentieth century geographical thought, school textbooks pointed out that the natives of such a weird and primitive place as Australia were 'naturally the most degraded in the world' (Lyde, 1920: 108). Innumerable texts taught children that mankind (*sic*) was divided into three (occasionally five) 'great races', whose physical and moral characteristics were directly related to the colour of their skin. Capturing the tone of texts throughout the period, Barrington Ward (1879: 45) summarised the situation thus:

> The white race is the highest. White men are nearly all civilised, and many of the yellow race are not far behind. Some of the black men have civilised too, but by far the most of them are savages, or in that half-savage state we call barbarous.

Within this racialised framework of understanding, Australian aborigines were almost invariably compared unfavourably with the non-white populations of all other British imperial territories. Individually and collectively, the texts presented an almost entirely unequivocal narrative of a people they judged to be the world's most savage, uncivilised, dirty and barbaric. The alleged elements of aboriginal life upon which such judgements were most commonly based were lack of permanent shelters, failure to cultivate food crops, brutality to women, and habitual nakedness. Occasionally, analogy was drawn between aborigines' contemporary condition and that of primitive ancient Britons. But it was far more usual to describe their existence as animal-like. Children read about tribes (or packs) that wandered around the wilderness fighting, sleeping, and devouring insects, roots and beetles. Some texts did suggest that the Australian 'natives' possessed enviable hunting and tracking skills. But even these admirable qualities were attributed to their evolutionary proximity to their prey – as further proof of their unity with the animal kingdom. Typically, texts described aborigines' limbs as scrawny, and their faces and bodies as covered with animal-like fur. They were almost invariably referred to as 'Black-fellows', a gendered label that transcended the usual distinctions between human males and females. Most discussions ended by remarking that, like the native plants and animals, the 'race' was dying out and its eventual extinction was inevitable. Explanations for their dwindling numbers varied, but most authors suggested that the 'natives' had resisted all efforts to improve their condition and sustain their existence, rejecting the virtues of civilisation and appropriating its potentially fatal vices (such as alcohol) with child-like cussedness. Few texts suggested that the demise of Australia's indigenous population should be cause for remorse on the part of the British. Indeed, the only regret that was expressed concerned the resultant shortage of 'reliable cheap coloured labour'.

Colonial settlers

In common with much popular juvenile fiction, geography texts portrayed British settlers in Australia as hard-working, self-reliant individuals determined to achieve prosperity for themselves, Britain, and the empire through their honest labours. Children were encouraged to identify with these compatriots, and to aspire to their admirable personal qualities. Many of the texts explicitly prompted pupils to consider the possibility of emigrating in the future. Text after text explained that emigration relieved the pressure of population at home and increased domestic supplies of food and raw materials for manufacture. By migrating to the colonies, children were taught, they could settle throughout the world and remain British – citizens of 'the land of the free', subject to the most just legal system and most democratic government the world had ever known. Within broader textbook discourses of imperial migration, Australia's immigrants were generally represented as intensely patriotic, industrious and dutiful British citizens. Their lives were depicted as prosperous and civilised in city and countryside alike. In stark contrast to the weird and primitive country in which the 'black-fellows' roamed, the immigrant's Australia was portrayed as a land where a bare-footed child or badly dressed adult was a rare sight.

Such, then, was the rosy image of the lifestyle of white Australians which was presented to working-class children in England.

Images from the past in the present

En route to Australia in 1995, I opened my contemporary guidebook and was confronted with rhetoric all too familiar to me from antiquated and – I had assumed – long-obsolete, geography textbooks:

> Stuck off in a far corner of the globe, the giant island-continent is literally a world apart. In many ways, to visit Australia is to pass through something of a time-warp into a Land of Fantasy . . . The seasons are reversed, the trees shed their bark rather than their leaves, and the native flowers have little scent. And there are the strange animals – not just the well-known kangaroo, koala and platypus, but wombats, bandicoots and Tasmanian devils as well. Even the light is different (and) the continent's original inhabitants, the Aborigines, have the appearance of an ancient people. (Robinson, 1994: 12)

During my stay, I discovered that, like geography textbooks before them, tourist postcards played with notions of Australia's 'upside down-' and 'down under-ness'. Others were (photo)graphic representations of aborigines as, quite literally, naked 'black fellows' armed with primitive weaponry or consuming insects. Some drew, nostalgically, on Australia's colonial history. In many respects, then, these images (specifically intended for global circulation) were laden with familiar ideological messages – the same cultural and moral value-judgements as century-old geography textbooks. Furthermore, it became all too apparent that in the present these are more than mere homespun export commodities. Time and again, shallow and good-natured exchanges with taxi-drivers, tour leaders, shopkeepers and restaurant workers became vehicles for white Australians' (unsolicited) views regarding their aboriginal compatriots. I was variously informed that the aborigines are 'a waste of skin', 'ugly – and I mean really ugly', 'diseased', 'lazy', 'inveterate drunkards', 'greedy cheats', 'liars and thieves', and 'dangerous'. Observations such as these reveal the extent to which inherited images of peoples and cultures, and historical notions of their relative appeal and worth, remain imbued within contemporary societies. They provide foundation for collective geographical imaginations and continue to shape popular attitudes and senses of identity. I am not suggesting that school geography textbooks alone created these stereotypes. Their authors were, of course, influenced by the prevailing geographical, educational and cultural discourses to which their work, in turn, contributed. Nor were textbooks the only source of information about the world that pupils had at their disposal. Indeed, they were actively encouraged to draw upon an ever-increasing range of media (see below). Nevertheless, belonging to broader cultural discourses themselves, many of these latter merely echoed the views in textbooks rather than offering alternative perspectives on the world. Within these wider discourses, I suggest that school geography texts played a central role in the process of translating academic theories such as environmental determinism into popular, but implicitly accurate and authoritative, narratives. Finally, it is important to recognise that the texts discussed here exerted influence upon children throughout the empire. In Australia, for example, education remained dominated by materials published in Britain (or modelled on these) well into the 1960s.

Picturing the world

Halford Mackinder is perhaps the best known champion of visual imagery as a means of furthering geographical knowledge and understanding. For Mackinder (1911) and many of his contemporaries, the ability to visualise distant places and peoples with accuracy signified the very essence of the geographer's power. Many believed that vivid and realistic appeal to the eye was absolutely fundamental to the process of training children to visualise the appearance and circumstances of environments and cultures beyond their (usually very limited) personal experience. Only when this was achieved, Mackinder asserted, would Britons truly think, act and identify as an imperial populace. Not only were visual images said to convey information clearly, but the impressions they created were believed to be particularly memorable. Indeed, recent research confirms that geographical ideas communicated visually are usually much clearer and long-lasting than those expounded verbally (Wright, 1979). Although they are often regarded as objective and impartial, be they sketches, paintings, photographs or maps, pictorial representations never create a neutral vision of the real world. Each captures and conveys a particular version of visual 'truth'. Whether they are glanced at or scrutinised in detail, then, their content, omissions, composition and contextual framing are powerfully encoded media for ideological messages (Wood, 1992; Van Leeuwen and Selander, 1995; Ryan, 1997).

Visual images in textbooks

From the very inception of the British state's provision of schooling in 1870, Inspectors (HMIs) were intensely critical of teachers who attempted to impart geographical knowledge solely through unexplained definitions and statistics rote-learnt from factual gazetteers. Believing that the interest of children in the subject could not be captured and developed by verbal explanations alone, they urged that full use be made of pupils' visual faculties. So it soon became usual for the pages of popular school texts to be peppered with illustrations. Authors drew special attention to the number, variety and accuracy of those in their particular publications, invariably claiming these were integral to the book's meaning. It is instructive, therefore, to consider the kinds of images that were chosen most frequently and their relationship to the written texts.

The physical, commercial and political strength of Britain and the empire were among the most common themes of textbook illustrations throughout the period under discussion here. The map of the world on Mercator's projection, with British 'possessions' coloured in red, was an almost ubiquitous frontispiece. Locating the various places on this was said to give pupils a sense of personal power and make rhetoric of possession meaningful. Children

Figure 5.1 Textbook imagery of 'Savages' (*source*: Barrington Ward, 1879).

were instructed to study the red patches carefully until they understood the geographical 'truth' of the claim that theirs was an empire on which the sun never set, and to reflect upon its political and moral implications. It was, many authors suggested, remarkable that such a tiny group of islands dominated the globe: 'How small the British Isles are, when compared with the rest of the earth! Yet they form one of the most important countries on the globe, and are home to the greatest nation' (*Chambers's Twentieth Century Reader*, n.d.: 7). They portrayed this as an unprecedented phenomenon, entirely due to Britain's geographical location and its effects upon the character of the population. Written descriptions of the appearance and lifestyles of 'civilised' white Europeans (particularly the British) and those of non-white 'natives' of, say, Australia were reinforced by starkly contrasting illustrations. Figure 5.1 is a typical picture, intensifying notions of non-white peoples as an indistinguishable mass of near-naked bodies, fighting with bare hands or primitive weapons, dancing around menacingly, or writhing animal-like on the ground. Illustrations of 'civilised people', which often appeared alongside, left English school children in little doubt as to the very different behaviour expected from them as members of 'the most civilised race of all' (Barrington Ward, 1879: 44). These featured neatly dressed, orderly and disciplined family groups. Frequently, as in Figure 5.2, possession of geographical knowledge was represented as a central distinguishing feature of 'civilised' Britons.

Figure 5.2 Textbook imagery of 'Civilised people' (*source*: Barrington Ward, 1879).

There was as much to be read about the relationship between the British and the 'natives' of their overseas territories from textbooks' visual images as from their verbal explanations. Supporting the assumption that manual work was physically impossible for white settlers in the tropics, book after book contained drawings and photographs of 'natives' labouring under the command and supervision of their British 'masters'. Furthermore, English children learned that such arrangements were peculiar neither to the tropics nor to particularly strenuous activities. For example, a generic image of South Africa depicted negroes sorting diamonds under the watchful eye of a white overseer. This supported verbal assertions that all over the world 'others' performed tiresome jobs for Britons:

> For us the Chinaman gathers the leaves of the tea-plant; the Arab shakes the ripe berries of the coffee plant; the Negro cuts down the sugar cane, and picks the down of the cotton-tree; and hunters, trappers and farmers in many lands are ready to supply us with corn, skins, wool and furs. (Anon., 1883: 113–4)

In addition to commercial service to their homeland, visual images informed pupils that it was the naturally determined duty of the 'natives' of imperial possessions overseas to ensure the maximum comfort of British settlers. Innumerable texts included pictures of white colonists and explorers attended, whether at work or leisure, by non-white servants (often portrayed quite literally as their beasts of burden).

The assertion that certain peoples were 'naturally' suited to occupy specific geographical locations was among the most pervasive visually illustrated themes,

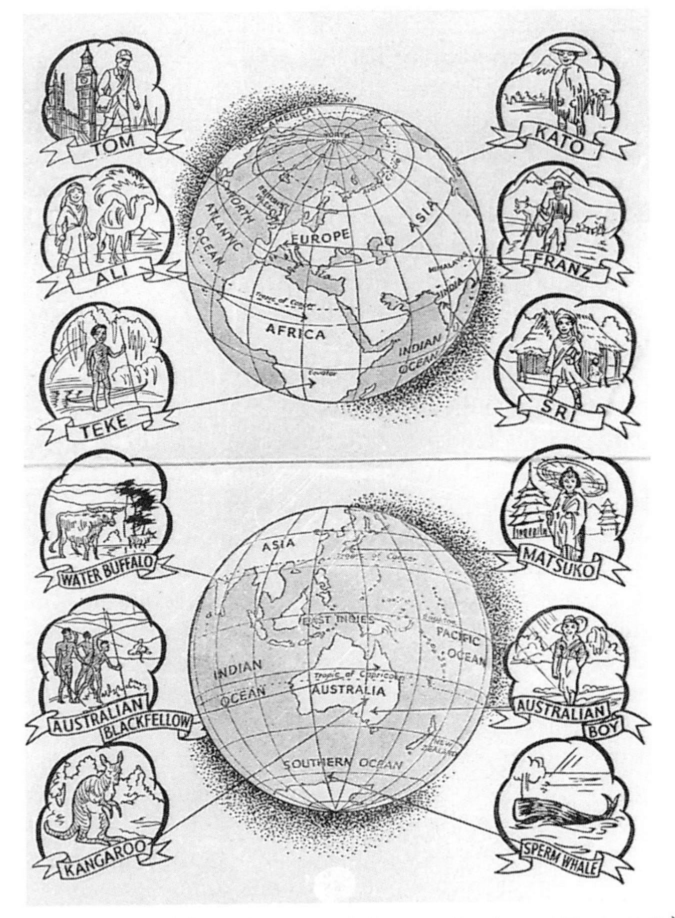

Figure 5.3 Children of many lands (*source*: Brooks and Finch, 1928).

exemplified by Figure 5.3. This was the frontispiece to the first in a progressive series of books, commonly used as class texts throughout (and beyond) the interwar period. It was therefore many English children's very first impression of geography, and its messages were reinforced each time they opened their textbook. In common with many other introductory texts (Ploszajska, 1996a), the narrative took pupils on an imaginary journey comparing the lives and appearances of children around the world. Characteristic of verbal and visual representations in other textbooks across the entire period, this image allocated various 'races' to their 'appropriate', spatially separate, locations and physical environments. Several enforced this lesson unequivocally, instructing pupils to copy out statements like 'Black men live in the south . . . Brown men live in the East . . . Yellow men live in the Far East . . . Red men live in the West' (Anon., 1887: 87, 102, 106 and 110). In this image, the notion that particular peoples (and their lifestyles) 'belonged' in particular places was conveyed by the physical settings in which the stereotypical children appeared. Their environmental surroundings could be 'read' as indicators of their cultural circumstances. The

British boy, for example, appeared before the Houses of Parliament, which served as an easily recognisable symbol for his nation's rational, just, stable and democratic political and social organisation. Britain, pupils learned from text after text, was 'the land of the free', whose citizens enjoyed greater liberty and even-handed justice than those of any other nation. In contrast, the North African boy was pictured against a backdrop of pyramids. This reflected suggestions in many textbooks that whilst this area had been a seat of cultural achievement historically, it had long-since reached stagnation and failed to progress further. Likewise, the portrayal of the central African boy surrounded by dense, entangled forest mirrored recurrent verbal and visual representations of Africa as the impenetrable, unknowable, 'Dark Continent'.

The image of indigenous Australian culture, though labelled in the singular, depicted three (apparently mature) individuals hunting in a group, almost an animal-like pack. This reflected the tendency of textbooks to bracket indigenous Australians with the animal kingdom rather than with humankind. The white 'Australia Boy', by contrast, stood confident and alone in an altogether less bleak and more domesticated part of this vast continent. Off-set against the Other (and Othering) Australian image, this illustrated the argument of numerous authors that the native inhabitants of warm climates became languid and sluggish, making no effort to exploit the natural potentials of their environment. Only white men (*sic*), they suggested, had the intelligence, determination and strength of character to dominate nature. Dressed like a Boy Scout, the white Australian child represented the exemplary imperial pioneer who, the image suggested, had learnt to command nature and tame his environment. It is interesting that, apart from the 'Australian Blackfellow', this was the only child in the composite image not to be assigned a name. Perhaps this was intended to encourage British boys to imagine themselves in his place, to ascribe their own name onto this character, and to assume some of his imagined attributes themselves.

In its totality, this composite image illustrates several themes recurrent in geography textbooks throughout the late nineteenth and early twentieth centuries. Most starkly, it endorsed the view that white people were civilised, intelligent, resourceful and progressive, and that Others were uncivilised and stupid. Moreover, because the illustration was an uncoloured line drawing, gradations of skin colour were indistinct. The peoples of the world were therefore dichotomised as, quite literally, black or white. Yet, like most verbal accounts, it conveyed a decidedly harmonious overall impression. All the children were smiling, apparently happy with their assigned position in the global order of things. So British children could be satisfied with their superiority and, at the same time, be assured that this was unlikely to be challenged either from within, or beyond, their imperial territories.

Other sources of visual imagery

By the interwar period, when the representation we have just reflected upon in some detail was produced and first consumed, geography pupils were collecting visual images of the world from a wide range of diverse sources. This was just one of many ways through which teachers endeavoured to involve children actively in the acquisition of geographical knowledge. Collecting geographical pictures from newspapers, advertisements, journals, postcards, cigarette cards, stamps and travel and immigration brochures, was also said to convey the subject's reality and everyday relevance. Practices of this sort were especially well suited to prevailing educational policies and thought during the interwar period but, as we have seen, the importance of the visual to geographical education was established long before. From the 1870s, educationalists and geographers alike urged teachers to illustrate their geography lessons with appropriate visual images wherever possible. Lantern-slides were deemed an especially effective medium. Indeed, the perceived urgency of securing sufficient and suitable slides of the British empire inspired the foundation of the Geographical Association (GA).

Although the GA was concerned with broader issues regarding the status of school geography from the outset, lantern-slide provisions remained central to its early activities. The very first image it acquired was a map of the world on Mercator's projection indicating, as was conventional, all of Britain's imperial possessions in red. Maps showing trade routes and the global distribution of ethnographic 'types' soon joined this, and within six months the GA had collected sets depicting Africa, Asia, Europe, South America and North America. Thereafter, its lantern-slide library grew steadily, in terms both of its content and membership. Subscribing schools received regularly updated catalogues of the visual resources at their disposal. The West African Collection was typical, containing 40 images of various physical and human features of the region. The 1903 catalogue (retained at the Association's Sheffield headquarters) provides a brief description of each view, drawing attention to particular features (Table 5.1). Like the verbal and visual imagery in many contemporary textbooks, these captions provided English school children with a representation of West Africa as a region where primitive, chaotic, superstitious native peoples and rational, industrious, civilised Europeans co-existed to mutual advantage, albeit in starkly contrasting and strictly segregated domestic circumstances. In harmony with dominant attitudes of the time, then, the GA's lantern-slides presented an imperialist vision of Britain's overseas territories.

Through to the interwar period and beyond, lantern-slides were regarded as among the most useful educational aids at the disposal of geography teachers, especially in their efforts to 'increase the possibilities of bringing the classroom into contact with the outside world' (Dempster, 1939: 94). There was general agreement that slides stimulated pupils' interest and rendered lessons vivid, realistic and memorable. In addition, teachers collected geographical pictures from an increasing range of sources. Advertisements and promotional posters supplied to schools free of charge by shipping companies were considered especially suitable materials. But posters issued by the Empire Marketing Board (EMB) were the most prevalent large-scale illustrations in geography lessons during the interwar period. Between 1926 and 1933, the EMB published some 800 vivid and eye-catching posters designed to influence consumer choice in

Table 5.1 Some descriptions drawn from the Geographical Association West African Lantern-slide Collection Catalogue, 1903

Collection number	Description of view
2	Freetown, Sierra Leone. This is a well laid out town with concrete or brick buildings, well drained streets, and one of the principal ports of West Africa. The mountains at the back are thickly wooded, though to the right may be seen some of the bungalows of the European town or hill station.
8	Native village scene. Note the straggling character of the huts. The village is in a forest clearing; the men in the foreground are wearing goods from Lancashire. Goats are found everywhere.
20	There are still throughout the forest region many native priests or medicine men as they are sometimes called, and here is one sitting in the fetish hut. Notice what one may call the idol and the drum.
32	Before roads and railways, and even now in many parts, the European travels by hammock. This is one kind of hammock. It will be noticed that it is carried on the heads of four Africans.

Source: Geographical Association, Sheffield.

favour of empire produce. In the interests of long-term popular effects, the Board explicitly targeted children and issued specially reduced copies of its posters free to schools. A much publicised geographical teaching resource, by 1933 some 27,000 schools (the vast majority, ranging from state infants' schools to Eton) in Britain received these (MacKenzie, 1984; Constantine, 1986). They included stylised maps of imperial trade and transport routes. Another common theme was the production of raw materials and foodstuffs for British domestic supply and consumption. These presented school children in Britain with views of, for example, Sudanese cotton pickers, a tobacco plantation in Southern Rhodesia, and fruit farming in South Africa. The posters were said to convey a realistic sense of the lives, homes and work of the peoples around the empire whose efforts provided Britain with everyday necessities and luxuries. Other posters featured shops crammed with imperial produce labelled by country of origin. These enforced in children's minds the links between their own local grocers and the far-flung reaches of the empire. In sum, most geographers and educationalists agreed that the EMB posters provided realistic and evocative images which linked the parochial concerns of British school children with the outside world in meaningful ways. Many felt that they fostered a sympathetic understanding of other peoples and cultures.

Imag(in)ing 'our' place in the world

Throughout the period under discussion here, the study of visual images was an integral component of school geography lessons. Geographers and educationalists shared broad agreement that appeal to the eye was among the most effective means of capturing pupils' interest in the subject and conveying geographical ideas with clarity. Thus trained, it was said, children developed the habit of accurate first-hand observation and the ability to imagine peoples and places beyond their personal experience. Halford Mackinder and many of his late-nineteenth and early-twentieth-century contemporaries considered these faculties essential not only to geographers, but to British citizens more generally. As an imperial nation, they suggested, it was imperative that the populace be able to envision the conditions of their overseas possessions with clarity and realism. During the interwar years, greater emphasis was given to the promotion of international tolerance and understanding through visual images. Photographs, for example, were said to convey a sense of place that fostered notions of personal intimacy. Collecting up-to-date visual images from newspapers and food packaging was intended to encourage pupils' realisation of world interdependence. Throughout the 1920s and 1930s, practices of this sort, and some textbook scripts, encouraged a humanistic, egalitarian global outlook. For the most part, though, visual images continued to portray the world from a distinctly imperialist or Anglocentric perspective. As we have seen, in essence textbook illustrations altered little between the 1880s and 1940s. To be sure, by the end of the period, photographs had almost entirely replaced the crude woodblock prints of early publications. But their ideological messages remained rather more constant than this apparent transformation might suggest. The majority, whether photographs, line-drawings, maps or diagrams, continued to encapsulate assumptions of British supremacy. Let it be remembered that the particular educational value of visual images was said to lie in their power to convey geographical ideas in an especially memorable way. And, indeed, in popular memory (or imagination) 'the red bits on the map' remain a potent and pervasive symbol of school geography lessons in the past.

Notwithstanding the close associations between geography and the visual, many educationalists believed lessons that appealed to senses other than sight alone were more effective still. Geographical modelling (Ploszajska, 1996c) and fieldwork (Ploszajska, 1998) exemplified this principle. Concrete realities and embodied experiences rather than abstract theoretic notions were pivotal to these closely related activities. They demanded the full sensory, physical and imaginative involvement of pupils in the very creation of geographical knowledge and understanding. However, while mobilising muscular and mental activity and senses of not only sight but also sound, touch, smell and (occasionally) taste, concerns with the visual remained central in discussions of modelling and fieldwork. Both were judged extremely effective in developing the power to visualise distant places and peoples accurately. The perceived need to cultivate this among imperial citizens was reflected in the rhetoric with which the practices were discussed during the early decades of the period. Thus Victorian and Edwardian geographers and educationalists emphasised the value of models in providing children with accurate and tangible knowledge of Britain's territories overseas. At a more conceptual level, observing, handling, making and naming models of these places was said to make the

Figure 5.4 The Colonisation Game (*source*: Lewis, 1909).

notion of possession meaningful. By the interwar period, their contribution towards the creation of a more sympathetic global outlook was foregrounded. Constructing models of the world's diverse physical, economic and cultural environments, it was suggested, developed children's understanding and tolerance of human differences.

Discussions about fieldwork underwent a similar, though less pronounced, ideological shift over time. Rhetorical allusion to geography pupils in the field (be it rural or urban) as imperial explorers retained popularity as means of introducing children to geography as a subject of excitement, adventure, and practical application. Pupils were encouraged to see themselves as heroic explorers following in the footsteps of the discipline's founding fathers, a concept familiar to them from their many textbooks which used adventure stories as ciphers for geographical knowledge (Ploszajska, 1996b). Some teachers devised elaborate field activities that made the imperial dimensions of these discourses quite explicit. Pupils at Kentish Town Road School in London, for example, learned geography through a series of visits to Hampstead Heath where they participated in the 'Colonisation Game' (Figure 5.4). This entailed the imaginary conquest, exploration, settlement and development of a previously 'undiscovered' country (Lewis, 1909). Fieldwork of this sort offered highly valued practical training for responsible and proactive imperial citizenship.

During and after the First World War, however, there were increasing calls for geography lessons that promoted a more egalitarian global perspective.

Systematic local surveys became increasingly numerous and detailed. Many saw this as a way of strengthening both local and national patriotism, along with notions of the shared responsibilities of citizenship. Furthermore, even the most apparently parochial fieldwork could be harnessed to inspire a meaningful sense of global interdependence. For example, study of commodities in local shops highlighted the dependence of pupils upon peoples and environments the world over for their daily necessities and cherished luxuries. This fostered, many commentators believed, a sympathetic sense of common humanity and international tolerance.

Conclusion

This chapter commenced with a discursive overview of the recent historiographies of geography and empire, emphasising the ways in which geography and imperialism have come to be conceived of as intimately related sets of cultural practices and ideas. The consequent critical reappraisals of the interconnected popular discourses of geography and imperialism have improved our understandings of both. In particular, restoration of hitherto marginalised individuals and practices to the frame of enquiry has produced more nuanced accounts that accommodate multiple experiences and perspectives. Within this intellectual and theoretical context, the body of the chapter has explored the role of popular geographical education in the imaginative construction of empire. Whilst focusing primarily on textbooks, these have been viewed in relation to the wider cultural, education and geographical discourses by which they were informed and to which they, in turn, contributed. I have emphasised, for example, that school geography textbooks were just one among the many media through which ideas about, and attitudes towards, colonised landscapes and peoples were constructed and transmitted. Nevertheless, I have suggested that textbooks' verbal and visual imagery played a key role in the popularisation of supposedly accurate and objective narratives of empire. As we have seen, these exerted (and continue to exert) hierarchical influence and power both in and beyond Britain.

Throughout, it has been stressed that knowledge production is never value free. This is equally true of the 'geographical knowledge' under scrutiny here, and the historiographical 'knowledge' of geography and imperialism thus derived. Interpretation is always and necessarily context-bound and subjective. For the purposes of this chapter, we have examined school geography teaching and learning in the past almost exclusively through the lens of British imperialism. The empire was certainly a constant material and imaginative theme from 1870 right through to 1944. However, this was neither all-pervasive nor a singular discourse. While verbal and visual images in teaching materials tended, for the duration of the period, to devalue non-British cultures and peoples, an increasing variety of resources and activities were employed in order to encourage a more sympathetic global outlook. As one geography teaching manual published in 1923 put it:

> To be taken out of our narrow egoistic selves, to see and feel things as others see and feel them, to see the world and all its life and activities from the point of view of those many others in our own and other lands whose lives and work we can only know through imagination and sympathy, this indeed is the real education which geography can bring to both young and old, the education of the Head and the Heart concerning the peoples of the world. (Welpton, 1923: 78)

This increasingly common ideal was expressed with ever-greater urgency during the interwar period. Yet the concurrent popularity of resources produced by the Colonial Office, the Empire Marketing Board and the Imperial Institute ensured that distinctly imperialist perspectives remained central to much classroom practice. Indeed, examination questions set and answers given during the 1930s indicated a resurgence of interest in empire geography among both teachers and pupils. Some schools continued to teach geography of the empire in terms of the relative strategic or commercial value of various territories and the 'problems' presented by their governance and development. It was not uncommon, for example, for grammar school pupils to spend their final year studying the geography of Australia largely from the perspective of its 'retarded' development – a result, pupils were taught, of the lack of 'coloured labourers' (Shaw, 1934). Meanwhile, even during the Second World War, annual Empire Day celebrations and membership of overtly imperialist youth movements kept their status as citizens of 'the empire on which the sun never sets' to the fore of many children's consciousness.

Ultimately, popular geographies of empire were constructed and negotiated through a complex interplay of formal and informal materials and practices. In the context of schooling alone, individual teachers commanded considerable power either to reinforcing or to challenging hegemonic ideologies. In turn, parental attitudes and the particularities of his or her social, economic, physical and cultural environments rendered individual pupils variably susceptible to such efforts. What seems certain is that the precise role of school geography lessons within broader cultural discourses of imperialism varied so considerably as to negate generalisations. Indeed, part of the value of this kind of contextual study lies in its very insistence upon the accommodation of diverse experiences and perspectives, and the nuanced exploration of the multiple interconnected histories of geography and imperialism.

References

Anderson, K. (1995) Culture and nature at the Adelaide Zoo: at the frontiers of 'human' geography. *Transactions of the Institute of British Geographers*, **20**, 275–94.

Anon. (1883) *The World at Home, Standard III – England and Wales*, Nelson, London.

Anon. (1887) *Longman's New Geographical Readers – The First Reader for Standard I*, Longman, London.

Balchin, W. G. V. (1993) *The Geographical Association: the First Hundred Years*, Geographical Association, Sheffield.

Barrington Ward, M. J. (1879) *The Child's Geography for Use in Schools and Home Tuition*, Marcus Ward, London.

Bell, M., Butlin, R. and Heffernan, M. (eds) (1995) *Geography and Imperialism 1820–1940*, Manchester University Press, Manchester.

Bristow, J. (1991) *Empire Boys: Adventures in a Man's World*, HarperCollins, London.

Britton, S. G. (1980) The evolution of a colonial space-economy: the case of Fiji. *Journal of Historical Geography*, **6**, 251–74.

Brooks, L. and Finch, R. (1928) *The Columbus Regional Geographies, Book I – Children of Many Lands*, University of London Press, London.

Butlin, R. (1995) Historical geographies of the British empire, *c.* 1887–1925. In Bell, M., Butlin, R. and Heffernan, M. (eds) *Geography and Imperialism 1820–1940*, Manchester University Press, Manchester, 151–88.

Cameron, I. (1980) *To the Farthest Ends of the Earth, 150 Years of World Exploration: The History of the Royal Geographical Society, 1830–1980*, Macdonald, London.

Cameron, M. (1912) *The World in School Book II – Our Wonderful World*, Nisbet, London.

Chambers's Twentieth Century Geographical Reader, Book III – England and Wales, (n.d.), Chambers, London.

Chancellor, V. (1970) *History for their Masters: Opinion in the English History Textbook 1800–1914*, Adams and Dart, Bath.

Christopher, A. J. (1997) 'The Second City of Empire': colonial Dublin, 1911. *Journal of Historical Geography*, **22**, 151–63.

Constantine, S. (1986) *Buy and Build: The Advertising Posters of the Empire Marketing Board*, HMSO, London.

Davin, A. (1989) Imperialism and Motherhood, in Samuel, R. (ed.) *Patriotism: The Making and Unmaking of British National Identity – Book 1, History and Politics*, Routledge, London, 205–35.

Dempster, J. B. (1939) Training for citizenship through geography. In Association for Education in Citizenship (ed.) *Education for Citizenship in Elementary Schools*, Humphrey Milford, London, 92–107.

Domosh, M. (1990) Towards a feminist historiography of geography. *Transactions of the Institute of British Geographers*, **16**, 95–104.

Driver, F. (1992) Geography's empire: histories of geographical knowledge. *Environment and Planning D: Society and Space*, **10**, 23–40.

Driver, F. (1994) *Geography, Empire and Visualization*, Royal Holloway (Department of Geography Working Paper, No. 1).

Driver, F. (1995) Geographical traditions: rethinking the history of geography. *Transactions of the Institute of British Geographers*, **20**, 403–4.

Driver, F. and Rose, G. (1992) (eds) *Nature and Science: Essays in the History of Geographical Knowledge*, IBG, London (Historical Geography Research Series, No. 28).

Duncan, J. (1989) The power of place in Kandy, Sri Lanka, 1780–1980. In Agnew, J. and Duncan, J. (eds) *The Power of Place*, Routledge, London, 185–201.

Fairgrieve, J. and Young, E. (1923) *Human Geographies Book III – Euro-Asia*, George Philip, London.

Godlewska, A. and Smith, N. (eds) (1994) *Geography and Imperialism*, Blackwell, Oxford.

Green, D. G. (*c.* 1924) *Our Island Cousins*, Edward Arnold, Leeds.

Gruffudd, P. (1994) Back to the Land: historiography, rurality and the nation in interwar Wales. *Transactions of the Institute of British Geographers*, **19**, 61–77.

Gruffudd, P. (1996) The countryside as educator: schools, rurality and citizenship in interwar Wales. *Journal of Historical Geography*, **22**, 412–23.

Hall, S. (1992) The West and the Rest. In Hall, S. and Gieben, B. (eds) *Formations of Modernity*, Open University Press, London, 275–320.

Haraway, D. (1989) *Primate Visions*, Verso, London.

Hardingham, B. G. (1934) *Foundations of Geography I – Round the Globe*, Nelson, London.
Herbertson, F. D. and Herbertson, A. J. (1903) *Descriptive Geographies from Original Sources – Australia and Oceania*, A & C Black, London.
Hicks, D. (1980) Bias in Books: Not Recommended. *World Studies Journal*, **2**, 14–22.
Hicks, D. (1981) The contribution of geography to multicultural misunderstanding. *Teaching Geography*, **16**, 64–7.
Hooson, D. (1994) (ed.) *Geography and National Identity*, Blackwell, Oxford.
Hudson, B. (1977) The new geography and the new imperialism, 1870–1918. *Antipode*, **9**, 12–19.
Hyam, R. (1993) *Britain's Imperial Century, 1815–1914: A Study of Empire and Expansion*, 2nd edn, Macmillan, Basingstoke.
Lester, A. (1998) 'Otherness' and the frontiers of empire: the Eastern Cape Colony, 1806–*c*. 1850. *Journal of Historical Geography*, **24**, 2–19.
Lewis, G. G. (1909) *Typical School Journeys: A Series of Open-air Geography and Nature Studies*, Pitman, London.
Lilly, T. (1993) The black African in Southern Africa: images in British school geography books. In Mangan, J. (ed.) *The Imperial Curriculum: Racial Images in the British Colonial Experience*, Routledge, London, 40–53.
Livingstone, D. (1991) The moral discourse of climate: historical considerations on race, place and virtue. *Journal of Historical Geography*, **17**, 413–34.
Livingstone, D. (1992) *The Geographical Tradition: Episodes in the History of a Contested Enterprise*, Blackwell, Oxford.
Lutz, C. A. and Collins, J. L. (1993) *Reading 'National Geographic'*, University of Chicago Press, London and Chicago.
Lyde, L. W. (1907) The teaching of geography as a subject of commercial instruction. *The Geographical Teacher*, **4**, 163–8.
Lyde, L. W. (1920) *Junior Geography, Book IVc – The British Empire*, A & C Black, London.
Lyde, L. W. (1924) *Man and His Markets*, Macmillan, London.
Mackay, Rev. A. (n.d.) *Geography of the British Empire*, Blackwood, London.
MacKenzie, J. M. (1984) *Propaganda and Empire: The Manipulation of British Public Opinion, 1880–1960*, Manchester University Press, Manchester.
MacKenzie, J. M. (1986) (ed.) *Imperialism and Popular Culture*, Manchester University Press, Manchester.
MacKenzie, J. M. (1992) Geography and imperialism: British provincial geographical societies. In Driver, F. and Rose, G. (eds) *Nature and Science: Essays in the History of Geographical Knowledge*, IBG, London (Historical Geography Research Series, No. 28), 46–51.
Mackinder, H. J. (1911) The teaching of geography from an imperial point of view, and the use that could and should be made of visual instruction. *The Geographical Teacher*, **6**, 79–86.
Maddrell, A. (1996) Empire, emigration and school geography: changing discourses of imperial citizenship. *Journal of Historical Geography*, **22**, 373–87.
Mangan, J. A. (1988) (ed.) *Benefits Bestowed? Education and Imperialism*, Manchester University Press, Manchester.
Mangan, J. A. (1993) (ed.) *The Imperial Curriculum: Racial Images and Education in the British Colonial Experience*, Routledge, London.
Marsden, W. E. (1990) Rooting racism and the educational experience of childhood and youth in the nineteenth and twentieth centuries. *History of Education*, **19**, 333–53.
Marsden, W. E. (1996) Researching the history of geographical education. In Williams, M. (ed.) *Understanding Geographical and Environmental Education: The Role of Research*, Cassell, London, 264–73.

Matless, D. (1995a) Effects of history. *Transactions of the Institute of British Geographers*, **20**, 405–9.
Matless, D. (1995b) 'The Art of Right Living': landscape and citizenship 1918–1939. In Pile, S. and Thrift, N. (eds) *Mapping the Subject*, Routledge, London, 93–122.
Mitchell, T. (1988) *Colonizing Egypt*, Cambridge University Press, Cambridge.
Muir, T. S. (1924) *A Regional Survey of the British Isles*, Chambers, London.
Murché, V. T. (1902) *The Teacher's Manual of Object Lessons in Geography*, Macmillan, London.
Nash, C. (1996) Geo-centric education and anti-imperialism: theosophy, geography and citizenship in the writing of J. H. Cousins. *Journal of Historical Geography*, **22**, 399–411.
Peet, R. (1985) The origins of environmental determinism. *Annals of the Association of American Geographers*, **75**, 309–33.
Phillips, R. (1996) *Mapping Men and Empire: A Geography of Adventure*, Routledge, London.
Ploszajska, T. (1995) *Making all the difference in the world: geography's popular school texts 1870–1944*, Royal Holloway (Department of Geography Research Papers, General Series, No. 3).
Ploszajska, T. (1996a) 'Cloud Cuckoo Land?' Fact and fantasy in geographical readers, 1870–1944. *Paradigm*, **22**, 2–13.
Ploszajska, T. (1996b) *Geographical Education, Empire and Citizenship, 1870–1944*, unpublished PhD thesis, University of London.
Ploszajska, T. (1996c) Constructing the subject: geographical models in English schools, 1870–1944. *Journal of Historical Geography*, **22**, 288–98.
Ploszajska, T. (1998) Down to earth: geography fieldwork in English schools, 1870–1944. *Environment and Planning D: Society and Space*, **16**, 757–74.
Pratt, M. L. (1992) *Imperial Eyes: Travel Writing and Transculturation*, Routledge, London.
Radcliffe, S. (1996) Imaginative geographies, postcolonialism, and national identity: contemporary discourses of the nation in Ecuador. *Ecumene*, **3**, 23–42.
Robinson, C. (1994) *Odyssey Illustrated Guide to Australia*, Odyssey, London.
Rose, G. (1993) *Feminism and Geography: The Limits of Geographical Knowledge*, Polity, Cambridge.
Ryan, J. (1997) *Picturing Empire: Photography and the Visualization of the British Empire*, Reaktion, London.
Said, E. (1978) *Orientalism: Western Conceptions of the Orient*, Routledge, London.
Said, E. (1993) *Culture and Imperialism*, Vintage, Chatto and Windus, London.
Scargill, D. I. (1976) The RGS and the foundation of geography at Oxford. *Geographical Journal*, **142**, 438–61.
Scholes, R. (1989) *The Protocols of Reading*, Yale University Press, New Haven, Conn.
Shaw, H. B. (1934) A Course on the Geography of the British Empire. *The Journal of Education*, **66**, 76–7.
Smith, N. and Godlewska, A. (1994) Introduction: critical histories of geography. In Godlewska, A. and Smith, N. (eds) *Geography and Empire*, Blackwell, Oxford, 1–8.
Steel, R. W. (1984) *The Institute of British Geographers: The First Fifty Years*, IBG, London.
Stoddart, D. R. (1989) A hundred years of geography at Cambridge. *Geographical Journal*, **155**, 24–32.
Van Leeuwen, T. and Selander, S. (1995) Picturing 'our' heritage in the pedagogic text: layout and illustrations in an Australian and a Swedish history textbook. *Journal of Curriculum Studies*, **27**, 501–22.
Welpton, W. P. (1923) *The Teaching of Geography*, University Tutorial Press, London.
Wood, D. (1992) *The Power of Maps*, Routledge, London.
Wright, D. (1979) Visual images in geography texts: the case of Africa. *Geography*, **64**, 205–10.
Yates, M. T. (1893) *Arnold's Geography Readers Book V*, Edward Arnold, London.

Chapter 6

Historical geographies of the colonised world

Brenda S. A. Yeoh

Introduction

In this chapter, I argue that 'historical geographies of the colonised world' have been overshadowed by the 'historical geography of colonialism'. The latter, usually interpreted as the study of the processes and practices of European colonial domination, needs to be destabilised in order to accommodate what might be termed a 'politics of space' in the colonised world where people resisted, responded to and were affected by colonisation. This is not so much a call for a complete turnaround in perspective from one which considers the ideologies and practices of the colonial enterprise to that which explores the everyday worlds of the colonised peoples, but a move to examine the intersections between the two in fleshed out accounts of the processes of conflict, collusion and negotiation that animated spatial politics.

The chapter begins by tracing the developing critique of standard accounts of the historical geography of colonialism. I then look at alternative projects that draw on notions of the 'contact zone', a multiplicity of 'resistances' and the dynamics of 'spatial politics' as conceptual tools. After a brief account of the colonial enterprise in Singapore in the late nineteenth and early twentieth centuries, I use three different accounts of the politics of space situated in colonial Singapore to illustrate the ways the discourses and practices of domination mesh with those of resistance in colonised spaces.

'Historical geography of colonialism' vis-à-vis 'historical geographies of the colonised world'

Of late, historical geographical approaches to colonialism, which focus exclusively on or privilege the impress of European coloniser culture or capital in shaping the shared worlds of coloniser and colonised groups, have come under more critical scrutiny than before. Two particular streams of scholarship are beginning to converge, and possibly cross-cut.

Indigenous worlds

The first stream draws on the ongoing debate on the 'indigenisation' of academic discourse in the social sciences, a movement which gained momentum in the early 1970s when Third World 'indigenous' scholars, joined by a number of Western counterparts, 'raised their voice against the implantation of the social sciences perpetuating "captivity" of mind' beyond the colonial era (Atal, 1981: 189). The plea was for an articulation of 'grassroots' consciousness and a rejection of 'borrowed' consciousness, the development of alternative perspectives that provide an insider's view of reality and a valorisation of historical contexts and cultural specificities in analysing human societies. In this project, the most difficult task is to move beyond 'iconoclastic talk about the "domination" of alien models and theories' to the construction of alternative conceptual frameworks and metatheories that reflect indigenous worldviews and experiences (Atal, 1981: 195). It has been acknowledged, for example, that 'to have an academic discourse beyond "orientalism" and "occidentalism" is rather a tall order as long as we cannot break away from and become totally independent of colonial knowledge' (Shamsul, 1998: 2). As Cohn (1996: 4–5) has argued, the colonial project not only invaded and conquered territorial space but also involved the systematic colonisation of indigenous epistemological space, such as indigenous thought systems, reconstituting and replacing these using a wide corpus of colonial knowledge, policies and frameworks. With decolonisation, ex-colonies have regained (sometimes partial) political territory, but seldom the epistemological space. The aim to reclaim such epistemological space has provided the impetus for subaltern studies, a school of political and literary criticism which considers how the voice of colonial subjects can be represented and heard without distortion. Calls for 'resistance' to and 'alternative readings' of European texts as a means to demystify colonialist power have also been issued. For example, Zawiah Yahya (quoted in Alatas, 1995: 131) argues that the most effective way of deconstructing Western discourse is to use 'its very own tools of critical theory, . . . not only to dismantle colonialism's signifying system but also to articulate the silences of the native by liberating the suppressed in discourse'. Among historians and geographers, there have also been several re-readings of the history of European expansion on the world stage such as Wolf's (1982: x) attempt to 'abrogate the boundaries between Western and non-Western history' and reinstate the 'active histories of "primitives", peasantries, laborers, immigrants, and besieged minorities' and Blaut's (1993) proposed non-diffusionist model of the world which seeks to decentre Europe as the maker of world history and geography.

In the debates about geographical historiography, there are tentative signs of a repositioning from an engagement with 'geography and empire' or 'geography and imperialism' with the accent on the colonising power, to a consideration of the multiple historical geographies of the colonised world. The recent debate about recasting what passes as the Western geographical tradition as an 'irredeemably hybrid product' which 'relied upon and appropriated many elements of other local ("indigenous") geographical knowledges' is one such example (Sidaway, 1997: 76–7).

From colonial domination to the agency of the colonised

If the call for alternative indigenous knowledges is one strand of the turn towards a focus on the 'colonised world', the second impulse relates to the developing critique of mainstream theories of colonial societies and cities that privilege the forms and faces of dominance as the leitmotif of the colonial enterprise. As Anthony King (1992: 343) (whose own work has had a major impact on 'the study of colonial domination and dispossession') observes in an insightful epilogue to *Forms of Dominance* (AlSayyad, 1992):

> the focus on European colonialism, by occupying the historical space of those urban places and people of which it speaks, has marginalised and silenced two other sets of voices: the voices of resistance and the voices of, for want of a better term, 'the vernacular'.

The metanarrative of postcolonial and, *ipso facto*, colonial discourse, has failed to break with – and, in fact, extended the reach of – the enterprise and discourse of dominance and dispossession which 'should have been dead for half a century' (King, 1992: 343). Ironically, it is the critical but exclusive engagement with 'the colonizer's model of the world' (Blaut, 1993) that produces a narration of history from the point of view of those who 'claim history as their own', further marginalising 'the people to whom history has been denied' (Wolf, 1982: 23). Work in postcolonial studies and cultural geography which trace their genealogy to Said's notion of the Orient as a hegemonic and homogeneous discursive construction of the west divorced from lived material practice has the effect of writing out the colonised (Young, 1995). The continual emphasis on the policies and discourses of the colonial power in 'the constant framing and creation of natives as the "other" in order to facilitate subordination' (AlSayyad, 1992: 8) leaves little epistemological and empirical space for the insertion of conflict and collision, negotiation and dialogue between coloniser and the colonised. In other words, a critical perspective which dissects but still centres the colonial vision inevitably reproduces its dominance, its gaze and its effects, without drawing the colonised body out from the shadowlands to which it has been relegated. In these accounts that ignore the agencies, struggles and practices of the 'colonised', not only is a false impression of the relative 'costs' and 'benefits' of the advances of colonialism created, the colonial attitude of homogenising the 'colonised' as an inferior 'other' is further reproduced. Non-Western cultures and peoples remain in 'an interpretive position as the perpetual inferior' (Tanaka, 1993: ix).

Everyday worlds

I have argued elsewhere (Yeoh, 1996) that the historical geography of colonialism, commonly interpreted as the imposition of the dominant logic of the colonial political and cultural economy onto non-Western time and territory, needs to be destabilised. This must be achieved by a valorisation of everyday practice – what Thrift (1997) calls 'non-representational theory', which emphasises the performative, embedded and relational manifestations of everyday life. Such an approach takes seriously the hard work of rescuing the common people in colonised territories from the 'enormous condescension of posterity' in the fashion of grassroots historians such as E. P. Thompson and George Rude, who worked to restore everyday lives and actions to the 'very stuff of history'. It is often the case that simply not enough is known about colonised groups given the asymmetries in the historical record for anyone approaching the colonial period to hope to mount a revisionist stance. Yet new approaches and an expanded sense of what counts as source material can uncover previously overlooked evidence. This is, however, not a call for 'a thoroughgoing empiricist endeavour purged of *any* theoretical, conceptual, or interpretative moment' (Philo, 1992: 143), but a concerted effort to imaginatively mine the official archives and re-filter colonial discourse through 'other' lenses and at the same time widen the net to include hitherto ignored source materials produced by the everyday workings in the lifeworlds of the colonised.

Contact zone

Another important endeavour in order to move beyond 'the historical geography of colonialism' to include the multiple 'historical geographies of the colonised world' is to reconceptualise the 'contact zone' in terms of contest and complicity, conflict and collusion, and to tackle the unwritten history of resistance. The term 'contact zone' has been used by Pratt (1992: 7) (as an alternative to 'colonial frontier', a term grounded within a European expansionist perspective) as 'an attempt to invoke the spatial and temporal copresence of subjects previously separated by geographic and historical disjunctures, and whose trajectories now intersect'. A 'contact' perspective

> emphasizes how subjects are constituted in and by their relations to each other . . . not in terms of separateness or apartheid, but in terms of copresence, interaction, interlocking understandings and practices, often within radically asymmetrical relations of power. (Pratt, 1992: 7)

In fleshing out the dynamics of the 'contact zone', Pratt (1992: 192) attempts to identify the 'parodic, transculturating gestures' on the part of colonised subjects as they undertake to represent themselves in ways which engage with and ultimately reshape colonial discourse. Drawing on different conceptual tools but with cross-cutting aims, I (Yeoh, 1996) have tried to reconceptualise power relations in the colonial city of Singapore, suggesting that the colonial city can be analysed as a terrain of conflict and negotiation where the specific techniques of disciplinary power and the multiple resistances 'formed right at the point where relations of power are exercised' (Foucault, 1980a: 142) are played out. I point out that the 'agency' of the colonised is not only highly differentiated but often effective because, whether tacitly or discursively, it drew on coherent ideologies, institutional structures and schemes of legitimation that were independent of, and largely impenetrable to the colonial authorities. At the same time, however, it is also circumscribed and modified by the disciplinary techniques of colonial power. It is by treating the 'colonial' enterprise and the 'colonised' world as 'different but overlapping and curiously interdependent territories' (Said, 1988: viii) that one seeks to 'faithfully mirror the complex weave of competition, struggle and cooperation within the shifting physical and social landscapes' (Harvey, 1984: 7).

Resistance

Such a project requires a reconsideration of colonial power relations in general, and the notion of resistance in particular. As Pile and Keith (1997: xi) note, 'issues of "resistance" have only recently become foregrounded in geographers' discussion of power relationships, political identities and spaces, and radical politics'. Part of this interest in 'resistance' draws from and folds into recent approaches in cultural studies which 'invoke a "tropology of resistance and hybridity" in their analyses of subaltern actors traversing landscapes of culture, power, and social contestation (Moore, 1997: 87). Geographers' specific contributions to the debate has been primarily in terms of demonstrating the value of a spatial understanding of 'resistance', and in grounding geographic metaphors of space, place and positionality (currently all the rage in cultural studies) in situated practices and local contexts. While this is not the place to trace the complex contours of geographers' recent engagements with the notion of 'resistance', I will briefly summarise two broad strands of the debate which would be useful in informing a contextualised reading of the politics of space in the colonised world.

Everyday resistance

First, beyond dissecting the strategies and technologies of domination, the domain of 'resistance' and 'politics' must be expanded beyond 'heroic acts by heroic people or heroic organisations' (Thrift, 1997: 125) (without detracting from the power and poetics of such acts or to suggest that they are scissored out of fabric of everyday contexts) and reconfigured to include resistant postures, ploys, tactics and strategies woven into 'the practice of everyday life' (de Certeau, 1984). Scott's (1985) work, in particular, was highly influential in valorising 'weapons of the weak' as 'everyday forms of resistance', but also attracted the attention of critics like Abu-Lughod (1990: 42) who cautioned against the

> tendency to romanticize resistance, to read all forms of resistance as signs of the ineffectiveness of systems of power and the resilience and creativity of the human spirit in its refusal to be dominated.

While we need to guard against trivialising 'resistance' by discerning it in all situations everywhere, paying attention to forms of resistance beyond the most explicit and heroic manifestations, allows us to appreciate the fluid, unstable nature of power relations. This creates the conceptual and creative space to rewrite the colonised world on its own terms rather than accept the dualistic model mirrored in colonisers' accounts of either a passive, quiescent and ignorant people 'ripe' for reform or salvation under benign colonial rule, or a hotbed of violence, riots, insurrections and revolts boiling over the moment colonial authority, surveillance and control were relaxed. As Vidler (1978: 28) points out (albeit in a different context), it is often '[b]etween submission to the intolerable and outraged revolt against it' that ordinary people 'somehow defined a human existence within the walls and along the passage of their streets'.

Rhizomatic resistance

These everyday and ordinary forms of resistance can also be thought of as rhizomatic. As Routledge (1997: 69) writes:

> Resistances may be interpreted as fluid processes whose emergence and dissolution cannot be fixed as points in time [or space] . . . [They are] rhizomatic multiplicities of interactions, relations, and acts of becoming . . . Any resistance synthesizes a multiplicity of elements and relations without effacing their heterogeneity or hindering their potential for future rearrangement. As rhizomatic practices, resistances take diverse forms, they move in different dimensions, they create unexpected networks, connections, and possibilities. They may invent new trajectories and forms of existence, articulate alternative futures and possibilities, create autonomous zones as a strategy against particular dominating power relations.

Strategies of resistance may thus be 'assembled out of the materials and practices of everyday life' (Routledge, 1997: 69), tracing visible lineaments in physical space; they may also 'engage the colonised spaces of people's inner worlds' and effect the production of 'inner spaces' or 'alternative spatialities from those defined th[r]ough oppression and exploitation' (Pile, 1997: 3 and 17). Thinking of resistance as 'rhizomatic practices' (the metaphor appropriately insists on a certain 'grounding' of such practices and at the same time conceives of resistance as sprouting both 'above' and 'below ground') points to the contingent nature of power. At the same time, it allows us to transcend the dichotomy between treating resistant spaces as purely autonomous, 'uncolonised' spaces exterior to or dislocated from the spatial parameters of domination, or as purely 'underside' spaces of social life confined to and reacting against authorised spaces of domination in a 'strategic' fashion where 'each offensive from one side serves as leverage for a counter-offensive from the other' (Foucault, 1980b: 163–4). Treating resistance as 'rhizomatic' emphasises its creative and elusive nature, as a subjectivity which is '"polyphonic", plural, working in many discursive registers, many spaces, many times' (Thrift, 1997: 135). In the colonised world, it is important to not only recognise that there were a panoply of resistances inhabiting different spaces but also to underscore the way they connect, collide, diverge, transmute, sometimes in unexpected ways, and often moving 'in' and 'out' of spaces of domination.

If the above ideas are intended to go beyond being elegant but empty metaphors, it is important to ground them in the concrete spaces of everyday life (Lefebvre, 1991) in the colonised world. To do this, I will narrate three (fragmentary) accounts of 'resistance' which are mapped onto three different spaces in the colonised world of late nineteenth- and early twentieth-century

Singapore drawn (with some minor modifications) from my book *Contesting Space: Power Relations and the Urban Built Environment in Colonial Singapore* (Yeoh, 1996). They are, given the constraints of space, somewhat truncated, 'scissored out' accounts, selected primarily to give a sense of the uneven terrain of resistance in the colonised world. This provides the localised context, social materiality and the human actors with which to further work out the multiplicity of 'resistances' and the dynamics of 'spatial politics' which animated the colonised world. I begin with a brief account of the colonial project as established in the British colony.

The colonial project in Singapore

In the early nineteenth-century world of mercantile capitalism, European imperial powers were engaged in a race to secure control of important sea routes to further their trading interests in the Far East. It was in this context that a British trading post was established on the island of Singapore in 1819, marking the start of British colonial presence which was to last for almost a century and a half. By the last quarter of the nineteenth century, Singapore's status as the premier entrepôt port in the Far East was consolidated. It was also positioned as a bridgehead for the extension of British political control over the Malayan hinterland from 1874, which went hand-in-hand with Western capitalist penetration of the interior. Singapore's rapidly expanding economy, coupled with a liberal open-door policy on immigration, drew large numbers of immigrants, primarily from China, India and the Malay Archipelago. By the turn of the century, the population exceeded a quarter of a million people, a figure which took just thirty years to double to over half a million in 1931. According to colonial census statistics, the principal 'races' were, 'Europeans' (1.5 per cent), 'Eurasians' (1.2 per cent), 'Chinese' (75 per cent), 'Malays' (12 per cent), and 'Indians' (9 per cent).

In a port city like Singapore where prosperity was largely dependent on entrepôt trade and the energies of a continuous stream of Asian immigrants, the colonial state had to consider not just the question of European demography in the tropical colony but also address the more intractable problem of Asian morbidity and mortality. The campaign for sanitary reform to banish death and disease, already invigorated by the Victorian preoccupation with personal cleanliness and public hygiene, gained momentum as a major prong of the colonial project, and in particular, the municipal government. To the vision of a sanitised city was blended the image of a progressive, civilised city (Figures 6.1 and 6.2). Singapore was to exemplify modern principles of order and efficiency as befitted its status as a great commercial emporium in the British Empire. Imposing a semblance of order and discipline on the urban environment was particularly crucial to the colonial enterprise. A well-ordered city was not only an object which engendered colonial and civic pride but it also facilitated mobility and surveillance. By rendering colonial society 'picture-like and legible', it became 'available to political and economic calculation' (Mitchell, 1989: 45).

Figure 6.1 Contrasting views of Singapore in the 1930s I: this plate captures the colonial hub of the city, with Victoria Theatre (once the Town Hall) left centre, the Municipal Buildings to the right, and Fort Canning Hill in the background (*source*: National Archives of Singapore).

Figure 6.2 Contrasting views of Singapore in the 1930s II: Depiction of a street scene in Chinatown. Shophouses line both sides of the thoroughfare (South Bridge Road) which served as a main conduit for *jinrikishas*, mosquito buses and the trolley bus (*source*: National Archives of Singapore).

The twin projects of sanitising and ordering the city provided much of the ideological and technological impetus for a wide range of control and surveillance strategies, implemented primarily through the agency of the municipal authorities, the chief social architect responsible for shaping the urban environment and local affairs. These specific strategies, usually couched in the language of reform and improvement, included intrusion into the domestic spaces of everyday practices; the spatial rearrangement of built forms to 'open up' the closely textured city; the replacement of 'traditional' methods of water storage and sewage disposal with municipally controlled systems; the improvement of environmental legibility through the inscription of a code of place-names; the demarcation and control of public spaces; and the deletion of 'obsolete', 'traditional', 'sacred' spaces to make way for 'modern', 'rational' uses.

I have argued that the built spaces of the colonial city did not simply bear the impress of the colonial powers but were sites of control *and* resistance, simultaneously drawn upon by, on the one hand, dominant groups to secure conceptual or instrumental control, and, on the other, subordinate groups to resist exclusionary definitions and tactics, and to advance their own claims. This is illustrated in the following excerpts detailing acts of 'resistance' in colonial Singapore.

Evading disease control

Asian counter-strategies of dissimulation, evasion, and concealment were clearly demonstrated in their attempts to evade municipal disease control. In 1900, the municipal health officer warned that in dealing with cases of infectious disease, the sanitary staff had to be on the alert to all the dodges practised by the natives to conceal such cases. Although the failure to report the presence of a 'dangerous infectious disease' carried a fine of up to twenty-five dollars, such a penalty was far outweighed by the practical trouble incurred by occupants if a 'dangerous infectious disease' were traced to a particular dwelling. The resulting inconveniences included domiciliary visits by sanitary inspectors, isolation, the disinfection and destruction of contacts' belongings and occasionally the dwellings concerned, the imposition of quarantine regulations, all of which measures were perceived by the Asian population to be irksome interferences with their livelihood routines. The removal of victims from the home to an orderly, sanitised hospital regime was also highly unpopular. There were frequent complaints that hospitals were socially isolating, that there was no strict segregation of sex, race, and religion, that conditions were overcrowded and inferior to what 'the better class of Asiatics were accustomed to in their own houses', and that Asians were not allowed treatment 'according to their own customs' by their own medicine men.

The Asians were hence keenly conscious of the consequences of detection and avoided it at all costs. The municipal health officer confessed himself nonplussed by the persistent reluctance of the Asians to report cases of infectious disease and the 'cheerfulness' with which they paid their fines when caught. Concealment of infectious cases was rife, as evidenced by the large number of cases discovered only after death. Victims of infectious disease were frequently smuggled out to the outskirts of the town and attempts to trace the disease to its source were often frustrated by the reluctance of patients and friends alike to reveal addresses. The 'persistent efforts of Asiatics to conceal the existence of [infectious diseases], and the deliberate mis-statements made to put Sanitary Officers off the track' were not confined to 'the more ignorant classes of the population, but were freely indulged in by the more intelligent and educated classes of the Asiatic community' including those who worked as clerks and storekeepers in European offices. Strategies resorted to by the Asians to avoid detection were highly imaginative, as illustrated by the following catalogue:

> The existence of cases of infectious disease [were] carefully concealed, the patients surreptitiously removed in *jinrikishas* [two-wheeled vehicle for one or two passengers pulled by a coolie] or *gharries* [horse-drawn carriage] to the hospital or to sick receiving houses. Or they [might] be taken in a moribund condition or after death and deposited on the street or any convenient piece of vacant ground from which they [were] removed to the hospital or cemetery by the Police. Every device is resorted to, to prevent the authorities from tracing the houses from which such cases were removed, such as changing *jinrikishas* two or three times between the house and the hospital, giving false addresses, or declaring the patient had newly arrived in the town and had been picked up on the street or 5-foot-ways [the local name for verandahs, see later].

In 1909, the municipal commissioners attempted to counter Asian dislike of quarantine by paying each contact sent to the Quarantine Station at St John's Island fifteen cents for each day of quarantine. The experiment was, however, scrapped after a year on the basis of the health officer's report that it had failed to encourage the Asians to come forward more readily with information pertaining to the existence or source of infectious disease.

Deaths suspected to have resulted from 'dangerous infectious diseases' also entailed post-mortem examination which was highly disliked and seen as 'an intolerable interference with the religions and sentiments of different classes of the [Asian] [c]ommunity'. As such, they were frequently misreported as resulting from other innocuous causes. An investigation into the accuracy of death returns in 1896 revealed that no less than 87 cases of infectious diseases (mainly cholera and enteric fever) were within six weeks returned under other names. As the cause of death in a large number of cases (an average of 65 per cent of deaths each year in the first decade of the twentieth century) was not medically certified but dependent on a perfunctory inspection of the corpse and inquiries from friends, it was inevitable that a significant proportion of misinformation evaded detection. This, according to the municipal health officer, was the vital flaw in municipal disease detection and control for 'where the cause of death [was] obscure and there were no means for satisfactorily determining them [*sic*], there must be many missing links . . . in [the] chain connecting many of the cases with each other'. As such, the municipal health officer felt that the inaccuracy of death returns militated against any attempt to introduce 'special measures' aimed at particular diseases and that 'it [would] only be possible to proceed on general lines and introduce such measures as experience in other places ha[d] shown to be instrumental in benefiting and preserving the public health'.

It was clear from this that municipal surveillance could only be effective if it were predicated on an efficient turnover of information. By the purposeful withholding, corrupting, and falsifying of information and by setting a high premium on it, the Asians were not only instrumental in thwarting attempts to trace the origin and course of infectious disease, they also played a part in rendering it difficult for the Health Department to devise more stringent and precise control measures (all quotations from Yeoh, 1996: 121–3).

Contesting public space: the 'Verandah Riots'

In the late nineteenth century, tensions between the municipal authorities and Asian shopkeepers and traders over the use of verandahs (narrow, covered walkways extending along the front of rows of shophouses or tenement dwellings) were heightened by the passing of Municipal Ordinance [IX] of 1887 which, *inter alia*, empowered the municipal commissioners to remove any obstruction of verandahs, arcades, or streets which either hindered the work of street cleaning or caused inconvenience to the passage of the public. When the ordinance came into effect at the beginning of 1888, the municipal engineer sent in a minute requesting to know whether the commissioners intended to take any action on the matter of verandah obstruction. The commissioners resolved that one month's notice was to be given in the local English and vernacular journals as well as through placards placed throughout the town of the municipal intention to 'rigidly enforce the clearance of all open verandahs abutting on public streets in Town'. The municipal president, Dr T. I. Rowell, stressed the importance of verandah clearance on sanitary grounds and anticipated that under his direction, the blocked up state of verandahs in the town would be 'greatly improved' within twelve months.

Typical of municipal manoeuvres, the new powers to order Asian society and the urban environment were couched in the language of improvement. Dr Rowell was, however, opposed by several of the non-official commissioners and in particular Thomas Scott, 'a gentleman of great influence with the Chinese community', who argued that enforced verandah clearance amounted to a confiscation of vested private rights and challenged the president's power to proceed without the consent of the commissioners. The 'verandah question' was also strenuously debated in the press. *The Straits Times* applauded the municipal resolve to clear verandahs and anticipated that once the provisions of the Municipal Ordinance were 'strictly enforced without fear or favour', it would 'take effect beneficially on the street traffic, and be no inconsiderable boon to pedestrians'. Like the municipal commissioners, it construed of the 'verandah question' as a debate between 'public interest' and 'private ends' and as such, public order could only be restored if the verandah were strictly demarcated as public space and its boundary with private property made less permeable. The rival daily, the *Singapore Free Press*, took the opposing stand that owners and occupiers of houses had a claim on the five-foot-way through 'long usage' and that the proposed campaign against verandah obstruction was 'harsh and coercive'.

Whilst the press continued to debate the verandah issue, municipal inspectors under orders from Dr Rowell started to clear verandahs in the neighbourhood of Arab Street, Rochore Road and Clyde Terrace Market (in the Kampong Glam district) on 21 February 1888. At North Bridge Road, shopkeepers and traders reacted by closing their shops as a demonstration of protest against the municipal order to clear their verandahs. News of this soon spread and shopkeepers in the vicinity eastwards of Bras Basah Canal followed suit by putting up their shutters. The inspector-general of police, Colonel S. Dunlop reported that from the 'quiet manner' in which the Chinese shopkeepers reacted, it was evident that the action had been preconceived in concert. Contrary to the official stand that no unwarranted harshness or precipitancy was displayed by the inspectors in carrying out the law, one of the local dailies reported that the verandahs were forcibly cleared, property damaged, sunshades pulled down, and shopkeepers roughly turned away.

On the afternoon of the same day, the municipal commissioners convened an emergency meeting to evaluate the situation. The power of the president to order the clearance of verandahs without the consent of the commissioners was again called into question by some of the elected members of the Board and a motion stating that the requirement of the Municipal Ordinance would be considered satisfied as long as there was sufficient space for two persons to pass abreast along the verandah was passed under overwhelming pressure from the elected commissioners. The motion was tantamount to a revocation of the president's original order and an acceptance of a more liberal definition of what constituted public right of way along the five-foot-ways. The Governor, Sir Cecil Smith, was later to censure the municipal commissioners for their lack of support of their president and for 'practically yield[ing] to the opposition which the Chinese had shewn [*sic*] to even limited interference with their occupation of the . . . verandahs'.

The municipal resolution failed to quell growing tensions and the next day (22 February), open rioting broke out in the streets, not only in Kampong Glam where verandah clearance had proceeded but elsewhere throughout the principal streets of the town. All shops and markets remained shut and crowds of *samsengs* [roughs] gathered in the streets, intimidating anyone who attempted to re-open for business. Stone and brickbat throwing was the order of the day, tramcars were set upon and damage done to public conveyances. Members of the European and other communities were attacked and the police were called out in full force. Trade and traffic were brought to a standstill. In the course of the afternoon, the municipal commissioners met again and ordered the circulation of the resolution passed the previous day intimating that the commissioners did not intend to act harshly and that only a clear passage of three feet across the width of the verandah was required. On the morning of 24 February, inter-clan fighting broke out among the Cantonese and Hokkien coaling coolies at Tanjong Pagar and sporadic rioting continued throughout the town.

By noon, however, shops began to re-open for business and the rioting gradually petered out.

The riots which originated in a confrontation between municipal inspectors and shopkeepers over the right to use one particular element of public space – the verandahs – rapidly escalated into a full-scale tussle for control over public streets and spaces in many parts of town. What was originally an act of disobedience on the part of shopkeepers spatially confined to a few streets became amplified by the involvement of the rowdier elements of Chinese society to become what the press described as 'an organised resistance to the law of the land'. As with other instances of urban disturbance, friction or conflict between two relatively small groups quickly assume an aspect of public violence, involving larger networks and loyalties which underlie Asian society. For a few days the crowds were able to hold public places at ransom and to force the town to a standstill, albeit temporarily. Everyday activities were dislocated, and social spaces in the city momentarily reclaimed. Also, the riots were not quelled without the municipal commissioners being compelled to adopt a far less stringent definition of public right of way along verandahs than that originally contemplated, thereby conceding much of the ground to Asian shopkeepers and traders who refused to relinquish their claim on the verandahs.

Whilst the 'verandah riots' represented by far the most violent aspect of the conflict over the use of 'public' spaces, it was not an isolated event but had instead been fomented in the wake of increasing ill feeling towards the authorities generated by a succession of unpopular measures implemented during the second half of 1887 and the early months of 1888. In particular, the customary right conceded to the Chinese to use the verandahs and the sides of public streets to burn sacrificial papers, display feasts for the dead, and hold street-*wayangs* [street theatres] during the *sembahyang hantu* season [Festival of the Hungry Ghosts] in August/September was withdrawn by the government in retaliation for a homicidal attempt on William Pickering, the Protector of Chinese, by Choa Ah Siok, a Teochew carpenter. The attack was interpreted by the government not only as a dastardly crime against the person of the Protector who 'ha[d] been for so long the friend and the benefactor of the Chinese community', but more crucially, as a 'great offence against the public order', which called for public punishment of the entire community and required public expiation on the part of the community. The mode of punishment chosen, the withdrawal of the 'privilege' of using public space for communal celebrations, was calculated to reinforce the public nature of the crime. The government further alleged that the Chinese could give information leading to the identification of those who had instigated the assault if they had a mind to do so and by withholding information, the responsibility for the perpetration of the crime lay upon the entire community. Until the Chinese co-operated to restore public justice and order, they forfeited all right to appropriate public spaces for their own use.

The irony of the prohibition order following on the heels of the Queen's Golden Jubilee celebrations in June of the same year during which the municipal commissioners granted free permits to all the inhabitants of the town to erect poles for decorative purposes in the public thoroughfares, was not lost on some of the more articulate members of the Chinese community. During the Jubilee celebrations, the Chinese took 'a most prominent part', putting up 'lantern processions' with 'gigantic dragons' winding their way through the streets, 'gorgeous decorations and illuminations' and *wayangs* scattered throughout the town. The Chinese who wrote to the press remarked that neither 'the ink describing the grandeur of the displays in the streets and the loyalty of the people in celebrating Her Majesty's Jubilee' nor 'the perspiration on the brow of [their] heads for having toiled so hard to make the occasion a marked event' had dried before the right of the Chinese to use the same streets for their own celebrations was abruptly withdrawn within a space of two months. The *sembahyang hantu* passed without any major incident but as with the clearance of the verandahs six months later, the prohibition order impinged upon what the Chinese saw as their customary right to order and use the urban environment, and in particular its public spaces, in a manner consistent with their own purposes. Against this background of growing resentment towards the authorities, and by the suspension of belief about the enormities of which 'the Other' was capable, a minor fracas quickly escalated into a full-blown riot. Although the 'verandah riots' were triggered off by what appeared to be conflict over a specific type of urban space, the fact that it rapidly assumed larger dimensions was an indication of widespread fears that the customary rights of the Chinese over the public environment were increasingly threatened (all quotations from Yeoh, 1996: 250–3).

The debate over Chinese burial grounds

In an attempt to rationalise land use and facilitate urban planning in the central areas of the city, a bill authorising the licensing, regulation, and inspection of burial and burying grounds was introduced during the August 1887 Legislative Council sessions. The Burials Bill, as it was called, generated much agitation and concern among the Chinese community in particular. Seah Liang Seah, the Chinese member of the Council requested a postponement of the second reading to allow more time for consideration as the Bill 'seriously affected the interests of the Chinese community, mostly those of the respectable class'. In particular, he perceived that the Bill aimed at 'the suppression of private burial grounds', a measure which would 'much affect the much-cherished customs of the better class of the Chinese'. The Colonial Secretary, in granting the postponement, assured the Council that the Bill had been drawn up 'with full regard for the feelings of the Chinese', and that there were no intentions to interfere with their 'feelings and religious sentiment'. However, he made it clear that Chinese custom could only be tolerated 'within reasonable limits and without sacrificing the good of the community':

> it is not right that all other classes of the community should be sacrificed to the desires of one section [the Chinese], to secure, for instance, all the small hills, which are the only places suitable for healthy houses in these countries, and take them for ever, sometimes merely as a monument to the honour of a man's family and his own personal vanity.

From the perspective of the Chinese, the Burials Bill represented an attack on Chinese customary rituals and an erosion of Chinese control over their own sacred spaces. Whilst there were initially rumours of riots and threats of violence, the protest against the bill soon resolved itself through constitutional channels. This was mainly because the bill principally affected the wealthier and more prominent members of the Chinese community who could afford private burial grounds and who were chiefly responsible for reserving large parcels of land for use as family graves. Protest against the bill was hence led by leaders of the community who were familiar with the legislative system and conversant with their own rights and privileges under such a system. The means of protest adopted included letters to the press, public meetings to gather support and petitions to the Resident Councillors and the Governor. When the latter failed to arrest the progress of the bill through the Legislative Council, memorials pleading their cause were sent to Sir H. T. Holland, the Secretary of State for the Colonies in London.

In the main, it was argued that among the Chinese, ancestor worship, and in particular sepulchral veneration, would be incomplete if the liberty to select propitious burial sites according to the principles of geomancy were curtailed by government interference. The concept of control of burial places by an external agency was an alien notion among the Chinese. *Feng shui*, or Chinese geomancy, was considered central to the Chinese faith because it was believed, so the petitioners claimed, that it was possible to site the grave in relation to the configuration of the landscape and the vicinity of watercourses in such a way that benign influences were drawn from the earth and transmitted to the descendants of the deceased. Once sited according to geomantic principles, both the tomb and its sepulchral boundaries were considered 'inviolable' as any interference with them would spoil the efficacy of the *feng shui* and imperil the welfare and prosperity of living descendants. In short, burial grounds, by nature of being sacred sites, must be exempt from government control and external interference. The petitioners argued that, as in Roman law, 'the ground in which one, who had the right, buried a dead body, became *ipso facto* religious; it ceased to be private property; it would not be bought or sold, or transferred or used; it was for ever dedicated to the dead and reserved from all current usages; and it should be sacrosanct to the memory of the departed'. *Feng shui* considerations hence had to be carefully taken into account in siting a grave (or any other building) if harmony between society and the physical landscape were to be maintained.

On both sides, arguments were deftly marshalled to press their case home. The Chinese quoted their sages, drew out innumerable examples illustrating how deeply seated and inviolable the principles of geomancy were among the Chinese, and appealed to the Imperial Charter which provided for due respect of the religious sentiments of all races. The discourse was strategically couched in terms of religious idealism, because religion and sacred places associated with it could claim the privilege of being beyond the purview of a secular government. The Chinese were also quick to counter the charge that hillside burial grounds were insanitary and liable to contaminate the city's water supply.

They argued that pollution of hill streams was impossible because unlike those of Malays and Europeans, Chinese graves were of considerable depth and lined with great quantities of quicklime. It was constantly reiterated that the grave of a wealthy Chinese was 'substantially built', 'planted round at great cost with shrubs', regularly visited and tended, and that the coffin was made of 'the hardest wood that [could] be obtained and well-lacquered' so that it was 'perfectly tight and waterproof'. The memorial sent to the Colonial Secretary of State summarised these arguments and put forward two main requests. First, notwithstanding the assurances given by the local government that their religious customs would be respected, the memorialists feared that the Burials Ordinance opened up the possibility, and even the likely prospect, that their burial grounds would be 'seized and turned to other purposes'. Their foremost request was that existing burial grounds would be protected from such 'contemplated desecration'. Second, they requested the liberty to choose their own burial sites outside a radius of two miles from the town proper in accordance with 'their religious faith in ancestor worship' and 'as directed in the teachings of their sages Confucius and Mencius'.

In their turn, the local government and the English language press cited land scarcity, sanitation, the public good, and attempted to counter the Chinese argument by showing that the much vaunted religious idiosyncrasies of the Chinese were by no means respected by the Mandarins in China who '[did] not scruple to appropriate with little delicacy of feeling burial grounds, shrines, etc. when required for Government purposes'. The Colonial Secretary, in defending the Burials Bill, cited evidence demonstrating that in China ancestral temples were summarily demolished to make way for government projects and claimed that 'the Chinese had no real feeling against removing graves'. In fact, he had been 'reliably informed' by the Consul-General of Shanghai that Chinese people 'were in the habit of placing the remains of their ancestors in urns in order that they might at any convenient season remove them from one place to another'. In short, the wealthy Chinese had no 'real' grievance and were simply pandering to vested interest under the guise of religious sentiment instead of protecting public health and advancing public good. Their request to be allowed to select their own burial sites outside a two-mile radius of the town was also dismissed by the Governor as tantamount to asking for an 'illegal privilege'. The protector of Chinese, Pickering, further reduced the issue to a question of choosing between two sets of priorities: 'are [Chinese] customs connected with the burial and worship of the dead compatible with the sanitary welfare of the living general community, and is the practice of buying land and appropriating some of the best sites as private mausolea conducive to the interest of the Government as regards the reasonable development of the Land Revenue, and the progress and prosperity of the Colony?' In his mind, the answer was unassailably clear: the claims of the living must prevail over those of the dead, customs should not be allowed to stand in the way of urban progress and land revenue, and private benefits should be subservient to 'that of the superior law, the welfare of the general public'. The burials question was thus polarised into a choice between diametrically opposed priorities such as those between the living and the dead, the progressive and the customary, and the public and the private.

On receiving the Chinese petitions, Holland, the Secretary of State for the Colonies, decided that the coming into operation of the Burials Ordinance should be postponed and instructed the Governor to repeal the ordinance and re-enact it with modifications to take into account Chinese sentiments. The ordinance was duly repealed but no further legislative action was taken on the burials question during the next eight years (all quotations from Yeoh, 1996: 289–93).

Historical geographies and the politics of colonised space

The aim of the above accounts is not so much to equate the 'historical geographies of the colonised world' with the 'historical geographies of resistance' but to argue that while colonialism produces its own space (Lefebvre, 1991), it does not do so without the processes of conflict and compromise between those who control space and those who live in it, both of whom must be seen as 'participants in the same historical trajectory'. Close examination of the 'contact zone, the social spaces where disparate cultures meet, clash, and grapple with each other, often in highly asymmetrical relations of domination and subordination' (Pratt, 1992: 4) opens up the possibility of constructing historico-geographical accounts of the colonised world which move away from depicting it as a passive, flattened out world, stamped upon by more powerful others, and fashioned solely in the image of colonialism.

These intersecting contact points between the different facets of the colonial world, and the multiple constituents of the colonised world are played out in a variety of ways and occupied different spaces, threaded into the fabric of everyday life and occasionally culminating into flashpoints of crisis proportions. A Chinese coolie immigrant who arrived in colonial Singapore in the early twentieth century was likely to encounter colonial and municipal authority as wielded by its minions in many guises, from the sanitary inspector who ordered the pulling down of his cubicle in a coolie lodging house, the municipal peon who served a limewash notice on the chief tenant of his house, the registrar of vehicles who issued his licence to pull a *jinrikisha*, the police officer who arrested him for obstruction of public places, to the municipal apothecary who inspected his corpse for signs of infectious disease and the burial grounds inspector who ensured his burial in a municipally sanctified plot. The view from the other side of the looking glass is an equally complex one. The colonial authorities had to contend with the continuous, clandestine strategies of evasion and non-compliance of myriad individuals, the voice of influential Chinese leaders in advocating alternative discourses, and the 'hidden' (from the colonial gaze) networks in Chinese society that occasionally drew the masses together in violent backlashes. While the 'agency' of the individual coolie, the leader and the masses was seldom concerted and never seamless, it drew sustenance from a clan-centred cultural economy governed by its own ideologies about life, work, health and death and visibly inscribed in the landscape in the form of clan-owned temples, medical halls, sick-receiving houses, burial grounds, schools, lodging houses, shops and business networks.

As Arnold (1987: 56) has argued, it is only through an awareness of the dialectical nature of the encounters between the colonised people and the colonial state that it is possible to avoid assumptions of mass 'passivity' and 'fatalism'. Historical geographies of the colonised world need to come to grips with these dialectical encounters, not only by dissecting the norms and forms of colonial dominance but also through a variegated account of the strategies and spaces of resistance. As the various excerpted accounts related above show, the colonised drew on multiple strategies to resist exclusionary definitions, to fend off the intrusive mechanisms of bureaucratic power and to advance their own claims. English-educated leaders of the Chinese community, for example, were able to defend the 'sacro-sanctity' of Chinese burial grounds though representations on various government committees, petitions, memorials and letters to the press. By drawing on alternative discourses on environmental management as embodied in Chinese geomancy, they sought to challenge Western urban planning ideas and sanitary science. As colonial authorities could not effectively govern without incorporating a segment of the colonised body into the ruling power structure, the views of those who acted as social brokers commanded sufficient weight for them to be incorporated into decision-making bodies in the colonial government. Policies were occasionally held back to accommodate expressed views (as in the case of the 1887 Burials Bill), although where policy changes were made, there was seldom any real acknowledgement of the legitimacy of alternative Asian discourses.

While the debate provoked by the introduction of burials legislation represented the better publicised, better ventilated aspects of the conflict over the rights of religious practice and the significance of burial grounds, it was but one of multiple 'contact zones' in which the politics over burial space were played out. The debates in discursive space afforded by legislative channels and the media were matched on the 'ground' by the small but pervasive, clandestine acts of illegal burial and exhumation among the Chinese labouring classes who were equally concerned with protecting their rights to burial even if they were unable to afford the finer points of *feng shui* insisted upon by their wealthier compatriots. Indeed, it was often at the level of daily practices and within the spaces of everyday life that the effects of power were renegotiated or countered. This is clearly demonstrated in the catalogue of colonised people's strategies to evade surveillance measures to detect and arrest infectious disease in the city. In Foucault's (1979: 139) terms, just as disciplinary institutions employ a panoply of meticulous, minute techniques embodying the 'microphysics' of power, the subjects of discipline are themselves capable of 'small acts of cunning endowed with a great power of diffusion'. While the strategies used by the Asian people were mainly self-help measures incapable of (nor were they directed at) upsetting the larger symbolic order of dominance and dependence prescribed for the colonised world, they were also not bereft of effects. They were often widespread enough to thwart the execution of particular policies and to inflect colonial control in everyday life. It should also be noted that these strategies were not purely idiosyncratic, obstructionist or irrational (as the colonial authorities often depicted them to be). Instead, the lack of 'co-operation' on the part of the people in implementing disease control in the city, for example, stemmed partly from the fact that they subscribed to different notions of disease and contagion, and what constituted appropriate measures of healing based on their own systems of medical care. It is in this context that colonial campaigns to sanitise the city were often perceived to be political, cultural and moral impositions to be 'passively resisted' (a phrase used by the authorities themselves to describe the people's reactions) whenever and wherever possible.

It should also be noted that while the spatial practices of resistance might take visible form through counter-strategies within colonised spaces (that is, the entire catalogue of evasion and non-compliance tactics to counter disease surveillance) reacting against the machinery of domination, they often drew on the resources of other, uncolonised, inner spaces 'outside' of the prevailing power relations (that is, an 'ethnic' medical geography which eluded colonial logic and control). In other words, the colonised world had its own spatialities of power which were not necessarily circumscribed or apprehended by the geographies of colonialism, underlining the point made by Pile (1997: 14) that 'geographies of resistance do not necessarily mirror geographies of domination, as an upside-down or back-to-front or face-down map of the world'.

While revolutionary uprisings did not loom very large throughout the history of colonial Singapore, the 'verandah riots' of February 1888 demonstrated how everyday rhythms might be disrupted by the culmination of tensions located at different sites in the colonised body and over a span of time. Ostensibly started by a localised confrontation between a group of shopkeepers and municipal inspectors clearing the verandahs of obstructions in one part of town, the involvement of other traders, hawkers, *rikisha* coolies, gangland elements, clan groups and the police galvanised the outbreak into full-scale rioting, a crisis event which also drew on the fissures within the municipal body itself as well as the rivalry between the two English dailies. As Routledge (1997) has observed, an act of resistance synthesises, albeit temporarily, 'a multiplicity of elements and relations'.

Not only were acts of resistance in the colonised world fluid processes which run the gamut of different forms, they were also interconnected in unexpected ways as 'rhizomatic practices'. Just as large-scale, violent clashes might grow out of an accumulation of small run-ins on a daily basis, flashpoints might also subside into everyday forays. The termination of the verandah riots did not signify a satisfactory conclusion to the question of how 'public space' should be defined and used; instead, the contest over the verandahs was reabsorbed into the daily arena as a site subject constantly to municipal reform on the one hand, and recolonisation by shopkeepers, traders and hawkers on the other. The contingent nature of resistance should also be noted, its emergence and dissolution inextricably linked to the specific context. The withdrawal of customary privileges to use public spaces for communal celebrations during the *sembahyang hantu* season did not provoke a visible response from the Chinese community immediately but clearly surfaced several months later in the foment building up to the verandah riots.

For some time now, the post-colonial critique has argued that colonialism was neither monolithic nor unchanging. In the same way that narratives of colonialism have been subject to closer scrutiny and refinement, accounts of the colonised world must also avoid homogenising and essentialising the 'colonised' as a category. Neither colonialism nor the colonised were unitary entities immune to the influence of the other. Neither 'dominance' nor 'resistance' forms a closed, complete circuit in itself. As Routledge (1997: 70) points out, 'practices of resistance cannot be separated from practices of domination, they are always entangled in some configuration. As such they are hybrid practices, one always bears at least a trace of the other, that contaminates or subverts it.' Historical geographies of the colonised world will be enriched by taking fuller account of the interweaving of the discourses and practices of domination with those of resistance as these take shape, not through abstract theories, but in the social materiality of colonised spaces.

References

Abu-Lughod, L. (1990) The romance of resistance: tracing transformations of power through Bedouin women. *American Ethnologist*, **17**, 41–55.

Alatas, S. F. (1995) The theme of 'relevance' in Third World human sciences. *Singapore Journal of Tropical Geography*, **16**, 123–40.

AlSayyad, N. (1992) Urbanism and the dominance equation: Reflections on colonialism and national identity. In AlSayyad, N. (ed.) *Forms of Dominance: On the Architecture and Urbanism of the Colonial Enterprise*, Avebury, Aldershot, 1–26.

Arnold, D. (1987) Touching the body: Perspectives on the Indian plague, 1896–1900. In Guha, R. and Spivak, G. C. (eds) *Selected Subaltern Studies*, Oxford University Press, New York, 55–90.

Atal, Y. (1981) The call for indigenization. *International Social Science Journal*, **33**, 189–97.

Blaut, J. M. (1993) *The Coloniser's View of the World: Geographical Diffusion and Eurocentric History*, Guilford Press, New York.

Cohn, B. (1996) *Colonialism and its Forms of Knowledge: The British in India*, Princeton University Press, Princeton, N.J.

de Certeau, M. (1984) *The Practice of Everyday Life*, University of California Press, Berkeley.

Foucault, M. (1979) *Discipline and Punish: The Birth of the Prison*, trans. Alan Sheridan, Penguin, London.

Foucault, M. (1980a) Power and strategies. In Gordon, C. (ed.) *Michel Foucault: Power/Knowledge, Selected Interviews and Other Writings, 1972–1977*, Harvester Press, Brighton, 134–45.

Foucault, M. (1980b) The politics of health in the eighteenth century. In Gordon, C. (ed.) *Michel Foucault: Power/Knowledge, Selected Interviews and Other Writings, 1972–1977*, Harvester Press, Brighton, 166–82.

Harvey, D. (1984) On the history and present condition of geography: An historical materialist manifesto. *Professional Geographer*, **36**, 1–11.

Lefebvre, H. (1991) *The Production of Space*, trans. Donald Nicholson-Smith, Basil Blackwell, Oxford.

King, A. D. (1992) Rethinking colonialism: an epilogue. In AlSayyad, N. (ed.) *Forms of Dominance: On the Architecture and Urbanism of the Colonial Enterprise*, Avebury, Aldershot, 339–55.

Mitchell, T. (1989) *Colonising Egypt,* American University of Cairo Press, Cairo.

Moore, D. S. (1997) Remapping resistance: ground for struggle and the politics of place. In Pile, S. and Keith, M. (eds) *Geographies of Resistance*, Routledge, London, 87–106.

Philo, C. (1992) Foucault's geography. *Environment and Planning D: Society and Space*, **10**, 132–61.

Pile, S. (1997) Introduction: opposition, political identities and spaces of resistance. In Pile, S. and Keith, M. (eds) *Geographies of Resistance*, Routledge, London, 1–32.

Pile, S. and Keith, M. (1997) Preface. In Pile, S. and Keith, M. (eds) *Geographies of Resistance*, Routledge, London, xi–xiv.

Pratt, M. L. (1992) *Imperial Eyes: Travel Writing and Transculturation*, Routledge, London.

Routledge, P. (1997) A spatiality of resistances: theory and practice in Nepal's revolution of 1990. In Pile, S. and Keith, M. (eds) *Geographies of Resistance*, Routledge, London, 68–86.

Said, E. W. (1988) Forward to subaltern studies. In Guha, R. and Spivak, G. C. (eds) *Selected Subaltern Studies*, Oxford University Press, New York, v–x.

Scott, J. C. (1985) *Weapons of the Weak: Everyday Forms of Peasant Resistance*, Yale University Press, New Haven, Conn.

Shamsul, A. B. (1998) Arguments and discourses in Malaysian studies: in search of alternatives. Paper presented at the International Workshop on Alternative Discourses in the Social Sciences and Humanities: Beyond Orientalism and Occidentalism, National University of Singapore, 30 May–1 June 1998, Singapore.

Sidaway, J. D. (1997) The (re)making of the western 'geographical tradition': some missing links. *Area*, **29**, 72–80.

Tanaka, S. (1993) *Japan's Orient: Rendering Pasts into History,* University of California Press, Berkeley.

Thrift, N. (1997) The still point: resistance, expressive embodiment and dance. In Pile, S. and Keith, M. (eds) *Geographies of Resistance*, Routledge, London, 124–51.

Vidler, A. (1978) The scenes of the street: transformation in ideal and reality, 1750–1871. In Anderson, S. (ed.) *On Streets*, MIT Press, Cambridge, Mass., 28–111.

Young, R. J. C. (1995) Foucault on race and colonialism. *New Formations*, **25**, 57–65.

Wolf, E. R. (1982) *Europe and the People without History,* University of California Press, Berkeley.

Yeoh, B. S. A. (1996) *Contesting Space: Power Relations and the Urban Built Environment in Colonial Singapore*, Oxford University Press, Kuala Lumpur.

PART III

SPATIAL CONTEXTS

Chapter 7

Historical geographies of the environment

J. M. Powell

Introduction

It seems essentially true that the historiography of historical geography reveals as much continuity as change, and certainly that applies with as much force to studies of the environment as to most of our other leading preoccupations. It makes good sense, therefore, to commence this chapter with an overview of a brief selection of work spanning almost a century of scholarly engagement before considering, in the second part, more recent developments in the evolution of a challenging academic frontier. An important qualification must be made at the outset; however convenient and practical this bipartite division, the two dimensions are not mutually exclusive, while they are rendered more complex by the multidisciplinary nature of what might be defined as historical geographies of the environment. Further the comparatively recent emergence, as an academic field of enquiry, of 'environmental history' is opening up fresh disciplinary needs and opportunities.

Changing the face of the earth

One of historical geography's most venerable traditions focuses on *reconstructions* of past physical environments. The discipline has always displayed both 'pure' and 'applied' motivations and potentialities, together with a range of intra- and inter-disciplinary engagements (*cf.* Baker, 1972; for an Asian perspective, see Kinda, 1997). According to some helpful – if contestable – arguments, this interest can be traced to a succession of factors. These include: the late eighteenth-century apprehensions of European intellectuals concerning the ecological reverberations of Alpine deforestation; George Perkins Marsh's *Man and Nature* (1864; most conveniently in Lowenthal, 1965), which outlined the environmental impacts of human settlement; a suite of American and other Western adaptations of late nineteenth-century German conceptualisations of *urlandschaft*; and early twentieth-century British syntheses, which essayed the importance of understanding the relationships between changing environmental conditions and the evolution of prehistoric settlement (see, in particular, Butlin, 1993). The importance of the human factor was particularly well explored, from the 1920s, in the work of the German-born American geographer, Carl Sauer, and

Table 7.1 Ecosystems converted to cropping, 1860–1978 (million ha)

	Forests/woodland	Savannah/grassland	Wetland	Desert	Total
'Core'					
Europe	8.1	10.2	3.4	–	21.7
USSR	57.6	90.1	–	3.2	150.9
Canada/USA	64.1	95.6	–	–	159.7
'Periphery'					
Africa	46.9	56.4	–	3.1	106.4
Asia	122.0	87.4	17.7	7.3	234.4
Latin Am.	64.1	56.2	0.1	3.1	123.5
Oceania	36.2	18.4	–	0.5	55.1
World total	242.8	296.8	21.2	17.2	851.7

Source: adapted from Williams (1996). Note that, in each paired case, the (roughly estimated) quantities cited for woodland and savannah systems are usually significantly lower than for forests and grasslands respectively; Asia and Africa provide important exceptions.

his followers at Berkeley. Building upon the established German preference for the analysis of sequential landscape change, this school of thought expounded the notion of the 'cultural landscape' as the didactic interface between the human and natural *milieux*. Among its other preoccupations, the Berkeley School was deeply committed to the inter-disciplinary pursuit of the links between vegetation change and the activities of prehistoric and pre-modern communities.

Variously co-opted, this rich legacy found unmistakable resonances in a more broadly-based 'environmentalism' which began to sweep through the Western world in the 1960s. The new movement's intellectual, scientific and policy wings placed a high premium on reliable inventories, which bench-marked environmental change. Historical–geographical studies made a vibrant contribution to this demand (see, for example, Turner *et al.*, 1990; Heathcote, 1995; Williams, 1996). Two examples are illustrated in Tables 7.1 and 7.2. The former draws attention to an apparent global correlation between the expansion of arable land and the conversion of vast natural ecosystems, meanwhile underlining the massive adjustments beyond a European 'core'. Table 7.2 indicates a (possibly hitherto unsuspected) likelihood of significant differences in the extent of environmental change between two 'New World' countries which are sometimes casually paired.

While these data serve as confirmation of the retention of a pivotal concern within historical geography, they also supply thematic leads to structure the first part of the chapter. In part, this can be interpreted as a prolongation of some rather unequivocal Marshian representations of transformative engagements with Nature's fundamental landscape assemblages – in so far as the discussion is mainly concentrated on the most visible physiognomy or superstructure of the surface of the earth. But inevitably, the discussion also turns towards interpretations of the large array of experiences subsumed within (or

Table 7.2 Environmental impacts of non-indigenous settlement, Australia and USA

Nature of impact by 1990	**% of total area**	
	Australia	*USA*
1. Pre-European ecosystems relatively intact	33.9	21.4
2. Ecosystems modified in detail		
– forests and woodlands thinned or replaced by exotics	5.3	28.9
– grasslands affected by domestic animal grazing	54.4	27.5
3. New ecosystems or land cover		
– cultivated and orchards	6.3	20.6
– urban areas	0.1	1.6
	100.0	100.0

Source: adapted from Heathcote (1995).

marked by) the stark tabular data, interpretations which prompt reflections on the ambiguous nature of 'core' and 'periphery'. The second part of the chapter moves beyond this introductory briefing to examine the elements of contestation and co-operation resulting from the recent expansion of research and teaching in 'environmental history', but also observes, *en route*, the opportunities for a very convenient and fruitful liaison with physical geographers. Figure 7.1, for example, illustrates some changes in forest cover in North America and Australia during the modern era. On the North American maps, lighter shading gives a very approximate indication of less intensive cover. The separately defined and broader distribution for 1993 shows that the estimated area of 'old growth' or virgin forest was chiefly confined to Alaska and Canada. At this scale, however, the retention of a number of apparently very small patches in the USA tends to underrate the success of local, regional and national efforts to preserve highly significant remnants.

The woods

Marsh's *Man and Nature* has been credited with stimulating the growth of conservationist sentiment but, like any other great work, it crystallised a good deal of innovative contemporary thought – notably but not only in the Mediterranean region – and was heavily indebted to earlier commentaries, including those on the Greek and Roman empires (Hall, 1998). The centrepiece of this work was the chapter on 'The Woods', which occupies more than a third of Marsh's text. From its inception in modern academia, historical geography declared a kinship with *Man and Nature*. Research into the 'clearing of the woodland' helped build the foundations of 'classical' historical geography in inter-war and post-war Britain, and, thereafter, has continued to offer intrinsically interesting, standalone or collaborative tasks. One early theme, shared by archaeologists, palaeoecologists and other specialists as well as by historical

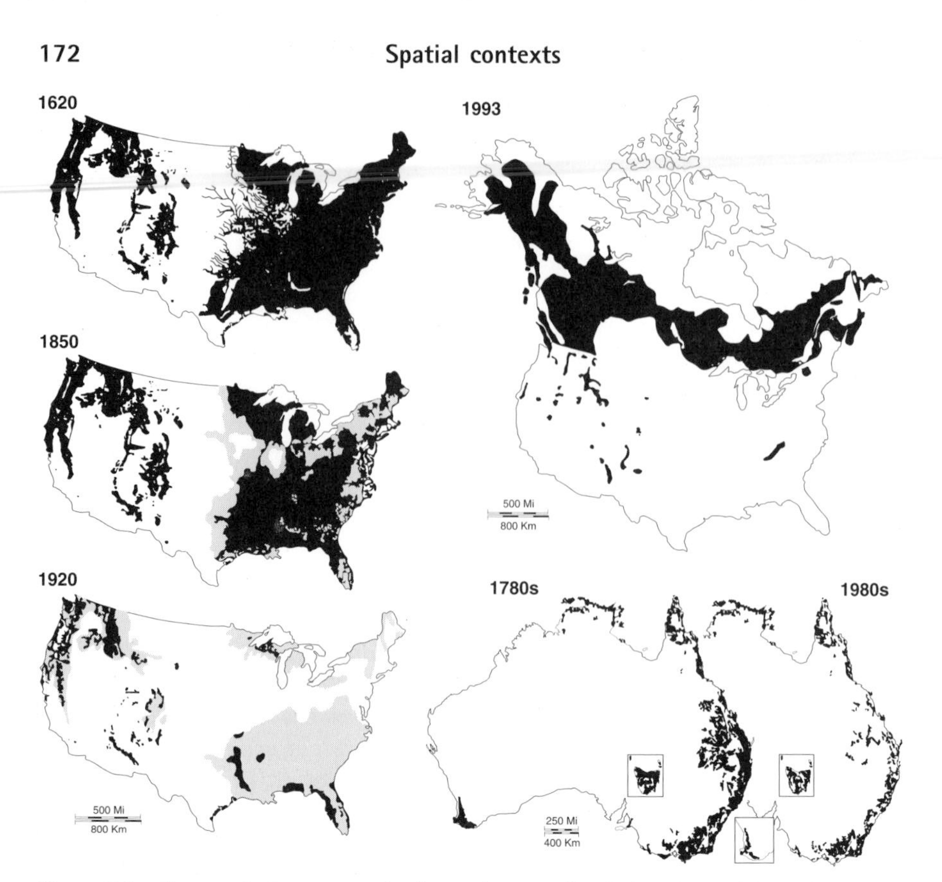

Figure 7.1 Changes in forest cover in the modern era, North America and Australia (after Greeley, 1925; Miller, 1990).

geographers, concerned the creation of impressive syntheses. These relied on a panoply of cartographical, documentary and physical evidence, which was often opaque and highly fragmented and demanded creative imagination as well as scholarly fortitude. Gradually, the meticulously contrived jigsaw was brought together in a series of national and global summations of the changes in forest and woodland cover, from the pre-agricultural era until the present. This project is far from complete, but the existence of certain critical regional and temporal variations seems confirmed.

Generally, Williams (1989a) suggests that the world's forest cover has diminished by a little more than 15 per cent over the modern period, and woodland has been reduced by approximately 14 per cent. He attributes much of that change to the clearing of deciduous and coniferous forests in the temperate belts of Europe and North-eastern America, principally for the expansion of peasant farming and commercial agriculture. Widespread clearances have also been effected in the warm temperate, subtropical and tropical zones of Africa, Australia and South America. Over the longer term many prominent deforestation episodes have been identified. In Europe, these occurred, for instance, during the Neolithic and early Bronze Ages, in association with the timber trade around the eastern Mediterranean from the seventh century BP, between the eleventh and thirteenth centuries in connection with German colonisations of the Slav lands, and in concert with the rapid surges of population growth and agricultural and industrial expansion between the sixteenth and early twentieth centuries. Clearances of temperate woodland occurred in pre-European America from as early as 12,000 BP, and in the equatorial rainforests, from between 3000 and 9000 BP.

Despite the occasional appearance, especially in earlier examples of such studies, of a type of fact-finding antiquarianism, there can be no denying the accumulated significance of this body of work for the scientific and moral projects of today. The painstaking documentations of pioneering researchers have become the raw materials of their successors. They may be exposed to more refined dating techniques, say, or used to guide particular conservation programmes, which might otherwise be quixotically aimed at the restoration of conjured 'original' landscapes. Yet it should be stressed that science and morality have long found common ground in tackling the question of whether nature or humanity has exerted the greater influence on environmental change. That engrossing theme underlies much of this chapter, notably its second part, and it is unfortunate that its important historiographical connotations lie beyond the scope of this short discussion: generally, the question has supplied a primary stimulant for geographical thought down the ages; in addition, its pursuit represents one of the most enduring, if not exclusive, contributions of historical geography to the wider discipline.

Not only did historical geographers produce specialist works dealing with the changes to tree cover, but they also contributed to a wider public arena. In Britain, the agendas of natural and historic conservation trusts were assisted and focused by the personal engagements and research of H. C. Darby and others. In the Americas, Sauer emerged as a respected historical–geographical exponent of cultural ecology. He wrote persuasively on the operation of human agency in pre-European landscapes, and (with Darby) subsequently collaborated in the production of a long-lived inter-disciplinary revival of Marsh's treatise, *Man's Role in Changing the Face of the Earth* (Thomas, 1956); that acclaimed enterprise would be emulated by a later generation (Turner *et al.*, 1990). However arresting the changes in tree cover, and the usefulness of all but the narrowest of the early inventories for global, national, regional and local summaries, common sense also underscored the necessity for due contextualisation, thereby ensuring some attention to ecological, economic, political and social factors. Increasingly, such perspectives complemented the trained predelictions and disciplinary loyalties, which had committed many geographers to focus excessively on cartographic records and representations, comments which apply equally to studies of the other landscape assemblages discussed below.

The waters

George Perkins Marsh was anxious to relate humankind's direct interventions in the earth's hydrological systems. These were occasioned by the need for domestic and industrial water supply, irrigation and navigation. Marsh also reported on the less direct but undeniable hydrological consequences of the removal of tree cover. In more optimistic vein, he observed the desirable interventions, including afforestation, the drainage of lakes and swamps, and the 'rescue' of land from the sea. Pioneering historical geographers were drawn to all of these sub-themes, but the Old World's 'reclamation' saga had left an unusually vivid imprint on the landscape, and it exerted magnetic appeal.

Research into the transformation of *wetland* environments constituted another major thrust in the early evolution of historical geography. Perhaps a rough distinction could be drawn between the broad-brush or large-scale 'clearing of the woodland' type ventures, and more patient local and regional reconstructions of the changing landscapes of fen, marsh and swamp. If it existed at all, the difference was superficial and short-lived, but the common adoption of larger scales for forest and woodland studies suggests a little more readily the probability of links with 'national' trends and policies, and correlations with piecemeal indicators of climatic change in the prehistoric and historic eras. In the case of the wetlands, the closely textured frames of reference invited research into constituent issues, which included: technological handicaps, innovations and competing propositions; the emergence of adaptive administrative, financial and legal structures; the resultant sequence of landscape expressions; and the reverberating role of key natural events such as flood, drought and wildfire.

The ultimate fact – or, as Marsh would have put it, the 'achievement' – of reclamation was normally the guiding research focus, but that necessarily encompassed an investigation of contemporary opposition to the proposed schemes. Between the 1930s and 1970s, historical geographers produced numerous reclamation accounts covering every inhabited continent and most large countries, particularly in the world's temperate zones. (Standard bibliographies are available elsewhere; see, for example, Baker, 1972; Butlin, 1993.) Annotated lists commonly commence with 'classical' and later accounts of the transformation of the Fenlands of Eastern England, and incorporate Western European, Russian and various New World studies (Darby, 1940a, 1940b, 1983; Hewes and Frandson, 1952; French, 1964; Williams, 1974; *cf.* Butlin, 1993). It would be misleading to reach for the term 'triumphalist' when describing this approach, and rather unconvincing merely to link its standard preoccupation with technological inputs and the writer's (scarcely uni-dimensional) provenances of depression, war and post-war reconstruction. Yet its set reckoning on the selected product – an ordered, frequently geometricised scene, aggressively tailored for new production systems – could just as easily be staking out or consolidating a disciplinary claim, and delivering a kind of bespoke narrative.

Grasslands and diffusions

The documentation also extended to global transformations of natural grassland cover, and in this context commenced by favouring modifications of the coastal and interior plains of the Americas, Australia, Eurasia, New Zealand

and Southern Africa. This entailed the study of plant and animal domestications, which had assisted in the conversion of the grasslands to commercial cropping and grazing belts. Among other things, this work was aimed at augmenting Sauer's cultural–historical perspective on agricultural origins and dispersals (Sauer, 1952), and was part of a long-range interdisciplinary project embracing a wider global spectrum (*cf.* Donkin, 1977, 1985). Andrew Clark's (1949) related study of the sequential colonisation of New Zealand by people, animals and plants quietly endorsed a better appreciation of a form of biological imperialism, which had accompanied and indeed still underpinned European expansion: in essence, this modest historical geography anticipated by almost four decades a more celebrated global overview by the historian, Alfred Crosby (1986).

Research trajectories seldom submit to thumbnail sketches. Historical geography's main practitioners were normally teachers as well as researchers and, in that capacity, many displayed specialist interests in particular places or regions, in selected periods or eras. Those interests prepared them for a lively discourse, which fleshed out and tested the bold longitudinal perspectives with anchoring contextualisations, including the addition of large-scale 'cross-sectional' analyses and examinations of the reasons for very distinctive transformations in key regions. In these regards, the grasslands offered spectacular theatre. The dominant theme was human interaction with the natural environment, and it sustained a rich diversity of opinion. As Tables 7.1 and 7.2 demonstrate, global environmental change has accelerated over the modern era and the natural grasslands (and open savannahs) of the New World have been profoundly affected by cropland expansions. In addition, the enormous areas laid down to sown pasture for livestock grazing further underlined the importance of this New World experience. These comprehensive outcomes were not too far removed in time and space from many pioneering geographers and historians, who seemed convinced by the proposition that the transformations became more accessible and therefore more intelligible at the level of regional and local analysis, not least because that is where the human signature is discernible. North America's Great Plains became the New World's archetypal region for grasslands scholarship. From the 1960s, historical geographers became embroiled in an absorbing scholarly controversy over shifting contemporary perceptions of the environment of the Plains, and especially over the origins, fate and practical import of a 'Great American Desert' image. Acknowledging the echoes of J. K. Wright's (1947) appeal for the study of 'geosophy' – that is, the evolving of scientific, aesthetic and vernacular forms of geographical knowledge – Bowden (1969) argued that previous scholarship had reflected the fluctuating environmental conditions on the Plains during the lifetimes of the interpreters, the influence of governmental and popular concern for what was deemed to be 'problem' or 'marginal' territory, and the individualistic emphases contributed by a group of gifted historians. For instance, he explained that in the period 1880–1905, 'Romantic Plains' historians, partly responding to severe drought conditions, popularised the idea of an authentic desert made over to settlement by the genius and sacrifice of American pioneers. Later, the spatial and behavioural interests of the powerful Turner school of 'frontier' historians, assisted by the experience of unusually favourable seasons and some misinterpretations of basic ecological data, reduced the strength of the environmental factor. A third phase reflected the droughts and depressions of the 1930s and, perhaps above all, the passionate concern of the historian, Walter Prescott Webb to make up for the neglect of his home state of Texas. He portrayed this as an area of special environmental difficulty, which had been redeemed by the design and application of scientific, technical, legal and other innovations (Webb, 1931). Webb's contemporary and Great Plains compatriot, James C. Malin, had too good a grasp on local history and ecological facts and conceptualisations to over-react to short-term flutters in rainfall regimes, and was similarly unmoved by the notion of a real or supposed desert, but for a time he was too easily dismissed as a cranky and isolated maverick. Malin's works would be dusted off and eagerly circulated when 'environmentalism' came to the fore (Bell, 1972).

The major growth of historical–geographical studies of the conversion of New World grasslands commenced in the post-war era. From the outset, in New Zealand and, if to a lesser extent, Australia, the process was dominated by transfers of the classical British preference for assiduously documented reconstructions. None the less, it contrived its own responses to the new *milieux* (see, for example, Baker, 1972). There was a fairly general tendency to be caught up in the primacy of a development imperative which was still powerfully extant in the researchers' own national communities. Effectively rather than designedly, where those researchers were also immigrants or of immigrant stock (a central characteristic of the group), the project as a whole must have promoted some improved rapport with their adopted countries.

For our present purposes it is equally important to observe that, although blinkers came with this preoccupation as with any other, this work stands up surprisingly well to late twentieth-century criticism. That is arguably a reflection of the comparative efficiency shown by this *leitmotif* in capturing material which was admittedly more intensively explored during the subsequent explosion of inter-disciplinary research. Development has undoubtedly been a ubiquitous and penetrating obsession in the New World, but its inclusiveness may be understated. An even-handed reading of a selection of pioneering works on Australia, for example, suggests that it supplemented, underpinned and played essential counterpoise to many other currents. These included:

- fervently held political and social ideas;
- organisational and legislative reforms, including those in and for the giant resource management agencies;
- the foundations of enduring relationships between governments, institutions and people;
- ecological concepts, true and false;
- the process of environmental inventory and the definitions of 'resource';
- the increasingly pervasive role of science and technology;
- a complex suite of environmental attitudes;

- the mixed results of high- and low-toned 'acclimatisation' efforts;
- the spearheading of reverberating encounters with natural hazards, and between immigrants, indigenous peoples and indigenous terrains;
- a seemingly endless formation of reservation policies with cultural, environmental, political and other implications;
- the environmental and cultural bases of the puzzling aetiology of disease in the transformed worlds of animals, humans and plants (see Powell, 1996 for contextualising reviews; for further bibliographical detail, see Baker, 1972; Baker and Gregory, 1984).

Soil, climate and fire

The study of the nature and significance of changing soils, climates and fire regimes has been particularly influenced by factors that include: effluxions of time itself, which allowed for the accumulation of experiential cues; the interposition or harnessing of science and technology; and diffusions and refinements of key interpretations and strategies. During the first half of this century, historical geographers confronted the fact that the problematical vulnerabilities of soil and climate to human intervention created a complex dynamic, which seemed less amenable than other themes, largely because the most closely engaged disciplines had still to produce the appropriate fundamental approaches and associated data. The observation could be extended, with reservations, to the roles of fire, while, in all three cases, interwoven social and cultural factors combined to render their interpretation even more complex.

Yet these three elements have also been widely recognised as ineluctable ingredients in a geography which studies the earth as the 'home of humankind'. A number of writers who were inclined to Sauerian-style cultural–historical geography succeeded in stirring a lively controversy over the influence of anthropogenic fire and crop and livestock domestications in sweeping ecological transformations. Similarly, while climatologists, palynologists and the like made the most evident progress in recording climatic and vegetation changes *per se*, historical geographers inserted a plethora of local and regional studies, which tested and qualified the generalisations. Indeed, this set or sub-field remained agreeably open and democratic until the later twentieth century, and interdisciplinary and intra-disciplinary adaptations and borrowings were probably as frequently initiated by climatologists and historians as by historical geographers. Researchers concerned with sustained expansions and contractions in rural settlement, and with economic trends or demographic swings, were increasingly prepared to concede the likelihood of correlations with profound climatic change and smaller perturbations or fluctuations (*cf.* Le Roy Ladurie, 1972). Evidence of a so-called 'Little Ice Age' (extending for between 700 and 300 years from either the thirteenth, fifteenth or early sixteenth centuries) was diligently compiled in Europe; in addition, puzzling inter-continental, regional and temporal discrepancies were identified, as were indicators of advance–retreat sequences in notoriously 'marginal' farming country (for example, Lamb, 1972, 1977, 1995; Parry, 1981).

The centrality of contemporary appraisals of soils was well recognised by historical geographers with interests in either the Old or New Worlds. Perhaps the key point is that a relatively common emphasis emerged from the adoption of one or more research directions which involved:

- the projection into the past of current sophisticated evaluations of local variations in fertility and susceptibility, in which case soil could be seen as the principal explanatory variable on or across (say) a settlement frontier;
- reconstructions of contemporary learning sequences in a coming-to-terms with – that is, essentially modifying, exploiting or rejecting – the qualities and disadvantages of local edaphic environments;
- and (very closely allied to the previous option) enquiries into modes of vernacular and scientific ecological thought, connecting the capacities and opportunities available to human societies to manipulate soils in such a way as to effect definite ameliorations in climate.

The most obvious common thread may be the tendency to explain the production and maintenance of profound changes in regional land use patterns, zones or massive belts. A further connection with the development imperative is therefore very palpable. It should be noticed once again, however, that the most thorough analyses of 'progress' and 'learning' – and even of contemporary wishful thinking, for that matter – necessarily locate the seeds of future environmental strife in the supposed successes or confessed failures of empirical testing procedures, whether founded on populist or scientific ideas or on some compound of the two (*cf.* Kollmorgen and Kollmorgen, 1973; Williams, 1974; Powell, 1977).

Facing the change

The preceding sketch of those aspects of the historiography of historical geography concerned with environmental issues illustrates a persistent emphasis on the interpretation of environmental change. In most countries, that stress was prominently, if not exclusively, directed towards landscape reconstructions. While 'classical' historical geography typically concentrated on presenting characterisations of landscape elements as scientific evidence, even to the extent of accepting an intrinsic or stand-alone significance, there were also intimations of nostalgia, of a discomfort about the decay or removal of a distinctive urban and rural heritage. The academic origins of this sentiment may be traced to British and European developments during the 1930s and 1940s, but it has longer psychological roots. These reflect a human need to consult that heritage in order to derive and communicate accessible stories capable of expressing and promoting an elevated sense of local, regional and national identity, a kind of 'topophilia' (Tuan, 1974) for place-bondedness. Works in this *genre* were noticeably popular in post-war Britain, most notably the writings of W. G. Hoskins (1955) and other local and regional historians. Over the next three decades, the resilient allure of 'reading the landscape' established that a mood had been caught, although all too often, the bulk of the credit went to history rather than geography.

Environmental history and historical geography

Assuming there is substance to these introductory claims, then the atmosphere of post-Second World War reconstruction must have been a contributing factor. A similar line of reasoning locates a more pronounced contextual shift in the 1960s, best symbolised by the publication of Nash's *Wilderness and the American Mind* (1967). This was received as a highly readable intellectual response to a rejuvenated discipline of dissent across the United States and later throughout the Western World, and applauded for anchoring an emergent environmentalism in cultural terms, which proclaimed the ascendancy of 'preservationism' over 'wise use' conservationism. Nash built superbly upon the lead of other American historians (for example, Hays, 1957; Huth, 1957; Richardson, 1962), proffering what he described as a history of 'attitude and action toward the land', which viewed the phenomena of environmental change 'as evidence of man's values, ideals, ambitions, and fears' (Nash, 1972: 363). Except for an assertively American focus on the need for due recognition of regional nuances, the spatial focus in *Wilderness and the American Mind* was comparatively subdued; in contrast, the search for cultural, philosophical, political and psychological causation was usually paramount.

If very significant, this shift was scarcely a revolutionary break and the singularity of Nash's early work can be overdrawn. Those historical geographers who were already attracted by inter-disciplinary 'environmental perception' studies were already quite well attuned to the importance of art and literature in shaping contemporary appraisals and behaviours. With them, Nash's survey definitely struck a chord. Nevertheless, geographers might have done more with his discussion of the innovative creation of national parks as an outgrowth of fundamental social and cultural change, including the rapid evolution of environmental attitudes and individual and collective reactions to environmental deterioration. They should also have been better placed than most of their colleagues in history to explain the conservational and other roles of the great reservations in national and regional space-economies – and somewhat better equipped, too, to interpret the parks themselves as managed landscapes.

Curiously, *Wilderness and the American Mind* first appeared in the same year as the geographer C. J. Glacken's *Traces on the Rhodian Shore* (1967); Glacken's book was an exhibition of prodigious scholarship, which made fewer concessions to time and place, and (in consequence?) was greeted with much less fanfare. The extraordinary scope of Glacken's book was defiantly, sometimes speculatively, ambitious; a powerful antidote to the 'satisficing' enterprise of modern social science and marked by its uncompromising injection of humanities expertise. Sadly, this landmark survey of nature and culture in Western thought did not venture beyond the eighteenth century and it was Nash's book which won the wider readership. The wilderness monograph seemed more closely attuned, in both content and idiom, to the then current public disquietude, and the antecedents it had found were recognisable and highly prized. The stunning impact of Rachel Carson's *Silent Spring* (1962) had given a stirring *imprimatur* for countless related academic works, and Nash's treatise was one of several key texts which launched a wave of 'environmental awareness' programmes. It was also influential in the introduction of a purportedly new version of 'environmental history' into tertiary curricula in the United States, from whence it has spread internationally.

Despite pronounced local successes, notably in North America, environmental history, however, has yet to win a secure niche in history's project. Nevertheless, as with any other aggressive new field, its rapid growth exposed the vulnerabilities of established sub-disciplines. Opening exemplars included 'environmentalist' revisions of key episodes – the American 'Dust Bowl' phenomenon, for instance (Worster, 1979) – and a flourishing range of contemporary philosophical readings, biographical portraits and gender perspectives (for example, Worster, 1977; MacCormack and Strathern, 1980; Merchant, 1983). In addition, the processes, consequences and local reactions to the march of imperialism were revisited and represented with a much stronger environmental colouring (for example, Crosby, 1986; Anderson and Grove, 1987; Mackenzie, 1988; Gadgil and Guha, 1992; Osborne, 1994; Arnold and Guha, 1995; Arnold, 1996; Grove and Damodaran, 1996), and cultural–historical studies reviewed changing environmental ideas at the 'heart of empire' (Charlton, 1984; Thomas, 1984). Relationships with historical geography promised still greater intimacy where the emphasis switched from period to place. That is evident in the locational specificities favoured by a number of these historians, and most especially, perhaps, in Cronon's work on ecological change and settler–indigenous community relationships in early New England (1983), and also the urban-environmental history of Chicago (1991), together with Worster's studies (1985, 1992) of the American West. Similarly, Grove's *Green Imperialism* (1995) bravely disputed the affirmed lead of the USA in modern conservation thought with proofs of meticulously reconstructed anticipations in British imperial territories, notably in ravaged island environments.

Convergences, old and new

As *Modern Historical Geographies* demonstrates, the perceived worth and therefore the longevity of disciplines and sub-disciplines partly depends upon a capacity to adapt to autonomous (or internal) interests and pressures and external demands and potentialities. Environmental history's minor boom disturbed the habitat within which environmental work in historical geography was normally conducted, but much of the change can now be described as interactional. One good example is Worster's commendation (1990, 1991) of an 'agroecological perspective in history', recognising three primary levels of analysis: nature, the interactions of productive technologies with the environment, and human attitudes and values. This approach would require an examination of the dynamics of natural ecosystems over time, together with the impacts of the political economy of productive systems on the environment, and the kaleidoscope of belief systems which suffused an individual's or group's dialogue with nature and helped create a 'second nature' – in the form of modified or artificial ecosystems (*agroecosystems*). Worster's conceptualisation envisaged studies of the ecological changes leading to the emergence of commercial farming belts, the effects of climatic swings, floods and drought, pests, plagues and the like, and the construction of bridges to the natural sciences. 'Wherever the two spheres, the natural and the cultural, confront or interact with one another, environmental history finds its essential themes' (Worster, 1990: 1095). The same nexus provides a core perspective for historical geography. For example, unarguably, historical geographies of the New World have contributed to the understanding of the evolution of rural production systems over immense tracts of country. Again, 'natural hazards' studies were first written into geography's modern repertoire more than thirty years ago and, as we have seen, they have recruited specialist historical geographers; yet it is also true that, until the 1990s, the contingent aim of achieving and disseminating a more comprehensive appreciation of ecological change *per se* was given top priority only infrequently.

There is no glaring or defining difference between environmental history and historical geography in the weight given to the 'society' component of the nature–society equation. On the whole, however, historians have been more conspicuous in chronicling the early growth of conservation movements. The most recent developments are well treated by human geographers who do not espouse avowedly historical interests, while geographers have shown greater resistance to the type of 'narrative impulse' which proclaims an awareness that authors are 'moral agents and political actors' (Cronon, 1992: 1370). Traditionally, historical geographers have been schooled to give a wide berth to morality tales and often, because their narrations have been consciously designed as multiple-purpose exercises, their work is less given to categorisation as 'environmentalist' literature. In contrast, the 'new' environmental history delivers studies on, say, the startling ubiquity of fire over the course of human history (Pyne, 1982, 1991, 1997a, 1997b), the origins and ecological consequences of the 'plague' of sheep which accompanied the Spanish conquest of Mexico (Melville, 1994), and the diffusion of the commercial rubber plant as an index of economic/ecological imperialism (Dean, 1987).

The most painstakingly researched of the related historical–geographical monographs include Williams's compendium (1989b) on American forests and Jordan's exposition (1993) on North American cattle-ranching frontiers. Neither chooses to make casual concessions to the environmentalist band-waggoners, while each consults and elaborates upon a very rich body of regional and national material to deliver (in a liberating and provocative fashion, markedly so in the case of Jordan) astute local and global contextualisations. For Australia and New Zealand, historical geographers remained more at ease with 'resource appraisal' and 'environmental perception', but have gradually extended those interests into an 'applied' series on the history of forest, soil and water management, which sketched the antecedents of current environmental concerns, policy directions and connected institutional arrangements (Roche 1987, 1994; Powell, 1989, 1991a, 1991b, 1993, 1998). Comparable convergences include, for example:

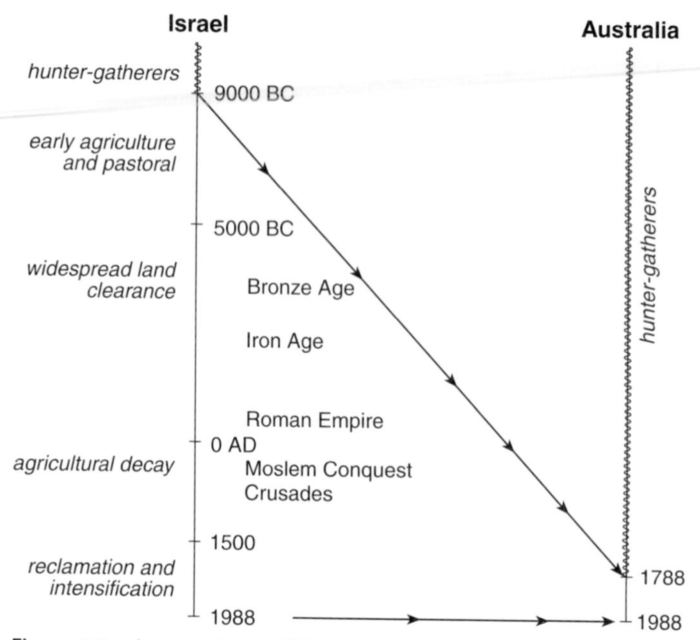

Figure 7.2 A comparison of 'Old' and 'New' World settlement sequences.

- environmental history's evolving focus on changes in ethical evaluations in the context of the emergence of animal rights issues and wildlife management (Dunlap, 1988, 1997); for a selection of Africanists' views, see Mackenzie, 1988, 1990; Beinart and Coates, 1995; Griffiths and Robin, 1997);
- reinvigorated inter-disciplinary emphases on the history of the actual and perceived relationships between changing environments and health/disease conditions in the human and animal realms (for example, Riley, 1987; Arnold, 1988; MacLeod and Lewis, 1988; Cranefield, 1991; Wilkinson, 1992; Beinart, 1997; and *cf.* Thompson, 1969, 1970; Powell, 1977: 119–43).

Explorations of the connections between each of these broad themes and the historiography of academic and non-academic geographical thought – an established but now oddly under-researched branch of historical geography – offer major opportunities (Livingstone, 1992, 1999). In another respect, contemporary debates on environmental change benefit from the illumination of cultural and cultural–ecological perspectives, not least in the study of primitive, peasant or 'pre-literate' communities (for one recent overview, see Mathewson, 1998; *cf.* Denevan *et al.*, 1987). Contemporary 'natural history' accounts simultaneously record environmental change and influence environmental adaptations and conservation ideas. New World examples once again strengthen the observation: Australian works include Finney (1993) and Griffiths (1996); a good overview is available in Jardine *et al.* (1996). Figures 7.2–7.5 provide

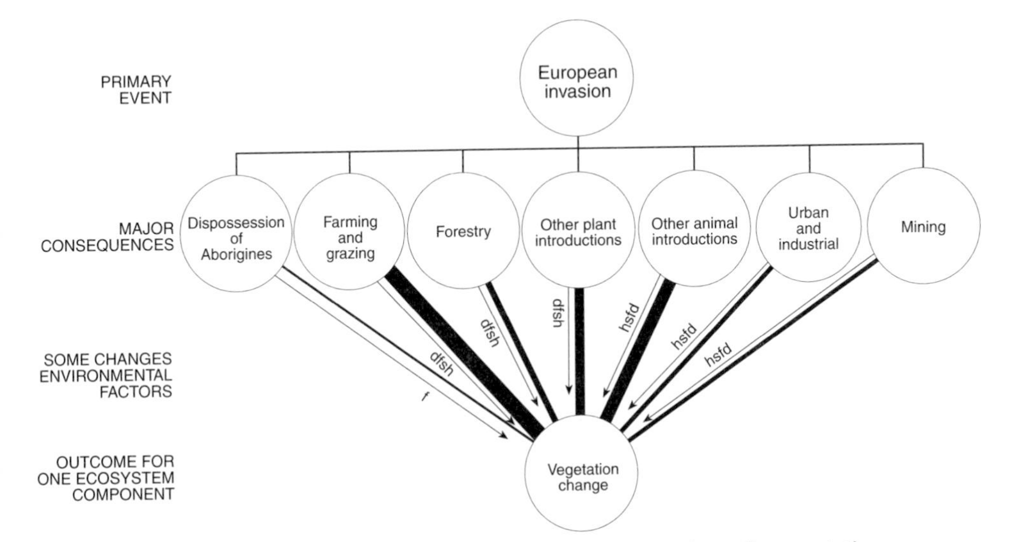

Figure 7.3 A representation of the major 'European' impacts on Australian vegetation (adapted from Adamson and Fox, 1982). Lines of varying thickness are used to suggest the range of impact. Lower case letters along each line distinguish principal cause of vegetation change as follows: **d**, direct result of the nominated human activity: **f**, changes in fire regime: **s**, alterations of soil properties: **h**, proximate/distant but related alterations to hydrological regimes.

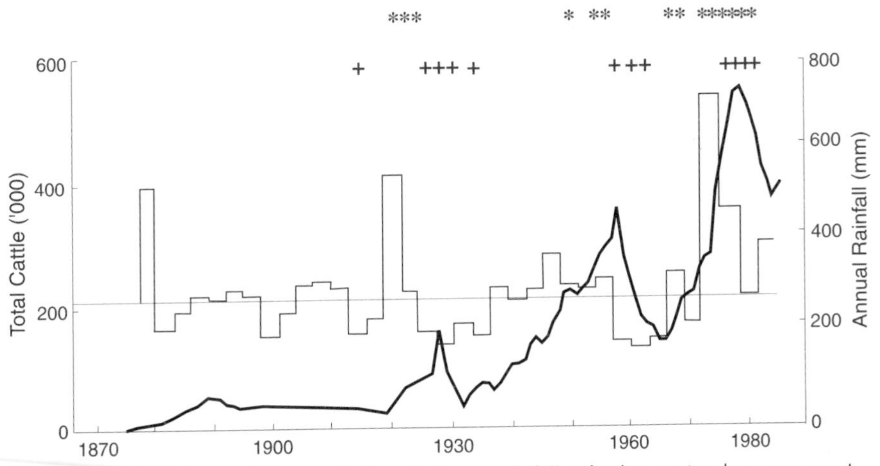

Figure 7.4 A background for correlations between rainfall episodes, pastoral management outcomes and associated environmental events; Alice Springs district (after Friedel *et al.*, 1990). Rainfall data (by 3-year means) are shown by the part-columnar line; the simple line graph indicates cattle numbers; asterisks at the top pick out extensive wildfires; beneath the latter, crosses identify occurrences of rabbit plague.

Figure 7.5 A documentation of the clearing of a district in the central wheat belt of Western Australia (after Arnold and Weeldenberg, 1991; Saunders, 1994).

specific examples of continuing inter-disciplinary interaction in Australia. The compressed time frame within which European settlement has wrought comprehensive environmental modifications in Australia is apparent from Figure 7.2. In contrast the landscapes of the Middle East experienced major transformations in the distant past, while the environmental impacts of modern technological communities have been – in relative terms – less significant.

Figure 7.3 is an interpretation of the major impacts made by the invading culture on Australia's indigenous vegetation, while Figure 7.4 addresses the correlations between rainfall episodes, pastoral management and associated environmental events in the Alice Springs district in the country's desert centre. Finally, Figure 7.5 records clearing activities in part of Western Australia's wheat belt. The maps underline the widespread reduction and fragmentation which accompanied the extension of the modern wheat-farming frontier. These effects contributed to a dynamic species loss over the past century, amounting to approximately 50 per cent of the district's indigenous mammals and 10 per cent of its birds.

It might be supposed that academic geography itself provides by far the most comfortable convergence. Yet despite periodical exhortations and a few well-cited early exemplars (for example, Lambert *et al.*, 1960), this meeting place was unaccountably bypassed until recently. Physical geographers and their fellow travellers in the sciences have traditionally contributed geomorphological and palaeoecological accounts of changing prehistoric landscapes. Some of their New World interpretations have either been taken up or were from the outset aimed at the public debates on the 'dispossession' of indigenous communities, partly through their descriptions of the landscape-forming interactions comprised in the rich word 'possession'; others have recovered the hydrological signatures and erosion scars left by the invading immigrant cultures in the historic era. (Some Australian examples are cited in Powell, 1996; see also Trimble, 1974, 1982, 1998; a useful related bibliography appears in Butzer, 1992.) Collaborative efforts in this field have largely been confined to the 'physical' sector of geography, where teamwork models are better developed. Latterly, however, a battery of improved dating techniques has been encouraging scientists to revive the dialogue with the *historic* past, and researchers trained in the humanities and social sciences are more obliged than ever to cross the bridge.

A promising guide is provided in the collection of essays edited by Butlin and Roberts (1995; see also Powell, 1997). It advocates a blending of 'human' and 'natural' archival materials at various scales, addresses a generous slice of time and marshals sub-disciplines within geography in a coverage that embraces a mixture of pure and applied evidence. This material: elucidates local and regional instances of environmental change; assigns the product to human and natural agencies; explains the formation of conservationally prominent (including 'relict') vegetation types; tracks fauna and flora migrations; and throws fresh light on the transformation of cultural landscapes. The detail picks out engrossing sub-themes: improvements in – and improved applications of – palynological and lake-sediment analyses; renewed connections with archaeology; an enterprising use of insect indicators; and speculations on the 'ecosystem shocks' which accompanied the Spanish and British invasions of the Americas and the Caribbean.

Taken as a whole, Butlin and Roberts's collection demonstrates that today's preoccupations with 'sustainability' are hollow indeed without an appreciation of real and putative *baseline* conditions and a grasp of ecological trajectories.

The book's northern hemisphere orientation and British Isles/North Atlantic emphasis is a little confining, and the bias towards very early periods necessarily advantages the natural scientists and proscribes the reach of radical modifications to research procedures. It could also be urged that the most profound ecological insights often require *global* data, that the southern hemisphere experience does rather more than endorse northern viewpoints, and that the rewards for similar collaborative work in the New World countries of both hemispheres include invaluable options for commentaries on the impact of government policies for settlement expansion, resource appraisal and environmental management during the past 200 years or so of tumultuous change. In Australia, physical geographers and ecologists have been moving briskly into the openings left by a perilously small band of historical geographers. Using an assortment of physical, manuscript and printed evidence for the nineteenth and twentieth centuries, these researchers have been explaining the effects of European settlement expansion on Australian mammals, and are investigating a similar mix of evidence for sample Outback properties to interpret the differential environmental costs of autonomous climatic change and pastoral management strategies (for example, Friedel *et al.*, 1990; Morton, 1990; Pickard, 1990; Lunney *et al.*, 1994; Lunney and Dawson, 1995).

Prospective

The topic under review affords one representative manifestation of the pace and direction of change in modern academia. Much of its distinctiveness resides in the fact that it seems redolent with civic purpose; in addition, however, it appeals strongly to those responsive and adaptable individuals who are prepared to test the margins of their specialist trades. As we have seen, historical geography made a deliberate investment in the investigation of environmental change long before the stimulus of the environmental movement intensified the kinds of converging inter-disciplinary research discussed here. That investment was both sound and ethical, but it would yield better returns for geography and for society at large if it were complemented and amplified – or otherwise applied in conjunction with – the insights of cultural and physical geographers, environmental historians, ecologists and others. If academic geography benefits from being as committed to the elevation of environmental sensibilities as it is to the appreciation of environmental processes, and as much to place-making as to the description and interpretation of places, then multifaceted historical geographies of environmental change surely provide a pivotal relevance. Modern environmental literature rushed in like a *tsunami* during the late 1960s and early 1970s. In the 1980s it withstood the braking effect of arguments which charged that words and stories were scarcely more than tools in the construction and advancement of interest groups, and was bolstered by the introduction of innovative teaching and research programmes in academic history. The best environmental stories address universals – our human needs for security, comfort, identity, the calming of conscience and its concomitant firm purpose of amendment – but also recognise locational integrities in so far as they have a very definite resonance in those national and regional contexts in which modern communities expect to participate in the making and breaking of government policy.

Historical geographers are better cued by training and the availability of proximate colleagues to set the rights and needs of the 'natural' environment alongside those of its human stewards, but while they continue to favour the latter emphasis, supporting environmental perspectives are often in evidence. In varying degrees, this is true of interpretations of environmental change within broader essays on the dispossession of indigenous communities (for example, Wishart, 1995) and the 're-settlement' of New World colonies (Powell, 1991b; Harris, 1997); and they have shown superbly the signal probity of scholarly encounters with contemporary recorders of the micro-detail of local transformations (Wynn, 1997). In other respects, historical geographers remain reluctant to follow their history colleagues into the presentation of 'moral tales', yet admit to a slightly disappointing ambivalence towards the remarkable successes of writers who have adopted materials and themes which tradition had assigned to the geographers' bailiwick – here a spirited biography of Francis Younghusband, the eccentric explorer (French, 1995) and a literary discourse on the sirens' call of the polar wilds for British publics (Spufford, 1996); there a well-realised account of the history of the fixing of longitude (Sobel, 1995), a connoisseur's elegantly evoked interfusions of the landscapes which exist within and outside the human mind (Schama, 1995), and so on. Such attitudes are understandable, but it would be a pity if they promoted an overly insular reaction. Rising demands for enhanced civic relevance in the academic arena promise continuing amendment to the transcendent environmental project, and historical geographers are well advised to acknowledge the necessity for a closer engagement with intra- and inter-disciplinary adjustments.

References

Adamson, D. A. and Fox, M. D. (1982) Change in Australasian vegetation since European settlement. In Smith, J. M. B. (ed.), *A History of Australasian Vegetation*, McGraw-Hill, Sydney, 109–46.

Anderson, D. and Grove, R. (eds) (1987) *Conservation in Africa: People, Policies and Practice*, Cambridge University Press, Cambridge.

Arnold, D. (ed.) (1988) *Imperial Medicine and Indigenous Societies*, St Martin's Press, Manchester.

Arnold, D. (1996) *The Problem of Nature: Environment, Culture and European Expansion*, Blackwell, Oxford.

Arnold, D. and Guha, R. (1995) *Nature, Culture, Imperialism: Essays on the Environmental History of South and Southeast Asia*, Oxford University Press, Delhi.

Atlas of Australian Natural Resources (1990), Australian Government Publishing, Canberra.

Arnold, G. W. and Weeldenberg, J. R. (1991) The distribution and characteristics of remnant vegetation in parts of the Kellerberrin, Tammin, Traying and Wyalkatchem Shires of Western Australia. *Technical Memorandum 33*, CSIRO Division of Wildlife and Ecology, Lyneham.

Baker, A. R. H. (ed.) (1972) *Progress in Historical Geography*, David and Charles, Newton Abbot.

Baker, A. R. H. and Gregory, D. (eds) (1984) *Explorations in Historical Geography*, Cambridge University Press, Cambridge.

Beaudry, M. C. (1993) *Documentary Archaeology in the New World*, Cambridge University Press, Cambridge.

Beinart, W. (1997) Vets, viruses and environmentalism at the Cape. In Griffiths, T. and Robin, L. (eds) (1997) *Ecology and Empire: Environmental History of Settler Societies*, Keele University Press, Edinburgh, 87–101.

Beinart, W. and Coates, P. (1995) *Environment and History: The Taming of Nature in the USA and South Africa*, Routledge, London.

Bell, R. G. (1972) James C. Malin and the grasslands of North America. *Agricultural History*, **46**, 412–24.

Bowden, M. (1969) The perception of the western interior of the United States, 1800–1870: a problem in historical geosophy. *Proceedings, Association of American Geographers*, **1**, 16–21.

Butlin, R. A. (1993) *Historical Geography: Through the Gates of Space and Time*, Arnold, London.

Butlin, R. A. and Roberts, N. (eds) (1995) *Ecological Relations in Historical Times*, Blackwell, Oxford.

Butzer, K. W. (ed.) (1992) *The Americas Before and After 1492: Current Geographical Research* (special issue, *Annals of the Association of American Geographers*, **82**), Association of American Geographers, Washington.

Carson, R. (1962) *Silent Spring*, Hamilton, London.

Charlton, D. G. (1984) *New Images of the Natural in France: A Study in European Cultural History, 1750–1800*, Cambridge University Press, Cambridge.

Clark, A. H. (1949) *The Invasion of New Zealand by People, Plants and Animals*, Rutgers University Press, New Brunswick.

Cranefield, P. F. (1991) *Science and Empire: East Coast Fever in Rhodesia and the Transvaal*, Cambridge University Press, Cambridge.

Cronon, W. (1983) *Changes in the Land: Indians, Colonists and the Ecology of New England*, Hill and Wang, New York.

Cronon, W. (1991) *Nature's Metropolis: Chicago and the Great West*, Norton, New York.

Cronon, W. (1992) A place for stories: nature, history and narrative. *Journal of American History*, **78**, 1347–76.

Crosby, A. (1986) *Ecological Imperialism: The Biological Expansion of Europe, 900–1900*, Cambridge University Press, Cambridge.

Darby, H. C. (1940a) *The Mediaeval Fenland*, Cambridge University Press, Cambridge.

Darby, H. C. (1940b) *The Draining of the Fens*, Cambridge University Press, Cambridge.

Darby, H. C. (1983) Historical geography in Britain, 1920–1980: continuity and change. *Transactions of the Institute of British Geographers*, New Series, **8**, 421–28.

Dean, W. (1987) *Brazil and the Struggle for Rubber: A Study in Environmental History*, Cambridge University Press, Cambridge.

Denevan, W. M., Mathewson, K. and Knapp, G. (eds) (1987) *Pre-Hispanic Agricultural Fields in the Andean Region*, BAR International Series 359, Oxford.

Donkin, R. A. (1977) Spanish red: an ethnographical study of cochineal and the opuntia cactus. *Transactions of the American Philosophical Society*, **67**, Part 5.

Donkin, R. A. (1985) The peccary – with observations in the introduction of pigs to the New World. *Transactions of the American Philosophical Society*, **75**, Part 5.

Dunlap, T. R. (1988) *Saving America's Wildlife*, Princeton University Press, Princeton, N.J.

Dunlap, T. R. (1997) Remaking the land: the acclimatisation movement and Anglo ideas of nature. *Journal of World History*, **8**, 303–19.

Finney, C. M. (1993) *Paradise Revealed: Natural History in Nineteenth-century Australia*, Museum of Victoria, Melbourne.

French, P. (1995) *Younghusband. The Last Great Imperial Adventurer*, Flamingo, London.
French, R. A. (1964) The reclamation of swamp in pre-revolutionary Russia. *Transactions of the Institute of British Geographers*, **34**, 175–88.
Friedel, M. H., Foran, B. D. and Stafford Smith, D. M. (1990) Where the creeks run dry or ten feet high: pastoral management in arid Australia. *Proceedings of the Ecological Society of Australia*, **16**, 185–94.
Gadgil, M. and Guha, R. (1992) *This Fissured Land*, Oxford University Press, Delhi
Gale, S., Haworth, R. J. and Pisanu, P. C. (1995) The ^{210}Pb chronology of late Holocene deposition in an Eastern Australian lake basin. *Quaternary Science Reviews*, **14**, 395–408.
Glacken, C. J. (1967) *Traces on the Rhodian Shore: Nature and Culture in Western Thought From Ancient Times to the End of the Eighteenth Century*, University of California Press, Berkeley and Los Angeles.
Greeley, W. B. (1925) The relation of geography to timber supply. *Economic Geography*, **1**, 1–11.
Griffiths, T. (1996) *Hunters and Collectors: The Antiquarian Imagination in Australia*, Cambridge University Press, Cambridge.
Griffiths, T. and Robin, L. (eds) (1997) *Ecology and Empire: Environmental History of Settler Societies*, Keele University Press, Edinburgh.
Grove, R. H. (1995) *Green Imperialism: Colonial Expansion, Tropical Island Edens and the Origins of Environmentalism, 1600–1860*, Cambridge University Press, Cambridge.
Grove, R. H. and Damodoran, V. (eds) (1996), *Essays on the Environmental History of South and Southeast Asia*, Oxford University Press, Delhi.
Hall, M. (1998) Restoring the countryside: George Perkins Marsh and the Italian land ethic 1861–1882. *Environment and History*, **4**, 91–103.
Harris, R. C. (1997) *The Resettlement of British Columbia: Essays on Colonialism and Geographical Change*, UBC Press, Vancouver.
Hays, S. P. (1957) *Conservation and the Gospel of Efficiency*, Harvard University Press, Cambridge, Mass.
Heathcote, R. L. (ed.) (1980) *Perception of Desertification*, United Nations University, Tokyo.
Heathcote, R. L. (1995) *Australia*, Longman, London.
Hewes, L. and Frandson, P. E. (1952) Occupying the wet prairie: the role of artificial drainage in Story County, Iowa. *Annals, Association of American Geographers*, **42**, 24–50.
Hoskins, W. G. (1955) *The Making of the English Landscape*, Hodder and Stoughton, London.
Huth, H. (1957) *Nature and the American: Three Centuries of Changing Attitudes*, University of California Press, Berkeley.
Jardine, N., Secord, J. A. and Spary, E. C. (eds) (1996) *Cultures of Natural History*, Cambridge University Press, Cambridge.
Jordan, T. G. (1993) *North American Cattle-ranching Frontiers: Origins, Diffusion and Differentiation*, University of New Mexico Press, Albuquerque.
Kinda, A. (1997) Some traditions and methodologies of Japanese historical geography. *Journal of Historical Geography*, **23**, 62–75.
Kollmorgen, W. and Kollmorgen, J. (1973) Landscape meteorology in the Plains area. *Annals, Association of American Geographers*, **63**, 424–41.
Lamb, H. H. (1972, 1977) *Climate, Present, Past and Future*, 2 vols, Methuen, London.
Lamb, H. H. (1995) *Climate, History and the Modern World*, revised edn, Routledge, London.
Lambert, A. M., Jennings, J. N., Smith, C. T., Green, C. and Hutchinson, J. N. (1960) *The Making of the Broads*, Royal Geographical Society, London.
Le Roy Ladurie, E. (1972) *Times of Feast, Times of Famine*, Allen and Unwin, London.
Livingstone, D. N. (1992) *The Geographical Tradition: Episodes in the History of a Contested Enterprise*, Blackwell, Oxford.

Livingstone, D. N. (1999) Tropical climate and moral hygiene: the anatomy of a Victorian debate. *British Journal for the History of Science*, **32**, 93–110.

Lowenthal, D. (ed.) (1965) *Man and Nature or, Physical Geography as Modified by Human Action*, Belknap Press, Cambridge, Mass.

Lunney, D. and Dawson, L. (eds) (1995) *Fading Fauna* (special issue, *Australian Zoologist*, **30**) Royal Zoological Society of New South Wales, Mosman, NSW.

Lunney, D., Hand, S., Reed, P. and Butcher, D. (eds) (1994) *Future of the Fauna of Western New South Wales* (*Transactions of the Royal Zoological Society of New South Wales*), Surrey Beatty, Chipping Norton, NSW.

MacCormack, C. P. and Strathern, M. (eds) (1980) *Nature, Culture and Gender*, Cambridge University Press, Cambridge.

Mackenzie, J. M. (1988) *The Empire of Hunting: Hunting, Conservation and British Imperialism*, Manchester University Press, Manchester.

Mackenzie, J. M. (1990) *Imperialism and the Natural World*, Manchester University Press, Manchester.

MacLeod, R. M. and Lewis, M. (1988) *Disease, Medicine and Empire*, Routledge, London.

Marsh, G. P. (1864) *Man and Nature or, Physical Geography as Modified by Human Action*, Scribner, New York.

Mathewson, K. (1998) Cultural landscapes and ecology, 1995–96: of oecumenics and nature(s). *Progress in Human Geography*, **22**, 115–28.

Melville, E. G. K. (1994) *A Plague of Sheep: Environmental Consequences of the Conquest of Mexico*, Cambridge University Press, Cambridge.

Merchant, C. (1983) *The Death of Nature: Women, Ecology and the Scientific Revolution*, Harper, San Francisco.

Miller, G. T. (1996) *Living in the Environment*, Wadsworth, Bellmont.

Morton, S. (1990) The impact of European settlement on the vertebrate animals of arid Australia: a conceptual model. *Proceedings of the Ecological Society of Australia*, **16**, 201–13.

Nash, R. F. (1967) *Wilderness and the American Mind*. Yale University Press, New Haven, Conn.

Nash, R. F. (1972) American environmental history: a new teaching frontier. *Pacific Historical Review*, **41**, 363–72.

Osborne, M. A. (1994) *Nature, the Exotic, and the Science of French Colonialism*, Indiana University Press, Bloomington, Ind.

Parry, M. (1981) Evaluating the impact of climatic change. In Delano Smith, C. and Parry, M. (eds), *Consequences of Climatic Change*, Department of Geography, University of Nottingham, Nottingham, 3–16.

Pickard, J. (1990) Analysis of stocking records from 1884 to 1988 during the subdivision of *Momba*, the largest property in semi-arid New South Wales. *Proceedings of the Ecological Society of Australia*, **16**, 245–53.

Powell, J. M. (1977) *Mirrors of the New World: Images and Image-makers in the Settlement Process*, Dawson-Archon, Folkestone and Hamden, Conn.

Powell, J. M. (1989) *Watering the Garden State: Water, Land and Community in Victoria, 1834–1988*, Allen and Unwin, Sydney.

Powell, J. M. (1991a) *Plains of Promise, Rivers of Destiny: Water Resources and the Development of Queensland, 1824–1990*, Boolarong Publications, Brisbane.

Powell, J. M. (1991b) *An Historical Geography of Modern Australia: The Restive Fringe*, Cambridge University Press, Cambridge.

Powell, J. M. (1993) *'MDB'. The Emergence of Bioregionalism in the Murray–Darling Basin*, Murray–Darling Basin Commission, Canberra.

Powell, J. M. (1996) Historical geography and environmental history: an Australian interface. *Journal of Historical Geography*, **22**, 253–73.

Powell, J. M. (1997) Interfusing aesthetics, ecology and history: disputing a non-convergent evolution. *Environment and History*, **3**, 117–25.
Powell, J. M. (1998) *Watering the Western Third: Water, Land and Community in Western Australia, 1826–1998*, Water and Rivers Commission, Perth.
Pyne, S. (1982) *Fire in America: A Cultural History of Wildland and Rural Fire*, Princeton University Press, Princeton, N.J.
Pyne, S. (1991) *Burning Bush: A Fire History of Australia*, Henry Holt, New York.
Pyne, S. (1997a) *Vestal Fire: An Environmental History, Told Through Fire, of Europe and Europe's Encounter with the World*, University of Washington Press, Seattle.
Pyne, S. (1997b) *World of Fire: The Culture of Fire on Earth*, University of Washington Press, Seattle.
Richardson, E. R. (1962) *The Politics of Conservation: Crusades and Controversies*, University of California Press, Berkeley.
Riley, J. C. (1987) *The Eighteenth-century Campaign to Avoid Disease*, Macmillan, Basingstoke.
Roche, M. M. (1987) *Forest Policy in New Zealand: An Historical Geography, 1840–1919*, Dumore Press, Palmerston North.
Roche, M. M. (1994) *Land and Water: Water and Soil Conservation and Central Government in New Zealand 1941–1988*, Department of Internal Affairs, Wellington.
Sauer, C. O. (1952) *Agricultural Origins and Dispersals*, American Geographical Society, New York.
Saunders, D. A. (1994) The effects of habitat reduction and fragmentation on the mammals and birds of the Western Australia central wheat belt: lessons for New South Wales. In Lunney, D., Hand, S., Reed, P. and Butcher, D. (eds) *Future of the Fauna of Western New South Wales* (*Transactions of the Royal Zoological Society of New South Wales*), Surrey Beatty, Chipping Norton, NSW, 99–105.
Schama, S. (1995) *Landscape and Memory*, HarperCollins, London.
Sobel, D. (1995) *Longitude*, Walker, New York.
Spufford, F. (1996) *I May be Some Time: Ice and the English Imagination*, Faber and Faber, London.
Thomas, K. (1984) *Man and the Natural World: Changing Attitudes in England, 1500–1800*, Penguin, Harmondsworth.
Thomas, W. L. (ed.) (1956) *Man's Role in Changing the Face of the Earth*, University of Chicago Press, Chicago.
Thompson, K. (1969) Insalubrious California: perception and reality. *Annals, Association of American Geographers*, **59**, 50–64.
Thompson, K. (1970) The Australian fever tree in California: eucalypts and malarial prophylaxis. *Annals, Association of American Geographers*, **60**, 230–44.
Trimble, S. W. (1974) *Man-induced Soil Erosion on the Southern Piedmont*, Soil Conservation Society of America, Washington.
Trimble, S. W. (1982) A sediment budget for Corn Creek Basin in the Driftless Area, Wisconsin, 1853–1977. *American Journal of Science*, **283**, 454–74.
Trimble, S. W. (1998) Dating fluvial processes from historical data and artifacts. *Catena*, **31**, 283–304.
Tuan, Yi-Fu (1974) *Topophilia*, Prentice-Hall, Englewood Cliffs, N.J.
Turner, B. L., Clark, W. C., Kates, R. W., Richards, J. F., Mathews, J. T. and Meyer, W. B. (eds) (1990) *The Earth as Transformed by Human Action*, Cambridge University Press, New York.
Webb, W. P. (1931) *The Great Plains*, Houghton Mifflin, Boston.
Wilkinson, L. (1992) *Animals and Disease: An Introduction to the History of Comparative Medicine*, Cambridge University Press, Cambridge.

Williams, M. (1974) *The Making of the South Australian Landscape: A Study in the Historical Geography of Australia*, Academic Press, London.

Williams, M. (1989a) Deforestation: past and present. *Progress in Human Geography*, **13**, 176–208.

Williams, M. (1989b) *Americans and Their Forests. A Historical Geography*, Cambridge University Press, Cambridge.

Williams, M. (1996) European expansion and land cover transformation. In Douglas, I., Huggett, R. and Robinson, M. (eds), *Companion Encyclopaedia of Geography: The Environment and Humankind*, Routledge, London, 182–205.

Wishart, D. J. (1995) *An Unspeakable Sadness: The Dispossession of the Nebraska Indians*, University of Nebraska Press, Lincoln.

Worster, D. (1977) *Nature's Economy: A History of Ecological Ideas*, Cambridge University Press, New York.

Worster, D. (1979) *Dust Bowl: The Southern Plains in the 1930s*, Oxford University Press, New York.

Worster, D. (1985) *Rivers of Empire: Water, Aridity and the Growth of the American West*, Oxford University Press, New York.

Worster, D. (1990) Transformations of the earth: toward an agroecological perspective in history. *Journal of American History*, **76**, 1087–1106.

Worster, D. (1992) *Under Western Skies: Nature and History in the American West*, Oxford University Press, Oxford.

Wright, J. K. (1947) *Terrae incognitae*: the place of the imagination in geography. *Annals, Association of American Geographers*, **37**, 1–15.

Wynn, G. (1997) Remapping Tutira: contours in the environmental history of New Zealand. *Journal of Historical Geography*, **23**, 418–46.

Chapter 8

Historical geographies of landscape

Susanne Seymour

Contested terrains: landscapes and landscape representations

The concept of 'landscape' has in recent years constituted a highly contested terrain of study and interpretation both within geography and beyond, in disciplines as diverse as art history, archaeology and anthropology (Bender, 1992). At a very basic level, debates have been structured around different approaches to the term 'landscape'. On the one hand, there is a group of approaches which define landscape solely as a physical entity, as 'real and tangible' land (Muir, 1998: 269). A contrasting set of approaches place more emphasis on the symbolic qualities of landscape, often considering the interplay between symbolic and material aspects, although some have also questioned this type of distinction (for example, Cosgrove and Daniels, 1988). Both approaches privilege a visual means of interpreting landscape although they do this in different ways.

The tradition of approaching landscape solely as a physical entity has been more important in the past in historical geographies, notably in the work of such figures as H. C. Darby, Carl Sauer and W. G. Hoskins, although it still claims a number of contemporary practitioners among geographers, including Muir (1998) and Hooke (1998). The historical geographies of landscape produced by such figures have focused on the physical landscape changes due to human and natural forces, tending to view landscape as a complex grouping of artefacts which can be surveyed and plotted or mapped. On the whole, such research has ignored or downplayed symbolic interpretations of landscape (Williams, 1989).

A concern with landscape representation is dominant in historical geographies which investigate the symbolic qualities of landscape and has been closely connected with the emergence of a new cultural geography (Duncan, 1995). Authors of such works have questioned a realist interpretation of landscape and have cast doubt on a division between objective and subjective approaches. The study of landscape has thus displayed explosive qualities in terms of the challenges it poses to the conventional boundaries of knowledge and disciplines (Bender, 1992: 3). This has brought the work of artists, landscape designers, musicians, novelists and poets within the realm of study, as well as encouraging reviews of the representational qualities of traditional geographical tools, most notably maps (see Harley, 1988; Alfrey and Daniels, 1990).

Such an approach to landscape thus registers a concern with the ways in which interpretations of the world are made and suggests there is no innocent, unmediated way of perceiving the world (Kinnaird *et al.*, 1997). All landscapes are taken to be representations, be they of earth, brick, verse, paint, ink or prose. That is not to argue that all landscape representations are the same: several authors have highlighted how differences in form influence the meanings which are conveyed and understood. There is a sense, however, in which a symbolic approach to landscape rests strongly on interpretation through the visual senses. Although several authors have discussed landscape as a text to be read (Duncan and Duncan, 1988; Daniels and Cosgrove, 1993), a landscape is more generally treated as something to be viewed, and often most associated with an open view organised through certain kinds of aesthetic conventions. A symbolic or iconographic approach to landscape recognises explicitly that there is a politics to representation. Landscape representations are situated: the view comes from somewhere, and both the organisation of landscapes on the ground, and in their representations, are and have been often tied to particular relationships of power between people. It was questions of class that were first explored by historical geographers interested in the politics of landscape.

Early historical geographies of this type focused on the influence of class relations in the production of landscape representations (Barrell, 1980; Cosgrove, 1985). Such interpretations have been criticised for reading landscapes only through the perspective of class relations (Rose, 1993) and restricting what qualifies as 'landscape' to a ruling class, Western and patriarchal 'way of seeing', such that the possibility of landscape subversions and other interpretations of landscape have been downplayed (Nash, 1994; Hirsch and O'Hanlon, 1995). None the less, the early work focusing on class relations has been important in raising the possibility of interpreting the symbolism of landscapes through an analysis of social relations.

More recently, greater attention has been given to the complexity of images and 'polyvisual' interpretations based on a series of social identities. As Kinnaird *et al.* (1997: 176) argue:

> Landscapes are not merely 'there' on the ground but are socially constructed within a complex and changing interplay of power relations, not least those between gender, class, race, sexual preference and other social differences.

This attention to wider dimensions of social relations and poststructural understanding of identity as culturally and historically contingent, has spawned a large amount of work which examines the interconnections between landscape and a series of interacting and shifting aspects of identity: national identity, gender, sexuality, colonialism, postcolonialism and race, as well as class (for example, Nash, 1994; Matless, 1995). Such work has embodied more fluid approaches, emphasising how landscapes may be both represented and understood in different ways. It has been stressed that a particular landscape might be subject to a range of interpretations depending on the social, political, economic and cultural position of the interpreter. Alternatively, a single landscape might be viewed simultaneously in a variety of ways (for example, Daniels and Seymour, 1990). It has been suggested that dominant understandings of image forms and particular landscape images are open to subversion (Nash, 1994; Seymour *et al.*, 1998). The understanding of what constitutes 'landscape' has been broadened beyond the confines of the open vistas characteristic of ruling class interpretations, to include a range of everyday landscapes in which senses of being, and senses of seeing, both help define what landscape is (for example, Hirsch and O'Hanlon, 1995). Work on landscape has broadened to consider the non-visual ways in which landscapes are experienced, and landscape is now considered in relation to a wider and more explicitly theorised suite of social identities including the traditional focus on class, as well as gender, nationhood and sexuality.

This chapter examines how landscape representations have been interpreted and re-presented in historical geographies and examines some studies of key landscape representations of the modern era: maps, landscape art, landscape photography, travel journals, poetry and music. It focuses on issues of landscape representation in the period *c.* 1660–1830, considering the relations between landscape and modern identities through the themes of property and labour. These motifs are then further explored through two case studies, the first considering the practices and cultural politics of English landscape improvement, the second addressing the aesthetics and politics of the picturesque in English plantations in the West Indies.

A landscape 'way of seeing'?

The dominant view of landscape as a ruling class 'way of seeing' in historical geographies situates the origins of the term in Western Europe in the sixteenth to eighteen centuries, a time when changes in relation to science, nature and religion were associated with emerging capitalism, colonialism and patriarchy (Bender, 1992: 1). A key historical geography text, which put forward this view of landscape, was Cosgrove's work (1993) on the emergence of the term 'landscape' in the fifteenth- and sixteenth-century Venetian Republic of north-eastern Italy. The generation of a landscape way of seeing, Cosgrove argues, was bound up with Renaissance theories of space and practical appropriation of territories by the urban merchant class, who were buying rural estates, and employing new theories and new geometric skills, such as linear (three-dimensional) perspective and improved surveying techniques to map and survey their lands. These new skills allowed more detailed mapping and they were also linked to the emergence of new styles of painting, incorporating expansive views and a high viewpoint and set to use in the paintings that landowners commissioned of their estates. The appropriation of space through land reclamation and drainage and landscape art was thereby intertwined. The 'landscape way of seeing' which emerged in landscape art, Cosgrove argues, was the exclusive preserve of a single social group, the urban merchant class (see also Kinnaird *et al.*, 1997). Such work forms part of the inter-disciplinary landscape school influenced by Marxist cultural theory, the focus of which has been the interpretation of landscape in the eighteenth century (see below). Within this school landscape itself has been interpreted as a 'visual ideology' of the ruling landed classes: a detached and privileged way of seeing, a symbol of their power and a celebration of their possessions which represents, naturalises and reproduces class relations. The term 'landscape' is reserved by Williams, for example, as a privileged way of seeing, confined to particular views. 'A working country,' he argues in his seminal text, *The Country and the City*, 'is hardly ever a landscape' (1973: 120). For Cosgrove (1985: 58) there is 'an inherent conservatism in the landscape idea, in its celebration of property and of an unchanging status quo, in its suppression of tension between groups in the landscape'.

The connections which Cosgrove established between the practices of surveying and mapping, and those of perspective in landscape art, raised some fundamental questions about the tools of geographical enquiry and geographical representation. But such an analysis of landscape, based principally on class divisions, has itself been criticised as too narrow a focus in a number of ways, including the perspectives of gender, sexuality and race. Geographers and others drawing on feminist and psychoanalytical thought have highlighted how many such Marxist-influenced discussions of landscape have ignored issues of gender, and even reinscribe landscapes with particular gender identities (Rose, 1993). They have emphasised the tradition of nature, and by implication rural landscapes, being viewed as feminine and a tradition of the female body as a landscape, pointing out how the organised and rational gaze of linear perspective might also be seen as a masculine gaze, with similarities to the distanced and objective mode of acquiring knowledge in conventional Western scientific inquiry. Some claimed, for example, that women were not able to see landscape in 'appropriate' ways in the eighteenth century (Barrell, 1990: 19). Similarly, in the nineteenth century, the category of landscape as a subject for English middle-class women was seen as inappropriate. As Deborah Cherry (1993: 118) writes, from the mid-nineteenth century,

> there were profound debates over landscape as a site and sight for women. Discussions focused on the propriety of women painting out of doors, the appropriateness of the artistic category and the suitability of rural scenes for women's visual representation.

Women were largely charged instead with flower painting. This 'gaze' has likewise been defined by class, race and sexuality. Edward Said's (1978) *Orientalism* has examined the 'othering' of 'the Orient' in landscape representations by Western travellers while Mitchell (1994a) has argued that landscape is an imperial way of seeing, becoming the predominant mode of representation at the height of a nation's colonial activity.

Alongside this work on the dominant discourses of landscape, there is a growing interest in the ways in which the conventional associations of landscape with masculine rational vision were negotiated in the past by those conventionally located outside the realm of reason – women. Complicating Mary Louise Pratt's (1992) gendered reading of Mary Kingsley's representations of landscape in *Travels in West Africa* (1982; first published 1897), Alison Blunt's (1994) account highlights how Kingsley draws on certain established masculine and imperial traits, such as an emphasis on the visual and the feminisation of

Figure 8.1 *Painting No. 488*, Marianne North (n.d.) (published with permission of the Board of Trustees of the Royal Botanic Garden, Kew).

landscape. However, these features are not fully developed. Kingsley's vision is circumscribed, walled in by mangroves or obscured by mists (although both seem to enhance the view for Kingsley), thereby placing her 'both inside and outside a masculine imperial tradition of exploration, conquest and surveillance' and illustrating 'the complexities and contradictions of a subject positionality' (Kinnaird *et al.*, 1997: 45). Thus, while criticising the male and colonial dimensions of landscape, Blunt and Rose (1994) have argued that in colonial situations, white women could look with almost the authority of white men specifically because the privileges of their racialised position outweighed the disadvantages of gender. Similarly, the paintings of Marianne North (1830–1890), who travelled extensively as a privileged English women along routes smoothed by the structure of British imperialism, both reproduce the associations of women with flower paintings and subvert them, painting neither conventional 'feminine' botanical images nor staying in the safe confines of the English garden (Figure 8.1). These cases highlight the constant and simultaneous negotiations of the meanings of landscape and dominant, shifting and resistance identities.

Recent work by anthropologists (Bender, 1992) and geographers (Morris, 1994; Cowell, 1997) has explored how landscape representation is not confined to privileged social groups. Nineteenth-century middle-class women, for example, as Morris points out, created garden landscapes. Others, particularly anthropologists have questioned the privileging of the visual sense in landscape interpretation and have considered landscape experience in ordinary landscapes and non-Western cultures. Morphy's (1992) work on aboriginal landscapes of memory and the research of Gow (1995) on the landscapes of kinship held by the Piro of Amazonian Peru are two examples. These accounts can tend towards an opposition of landscape as a 'way of seeing' with landscape as a 'way of being'. Matless (1995), in his work on English interwar landscapes, avoids such a distinction by adopting a more bodily investigation of landscape in which the visual is linked to the other senses. He focuses on preservationists' emphasis on 'being in the landscape' and practices of leisure, examining the way in which their representations of the experience of the scenic qualities of English rural landscape constructed various modes of 'good' and 'bad' citizenship. Loudness, litter, jazz or any other 'unsuitable' and 'intrusive' elements were defined as 'alien', along with the cultures and activities of urban, working class, Black and American people. Landscape, both in terms of its use and meanings in the past, and its theorisation within academic research, has always been the subject of contention and debate, caught up in struggle over the legitimacy of ownership and control. As some feminist geographers critique landscape others seek to reclaim its subversive potential (Nash, 1996).

Landscape representation, c. 1660–1850

In this section of the chapter, I consider some key ways in which landscape representation in the period from the end of the seventeenth century to the beginning of the nineteenth century has been approached in historical geographies. I will focus on two key themes:

- landscapes of property, using historical geographies of landed estates as examples;
- landscapes of labour.

As highlighted above, landscape was an important concept in Western European élite circles in the eighteenth century, particularly in England. This was witnessed in the emergence of landscape schools in painting, landscaping of landed estates and landscape tourism. According to Bender (1992: 2), landscape at this time was

> tightly tied to a particular 'way of seeing', a particular experience, whether in pictures, extolling nature or landscaping an estate [which naturalised a] deeply unequal, way of relating to the land and to other people.

Historical geographers have examined how various types of landscape representation were integral elements in the construction of modern identities in terms of nationhood, class, race and gender. This interest in landscape was particularly associated with rural areas and linked to new ideas about nature and property.

While most work by historical geographers of this period has been on rural landscapes, the most prominent exceptions to this include Daniels' (1993) work on Leeds and London, which examines the connections between town and country. In Britain, while the ownership of a country estate was the necessary entrée into the higher levels of society (patrician) in a way in which money alone was not, most of the great landowners also held urban properties. An appreciation of 'landscape' was an equally necessary entry requirement into the circles of a wider 'polite society', which developed in the later eighteenth century. In the assembly rooms of county towns or spas, in particular, lords and ladies mixed with country squires and their families, clergy, the military, bankers, merchants and other propertied professionals 'without the crippling respect for rank and hierarchy' which characterised earlier eras (Langford, 1989: 102). While tensions existed within polite society, a sense of shared interests, concerns and culture, including a taste for landscape, served to distinguish the polite from plebeian classes. Brewer (1997: 624) has argued that during the course of the eighteenth century, the visual quality of English people's relationship with nature increased as the country became increasingly urbanised and the proportion directly involved in agricultural or woodland activities declined. The number of printed books and images on sale to the public increased. In addition, interests in science and art were not separated out in Europe in the eighteenth century in the way in which they became divorced in the nineteenth century. Both urban botanical gardens and rural landed estates constituted sites of science and pleasure.

As Brewer (1997: 619) has emphasised, for Britain (and also Ireland and the wider context of Europe), there were many styles of representing rural 'landscape' in circulation, even amongst polite society, in the late eighteenth century:

> The countryside could be a place of arcadian rest, even indolence, the home of social harmony and virtuous self-sufficient work, a site of aesthetic pleasure, or a place in which to realise oneself through confronting 'nature'.

These ways of representing landscape could be mobilised by the same groups in different contexts. For example, in the country houses of the period, 'landscapes' of different genres were hung in different parts of the house and adjoining buildings: pastoral, mythological landscape paintings by artists such as Claude in a 'landscape room'; hunting scenes in the entrance hall; a decorated estate map in the estate office; picturesque watercolour landscapes in the drawing room. Likewise landscape parks were represented by their owners as multifunctional spaces: sites of leisure, scenic beauty, production, benevolence, seclusion and display.

However, concerns with the pictorial quality of landscape grew during the eighteenth century, emerging in the later years as 'the picturesque', a way of seeing landscape which dominated the early nineteenth century. Descriptions were given of places, highlighting how far they resembled paintings. Poetic descriptions of places were also constructed in ways which reflected the composition of a landscape painting (Barrell, 1988). But in the late eighteenth century, this term had very particular meanings. The key elements of picturesque imagery in written descriptions or visual imagery was that of a panoramic landscape, framed by trees or rocky outcrops, organised though alternative light and dark bands of topography, into a foreground, mid-ground and background. This structuring of the view encouraged the viewer to enjoy from an elevated position a vicarious sense of unimpeded movement across the landscape. This sense of the freedom of the eye to roam, Barrell argues, was linked to a sense of modern subjectivity, differentiated from nature, free from constraint, escaping confinement in visual excursions. This ability to see the landscape in certain ways was not neutral however. The faculty of seeing landscape in this ideal, distanced and abstract way was linked to the capacity to regard society in an impartial, objective and rational way. It also supported the 'natural' authority of upper-class men over other men and women in general. Thus the organisation of vision in landscape was deeply tied to the definition of upper class, white, male political and economic privilege. The ability to see landscape in particular ways became part of the construction of the position of the aristocracy and newly developing middle class, within the countryside, within state politics and thus also the position of the labouring poor.

Landscape in eighteenth-century England was, consequently, far from being merely a matter of taste. As Daniels and Seymour (1990: 488) have argued, representations of landscape constituted a 'highly complicated discourse in which a whole range of issues, which we might now discriminate as "economic", "political", "social" and "cultural" were encoded and negotiated'. Depictions of rural landscapes served to naturalise patrician authority, possession and power. For example, the Restoration poet, Abraham Cowley, represented a wooded landscape as comprising 'patrician trees so great and good', towering above 'the plebeian underwood'. Landscape was also gendered in particular ways in eighteenth-century Europe and the European colonies. In dominant strands of Enlightenment thought, women were represented as 'closer to nature' than men. This encouraged the sense of landscape as feminine, an object of nature in need of civilisation and 'improvement' by scientists, landscapers and artists, generally represented as masculine. For example, Mellor (1993: 85) argues that in Edmund Burke's highly influential *Philosophical Enquiry Into the Origins of Our Ideas of the Sublime and the Beautiful* of 1756, his idea of the sublime is associated with 'an experience of masculine empowerment', whereas its contrasting term, the beautiful, is associated with 'an experience of feminine nurturance, love and sensuous relaxation'. In turn, the representation of sites of colonial interest as female figures or 'virgin territory' provided a naturalising and legitimated discourse for the processes of European colonisation.

Landscapes of property

In Britain, the eighteenth century witnessed the growth of large estates and farms at the expense of smaller landholdings, through processes such as enclosure, drainage, and farm amalgamation. Until the Reform Act of 1832, the landed estate constituted the basis of social, political and economic authority in Britain. Such estates were not solely rural or even confined to Britain or Ireland. The century also witnessed the development or creation of estates in the European colonies. Larger landowners owned lands in a variety of regions, industrial and urban as well as rural, with a number having plantations overseas (Seymour *et al.*, 1998). Several studies have emphasised the importance of landscape representations in the promotion of the interests of aristocrats and landed gentry, examining Enlightenment changes beyond its radical metropolitan centres (for example, Williams, 1973; Barrell, 1980). This is not to give the impression that enlightened culture radiated from model estates through the networks of polite society. Not only were the flows always more complicated, but the discourse and practice of improvement were complicated and – at times – contentious. The very power and range of the term improvement, which extended from reading to statecraft, dispersed and destabilised its meaning in an increasingly complex society. During the Napoleonic Wars, a powerful anti-improvement literature emerged, initiated by defenders of the landed estate as a stable and stabilising domain. Enclosure was censured for greed and grinding the poor; model farms and gardens dismissed as useless showpieces (Daniels and Seymour, 1990; Daniels, 1993).

Sir Joseph Banks, President of the Royal Society, was a well-known figure who promoted the patrician version of improvement. His estates at Revesby, Lincolnshire and Overton, Derbyshire, his suburban gardens at Spring Grove, Isleworth, his house at 32 Soho Square (on the Portland estate), his management of the Royal Gardens at Kew, his direction from Kew of improvements in plantation agriculture overseas, formed the framework for his adventurous voyage with the *Endeavour*. Such estates were models to be publicised and to be emulated. Figure 8.2 is a representation of Croome Court in Herefordshire by the Welsh artist, Richard Wilson. Classically trained, Wilson produced landscapes in the style of Claude Lorraine (commonly known as Claude), a landscape artist of the Italian Renaissance much collected by young aristocratic Englishmen while on their Grand Tours in the early and middle years of the century. Croome, like many other English estates of the period, particularly those in the south and east, had recently been remodelled by Capability Brown. His work was characterised by landscape parks with huge lakes, sweeping lawns and large tracts of woodland, designs which were later condemned – particularly during the Napoleonic and Revolutionary Wars – for their arrogant display of exclusive property. Despite the placing of the house at the edge of the picture, everything in the scene appears subject to the control of the owner. The representation is of rural arcadias with little emphasis on connections to agricultural or sylvicultural production, despite the complex management systems that many indeed involved. Such scenes often featured landed families at leisure, viewing the scene (as here) or, conversely, were devoid of people. This image can be read as a landscape of luxury, surveillance and possession. Often the complexities of such representations has been ignored and landscape parks in the Brownian style have been simply read as sites of conspicuous consumption.

Figure 8.2 *Croome Court*, Richard Wilson (n.d.) (Trustees of the Croome Estate).

A development from this style was the open-air family portrait in which landed families were positioned in the scene, displaying the estate, often from the shade of a venerable oak tree (Rosenthal, 1982). As Daniels (1988) has pointed out, in England there was a complex symbolism associated with trees, dating from at least the Restoration (1660). Such representations appear more frequently to have included upwardly mobile families (see Daniels, 1988; Seymour, *et al.*, 1998). One such painting, Thomas Gainsborough's *Mr and Mrs Andrews* (*c.* 1749), has featured centrally in historical geographies of estate landscape representation (Figure 8.3). The debate was instigated by the art historian, Kenneth Clark's interpretation of the painting as depicting a couple in nature and Lawrence Gowing's idea that they were involved in 'the philosophic enjoyment' of the natural world (quoted in Berger, 1972: 106–7). Such interpretations prompted the Marxist art critic, John Berger to use the painting as an example in his radical review of the art world, *Ways of Seeing* (1972), based on the television series of the same name. The painting, Berger argued, illustrated the connections between the development of oil painting and the capitalist development of society and represented 'a way of seeing the world, which was ultimately determined by new attitudes to property and exchange, [which] found its visual expression in the oil painting' (Berger, 1972: 87). Oil painting, he argued had the particular quality of rendering the items it

Figure 8.3 *Mr and Mrs Andrews*, Thomas Gainsborough (c. 1749) (The National Galley, London).

depicted in a highly tangible and material manner. Thus, according to Berger, *Mr and Mrs Andrews* did not innocently depict a couple in philosophic enjoyment of nature but a couple enjoying their own estate as landowners. He argues that

> their proprietary attitude to what surrounds them is visible in their stance and in their expressions . . . [A]mong the pleasures that their portrait gave to Mr and Mrs Andrews, was the pleasure of seeing themselves depicted as landowners and this pleasure was enhanced by the ability of oil paint to render their land in all its substantiality. (Berger; 1972: 107–8)

Berger's *Ways of Seeing* came to constitute a founding text for the interdisciplinary landscape school influenced by Marxist cultural theory, which the early cultural geographers interested in landscape drew on and contributed to (see, for example, Cosgrove and Daniels, 1988). A concern with the material context of landscape art also informed Prince's (1988) analysis of the agrarian elements of Gainsborough's painting in which he identified straight rows of corn stubble, produced by the use of a seed drill rather than hand-sowing, a new-style five bar gate and 'improved' breeds of sheep. The painting, Prince argues, is informed by a practical knowledge of current agricultural practices in Suffolk, the place of Gainsborough's birth. Although Prince demonstrated that in its open display of cultivation, *Mr and Mrs Andrews* is an unusual landscape painting of its time, his analysis helped open up a debate about the display of improved agriculture in polite landscape art and landscape park design (see for example, Daniels, 1993; Seymour, 1993; Payne, 1993). This perspective, however, constituted another angle on the examination of class relations.

From a feminist perspective, Rose (1993) has re-viewed these class-based commentaries on the painting, using knowledge of eighteenth-century property relations, ideas of the modernist dualism, which opposes culture and nature and associates men with culture and women with nature, and the idea of the 'male gaze'. In contrast to Berger's account of the painting in which the couple were together viewed as landowners, Rose's account sets out to 'prise the couple . . . apart'. The iconography of the painting itself, she argues, highlights the different roles of husband and wife. Rose contrasts the active stance of Mr Andrews, ready to stride off to another part of his estate, with the static posture of his wife. Her sphere of activity, both physical and social, is, it appears, limited and certainly not the same as her husband's. Indeed, as Rose rightly points out, Mr and Mrs Andrews were not both landowners in legal title: 'only Mr Andrews owns the land' (Rose, 1993: 93). The gendered nature of property relations in the eighteenth century left all married women unable to own property in their own right, and any property they owned when single was usually transferred to their husband upon marriage. Mrs Andrews, Rose argues, has a different role. She is represented as part of nature, a static figure beneath an oak tree, symbolic of her expected role as wife and mother in extending and nurturing the family tree. This projected fertility is substantiated in the current fertility of the adjacent cornfield, piled with abundant stooks of corn.

Landscape, Rose argues, thus represents not only class relations but also gender relations. The landscape way of seeing is not only a privileged one in terms of class but also a masculine gaze. Such associations are only reinforced, she argues, in twentieth-century accounts of landscape. For example, in Berger's discussion of the energy of the visual image, Rose points out how his viewer experiences 'a going further than he could have achieved alone, towards a prey, a Madonna, a sexual pleasure, a landscape, a face, a different world' (Rose, 1993: 99). The landscape is thus re-projected as a feminised other, pursued by a voracious male spectator.

Despite the prominence of landscape as a key icon of eighteenth-century society, issues of gender and sexuality have been little explored. Bermingham (1993) has identified an institutionalised division defining education in the visual arts, whereby women were educated in the skills of drawing and painting and men in the skills of judging drawing and painting. Fabricant (1982: 223), argues that the early eighteenth-century landscape garden works with a feminised idea of landscape which is both maternal and erotic. The typical estate landscapes of this period, she argues,

> represented womblike enclosures specifically designed to allay the anxieties and satisfy the needs and desires (whether aesthetic, spiritual, or sensual) of their privileged owners, who were thus assured protection from the harsh, unpredictable world outside the garden and allowed to enjoy an environment shaped exactly according to their specifications and completely under their control.

Fabricant uses a description of Hagley Park from one of the most famous English poems of the eighteenth century, *The Seasons* by James Thomson, to illustrate how 'the feminized landscape opens herself up completely to the visual and perambulatory penetrations of both estate owner and poet, readily yielding up all of her treasures to the eye and other senses' (1982: 224). Thomson, imaging the park as a woman, describes how as he walks, 'she spreads/Unbounded beauty to the roving eye' and in the 'mingled wilderness of flowers. Fair-handed Spring unbosoms every grace.' Yet Fabricant's analysis focuses on a high culture approach, and like so much landscape history of the period, fails to relate literary ideas of early eighteenth-century landscape gardens to more practical issues of estate management and the day-to-day lives and living spaces of men and women.

Landscapes of labour

However, in the period from the mid-eighteenth to mid-nineteenth century, landscape provided more than just a measure of cultivated status for the upper and new middle class. It involved not only ideas of appropriate ways of seeing the land, but of seeing those who worked within it – the rural poor – and naturalising the social relations between those who worked the land and those who employed the labour of others. In John Barrell's (1972) famous and influential account of images of the rural poor in English paintings of this period, their representation is located within the social and economic changes of rural modernity, including the worsening condition's of the lowest strata of rural society, the growth of a new middle class and new forms of wealth creation at home, in commerce and industry, and new sources of income abroad in colonial endeavours. Drawing on accounts of the shift from a moral economy, in which the inequities of the social hierarchy were tempered by patrician benevolence and the existence of common rights, to a capitalist economy, Barrell traces the growing unease in the representation of the poor, with an increasingly class conscious, literate and disaffected workforce suffering from the erosion of common rights and, with increasing mechanisation and urbanisation of the textile trade, the decline of alternative sources of rural employment. The new economic aspirations of the aristocracy, interested in the improvement of their estates, and the potential profits from mining or manufacture, Barrell argues, necessitated a shift in the ways in which the poor and their work was understood. By the later half of the eighteenth century, the Pastoral tradition, which in literature and the visual arts had figured the relationship between the earth and humans as one of unproblematic harmony, with nature offering all its fruitfulness without demanding labour, was being replaced by the Georgic. Here, in contrast, nature was featured as a reluctant source, wild and hostile; only by hard and sober labour could land be made productive. This discourse which made hard work a moral imperative, and equated landscape beauty with ordered productivity, ideally dovetailed into the ideologies and practices of commercial capitalism, economic individualism and agricultural improvement.

Georgic images of the rural poor had to show them as a cheerful, sober, domestic peasantry, only momentary at rest from hard labour, and denied the individuality that was central to the concept of modern subjectivity. In landscape paintings, the rural poor, Barrell argued, were located on the dark side of the landscape:

> This division has the advantage of marking the differences in status and fortunes between rich and poor, while showing that the unity of the landscape and of the society it can be seen to represent, is dependent on the existence of both, which combine in a harmonious whole. As the landscape could not be structured without the natural contrasts of light and shade, so the society could not survive without social and economic distinctions which are thus also apparently natural. (Barrell, 1980: 22)

Although the Georgic constituted a dominant discourse, historical geographers and art historians have explored the ways in which English 'national' artists, such as John Constable and J. M. W. Turner informed by specific and situated experiences and knowledges followed, reformulated or resisted this formula in constructing relationships between national prosperity, local conditions and modern labour in landscape imagery (Helsinger, 1989, 1994, Howkins, 1992; Daniels, 1993; Lukacher, 1994).

The Georgic and the picturesque could also be combined. While, the picturesque was the style most popular with the growing middle classes of the late eighteenth century, it was not eschewed by patrician landowners or even explorers and scientists. Uvedale Price was one patrician estate owner who espoused the picturesque. He saw 'pictures in every tangled wood and thicket' (quoted in Daniels, 1988: 62) and admired the compositional picturesque for its quality of 'connection' in terms of scenic composition and in social terms. Price was also an active estate improver, extending and managing his woodlands through planting and felling, draining his land and rationalising his farms, creating several larger holdings but retaining a range from a few acres, on which a cow could be kept, upwards. This style of picturesque thus includes Georgic elements in a way in which William Gilpin's version did not. Whereas other polite styles of representation eschewed issues of labour and production, the Georgic celebrated them and, by implication, the role of polite patrons in British agricultural and mercantile expansion. Such Georgic representations were inspired in polite society by readings of Virgil's poem, *The Georgics*, in the context of social and economic disruptions linked to industrialisation, urbanisation, agrarian change and the wars with France and the potential fortunes that could be made from such rural developments in Britain or the colonies. Thus landscapes were praised for combining 'pleasure, patriotism and profit' or 'beauty and use'. In the remainder of the chapter, I move to consider the two case studies which exemplify the themes of property and labour.

Enclosed vision

Landscape representation played a key role in patrician versions of 'improvement' (Daniels and Seymour, 1990). Wild or 'unimproved' landscapes in eighteenth-century Britain could be appreciated. The antiquarian, Hayman Rooke, combined a statistical assessment of Sherwood Forest with an appreciation of its antiquity and trees, connecting their ancient beauty with 'heroic' figures from folklore and naval battles: 'I think we could not behold them without some degree of veneration' (cited in Watkins, 1998: 105). More often, however, common lands and Royal forests were commonly represented as 'wastes' or 'unimproved'. This view of these lands as unproductive helped justify the expansionist interests of larger landowners who generally supported: enclosures of lands held and used in common ownership or with customary rights; drainage schemes; the creation of enclosed fields or fenced plantations and individually owned pieces of land. A whole section of society thus lost customary rights. In 1793, for example, the Crown Commissioners in an assessment of Royal Forests found the remnant of Birkland and Bilhagh, where the inhabitants of neighbouring parishes had rights of common, as aspects which 'tend to obstruct Improvement, and lessen the Value of private Property, without bringing at present, any Profit to the Crown' (cited in Watkins, 1998: 95). Commons were appraised in terms of their potential as a resource, agricultural writers viewing unstinted commons as filthy blotches on the face of the country. Likewise the inhabitants of such 'wastes' were viewed as lazy or even criminal elements of society.

Accounts of the Clumber estate in Nottinghamshire emphasise how it was represented as without scenic quality before it had been landscaped and 'improved' in the later eighteenth century. Its owner, the second Duke of Newcastle, aspired to transform the 4,000 acres of Clumber Park into a 'garden', eventually creating a 2,000 acre farm inside it. This was mainly grassland but measures were taken to improve the quality of the sward, including rotations of clover and turnips and extensive water meadows to promote earlier grass growth. Improved sheep were bred by crossing Robert Bakewell's New Leicesters with the old forest breed and these were grazed in the park (Seymour, 1993).

A more famous site for agricultural and scientific improvement was the Holkham estate in Norfolk, owned by Thomas Coke. Figure 8.4 represents the park, landscaped by Capability Brown and Humphry Repton, stretching into the distance, with the new hall, built as a treasure house for Grand Tour spoils as a focal point and the church on a hillock to the left rising above extensive woodlands. The park, however, was also the site for Coke's extensive home farm, where Coke experimented with new breeds of animals and seeds and displayed his improving activities by holding agricultural shows, which were open to polite society from across Europe. In the foreground Coke inspects his Southdown sheep, taking notes to record the progress of his breeding and feeding trials.

By the end of the eighteenth century, the term, improvement, had broadened from its earlier associations with progressive farming practices to embrace a broad range of activities 'from music to manufacturing' (Daniels and Seymour, 1990: 488). However, ideas of what constituted 'improvement', were fiercely debated within polite society and contested from more radical quarters. In her novels, Jane Austen made use of metaphors of estate improvement to reflect upon her characters and discriminate their worth (Duckworth, 1971). At the same time, the effects of 'improvement' were keenly felt by those whose everyday working landscapes were sometimes dramatically transformed. The poetry of John Clare provides a sense of what this change might have meant (Barrell, 1972, 1988; Williams, 1973; Sales, 1983; Helsinger, 1987).

Figure 8.4 *Thomas Weaver Coke inspecting Southdowns* (anon.) (1806?) (Holkham Estate).

The processes through which the productivity of English agriculture was dramatically increased in the eighteenth and nineteenth centuries, especially Parliamentary enclosure, have long been explored by historical geographers. Their accounts of the practices which shaped the landscape can be fruitfully combined with an exploration of both the ideologies of improvement and, where possible, the impacts of those least able to determine the course of agricultural 'improvement'. While John Barrell's early work helped inspire historical geographers' interests in class, his more recent accounts of John Clare's (1793–1864) poetry now speak to contemporary interests in ideas of the body, nature and subjectivity. Clare, an agricultural worker, lived in Helpston, a village in what is now north Cambridgeshire (formerly Northamptonshire) which was undergoing enclosure. This complete reorganisation of the landscape into hedged fields was completed in 1820 when Clare was 27 years old. As Barrell argues, Clare's poetry explored the senses of openness and enclosure, and the relationships between subjectivity and place, as the geography of his locality was changing around him.

> Openness could mean for him the freedom of the open fields before the enclosure of Helpston, and so a landscape in which he was free to wander out in any direction, more or less, from the village, across the fallows, or along the balks and headlands of the fields under active cultivation. The division of the parish into

> small hedged fields, individually owned and cropped, and the stopping-up of old footpaths, curtailed that freedom, and Clare as well as the landscape was enclosed. (Barrell, 1988:123)

Yet unlike the sense of individual subjectivity, autonomy and freedom defined through separation from nature, Clare's poems spoke of an identity shaped through rather than in opposition to its context, 'constituted as a subject by the complex manifold of impressions that is the experience of being' in the landscape (Barrell, 1988: 127). Clare's identity was relational rather than individualistic, but in the face of the trauma of enclosure, also fragile. He died in 1864 in Northampton Asylum and his body was taken by train to Helpston for burial.

Colonial improvement

The desire for improvement was not confined to Britain or Europe. A significant number of British estate owners also had interests in the colonies. Some had connections through colonial or military service by a member of their family, or through investments or interests in colonial trade. Others commissioned artists, writers and cartographers to depict colonial landscapes. In later eighteenth-century Britain, Georgic discourses of landed estates increasingly integrated landscape aesthetics with estate management, mixing material and cultural concerns. Discourses of estate management encouraged owners to undertake estate 'improvements' which combined aesthetic, financial and patriotic imperatives (Daniels and Seymour, 1990). The process of polite colonisation proceeded not only by the acquisition of land overseas, but also in Britain through such practices as land enclosure and the increased surveillance of land and labour. Still, estate mapping, together with the topographical drawings of military campaigns, and Georgic literature and art produced by artists and writers helped extend British cultural authority into the colonies (Smith, 1985; Mitchell, 1994b). Caribbean property was accommodated by the adoption of conventional modes of representing and managing British landed estates, a process important in the assimilation of the islands as British colonies and in the integration of those with colonial interests into British élite society. Georgic discourses of landscape and estate management were modified to accommodate the physical situation of the tropics and the system of slavery in the British Caribbean, highlighting certain features and obscuring others. The islands were generally celebrated as a luxuriant tropical paradise of great productive capacity and financial value. Such a view screened out controversies over slave labour and conduct, the dangers of tropical climates, sexual exploits on plantations and the fragility of the 'West Indian' society and economy. Paintings commonly depict slaves engaged in 'cheerful toil', attending to the needs of their masters or in reassuring family groups (Seymour *et al.*, 1995). Through the medium of a Georgic poem, visitors attempted to transform the strange into the familiar, to play down the differences between the tropical slave plantation of the Caribbean and the British landed estate.

Figure 8.5 *Parham Hill House and Sugar Plantation, Antigua*, Thomas Hearne (1779) (The British Museum).

A comparison between views of Britain and of the colonies can be made using the work of the watercolourist Thomas Hearne. He produced views of estates in both the Caribbean and Herefordshire, where he completed a series of paintings for the picturesque author, Richard Payne Knight, and Sir George Cornewall, an estate owner of merchant descent. Hearne spent nearly four years in the Leeward Islands in the 1770s, under commission to the Governor, Sir Ralph Payne, producing a collection of around twenty large watercolours akin to the series of estate portraiture he later undertook in England. Upon his return Hearne worked up another estate sketch as a commission for its absentee owners, the Tudways of Wells in Somerset. Their estate, Parham Hill, was located on the eastern side of Antigua and in 1776 covered about 800 acres with a workforce of 533 slaves. Hearne's painting, *Parham Hill House and Sugar Plantation, Antigua* (1779) (Figure 8.5) includes some conventionally picturesque elements. The foreground includes a framing tree while the background exaggerates the height and rockiness of the Antiguan hills. Hearne's high view-point allows him to map the activity and production of the estate in considerable detail. Absentee owners, particularly those who had never seen their colonial estates, would be interested in such a record of their property and the prosperity it promised: substantial incomes were made, even from modest plantations, as sugar consumption grew rapidly in Britain (Sheridan, 1974; Drescher, 1977). Hearne depicts spring, the busiest time of the year on sugar plantations, and charts various stages in the production process. In the centre of the painting a field gang is hard at work cutting the ripe cane.

Nearby, another group of slaves load the cane they have cut onto a cart. A second cart, fully loaded, makes its way to the nearby sugar mill and boiler house, a semi-industrial site driven by wind power, nestled under a billowing cloud of steam produced by processing the cane into raw sugar. In the foreground, to the right, another cart is about to disappear round a bend in the road, transporting the barrels of raw sugar down to the local Parham Harbour to be shipped to Britain.

Hearne is likewise careful in his depiction of slaves. The British Caribbean islands had been under threat of invasion in a series of wars between the French, British and Spanish from 1689. Slave insurrections were also frequent. Twelve took place in Jamaica during the eighteenth century and although fewer occurred in the Leeward Islands, an island-wide slave insurrection had occurred in Antigua in 1735–6, and was followed by particularly bloody reprisals from the planters. Another slave-led revolt, albeit abortive, occurred in nearby St Kitts in 1778. Security was uppermost in the minds of plantations owners. At home there was also rising disquiet over the operation of the slave system and the treatment of slaves leading to calls for abolition of the slave trade (Goveia, 1965; Craton, 1982; Drescher, 1982). Using conventions of the picturesque style, Hearne reassuringly represents black people either hard at work in the fields, under the supervision of the overseer and overlooked by Parham House, or at rest in a sentimental domestic vignette of a family group in the foreground of the painting, a grouping which belies the social and economic position of slaves in the 1770s, discouraged from having families and frequently poorly dressed and fed (Hamshere, 1972; Ward, 1988).

Such Georgic representations obfuscate the tensions found on such estates and those between Britain and the colonies. Controversies raged in British polite society over flamboyant, fabulously rich and notoriously immoral West Indian nabobs, while the long-running debate over slavery not only posed a radical challenge to landed authority but also divided landed interests (Hamshere, 1972). But this accommodation was disrupted at certain times and places by debates over slavery in particular. An alternative view of both the English countryside and the system of slavery to Hearne's Georgic landscapes was put forward in an anti-slavery print, illustrating lines from *The Task* (1785) by the poet William Cowper (Figure 8.6). The image is hardly a landscape. Instead of a framing tree, there is only a stump, acting as a chain post. It is peopled, not by the estate owner but by a slave, positioned in the foreground, not obscured in labour in the middle distance. Nor does he form part of some reassuring domestic grouping but is chained, alone and on his knees. The slave, dignified despite his chains, and with none of the common marks of beatings visible, looks out to sea. The lines quoted from Cowper question when 'We have no slaves at home – why then abroad?' (Cowper, 1785). This appeal to Britain helps situate the print as one of a number produced by the Anti-Slavery Society and undercuts its radical edge. The black slave's hope for salvation is represented as emanating from the mainly white, often middle-class and evangelical anti-slavery groups of Britain and their supporters.

The colonies were sources of capital, raw materials and labour but they also fed a European taste for the 'exotic'. Landscape painters were frequently

Figure 8.6 Anti-slavery print: illustrating lines from William Cowper's *The Task* (n.d.).

recruited to such expeditions to work as topographers and to produce more developed images, such as oil paintings. One such artist was William Westall, employed on the voyage of the *Investigator* (1801–3). The expedition leader, Flinders, described how Westall 'took a view' of the coast five miles out from Port Philip in Australia which he felt 'will be an useful assistance in finding this extensive, but obscure port' (cited in Smith, 1985: 191). Westall was, however, disappointed by the pictorial potential of what he saw of the Australian coast and wrote to the expedition sponsor, Sir Joseph Banks, to ask his permission to go on to what was then Ceylon, a much more promising place for his skills he felt:

> the rich picturesque appearance of that Island; every part affording infinite variety, must produce many subjects to a painter extremely valuable. And as no painter has yet been there what I should acquire would be perfectly new, and probably interesting from the island being one of the richest in India and lately acquired. (Cited in Smith, 1985: 193–4)

In his study of European exploration of the South Pacific, Smith exaggerates the distinction between the approaches of artists and scientists. Likewise he fails to challenge Enlightenment models of science as objective and empirical and tends to overplay the role of philosophical ideas and religion rather than examining some of the imperial motives in terms of trade, expansion of territory and competition amongst European colonial powers. Banks had this wider view. The expeditions to the South Pacific were not just informed by objective scientific curiosity, a search for Eden, filled with noble savages or Christian zeal but had wider goals in terms of expanding empires, controlling resources, and displaying cultural power at home and to other competing powers. Of significance to both the painter, Westall, and perhaps more so his patron was the property as well as the pictorial value of the island. Westall was in search of an exotic picturesque which promised both.

Conclusion

Muir (1998) has recently been critical of what he regards as 'fragmented and unbalanced' accounts of landscape which dominate in geography. Viewing cultural geographical accounts of landscape as 'subjective' and focused on 'the imponderable subtleties of the human response', he advocates a return to what might be termed a 'scientific exploratory' approach to a 'real and tangible' landscape. Apparently blind to the significances of studies of subject positions, such as class, gender, race and sexuality (and the albeit complex interactions between them) in cultural understandings of landscape, Muir views subjectivity predominantly in terms of individual personality:

> the student of landscape whom one encounters surveying earthworks at the muddy site of a deserted medieval village will certainly be a very different personality type from the one seeking to infer meaning from an eighteenth-century landscape painting. Their approaches to landscape are different and their languages are mutually incomprehensible. (Muir, 1998: 265–6)

In some respects this critique follows the long tendency in studies of landscape to make distinctions between the 'real' material landscape and its symbolic representation. This has come from two distinct quarters, both from those who favour the empirical reconstruction of past landscapes over considering their meaning, and from those more recent researchers, who have treated the cultural representations as false or inauthentic and masks, which are a distraction from the harsh social reality of unequal class or gender relations – in images of rural labour for example. This notion of landscape images as deceptive screens, which hide the truth of the 'real' world (whether topography or social divisions) with its language of the authentic, and inauthentic, true and false, has been criticised by Matless (1992). Landscape representations, he argues, are both real in a material sense as objects, and real in the sense that they have material effects. Matless offers

> a way of considering representation, drawing in particular on the work of Foucault, which treats the image as neither significant of an essence nor reflective of a more basic reality; a way of operating which emphasises the image's substance, its substantive, power-sodden, moral and aesthetic make-up. Representations, images, fantasies are suggested here as being highly concrete stuff, not to be regarded as

> merely reflective or distortive of the world (through mirroring or distortion may be their declared aim), but as constitutive, as what the world is made of, really. (Matless, 1992: 44)

This sense of the work of images in the making of the world in a concrete sense, as distinct from simple reflections or distractions from the 'truth', is also highlighted by Daniels (1993: 8) who argues that:

> Landscape imagery is not merely a reflection of, or distraction from, more pressing social, economic or political issues; it is often a powerful mode of knowledge and social engagement.

This challenge to the convention of understanding landscape has also come from another direction. In referring to Cosgrove and Daniels (1988), the anthropologist Hirsch (1995) criticises historical geographies of landscape as too static in nature, meanwhile advocating an approach which treats landscape as a cultural process rather than an artefact. Bender (1992: 3) captures well the sense of energy which anthropologists in particular have emphasised: 'The landscape', she recounts, 'is never inert, people engage with it, re-work it, appropriate it and contest it. It is part of the way in which identities are created and disputed, whether as individual, group or nation-state.' This emphasis on process and on human action and agency in relation to landscapes, is also evident in W. J. T. Mitchell's suggestion that landscape is best understood, not as a noun, but as a verb. 'We should ask', he writes, 'not just what landscape "is" or "means" but what it does, how it works as a cultural practice. Landscape, we suggest, doesn't merely signify or symbolise power relations; it is an instrument of cultural power.' Landscape, he argues, is a dynamic medium,

> itself in motion from one place or time to another. In contrast to the usual treatment of landscape aesthetics in terms of fixed genres (sublime, beautiful, picturesque, pastoral), fixed media (literature, painting, photography) or fixed places treated as objects for visual contemplation or interpretation [we should] examine the way landscape circulates as a medium of exchange, a site of visual appropriation, the focus for the formation of identity. (Mitchell, 1994c: 1–2)

Whatever the differences in the metaphors and terminology adopted by these authors, all share a sense that landscape, identities and politics are both dynamic and entwined. From these perspectives, landscape does not simply mirror or distort 'underlying' social relations, but needs to be understood as enmeshed within the processes which shape how the world is organised, experienced and understood, rather than read as its end product. Despite the implied redundancy of considering landscape genres and landscape media implied by Mitchell's manifesto, much of historical geography's most fruitful work has been in bringing together a sensitivity to the constructive or formative nature of landscape and attention to the specific workings and meanings of these genres and media. As the accounts of landscape change and representation in this chapter show, historical geographies of landscape are also both attentive to the materiality of physical changes on the ground, and the substantive, power-laden and transformative nature of representation.

References

Alfrey, N. and Daniels, S. (eds) (1990) *Mapping the Landscape: Essays on Art and Cartography*, University of Nottingham, Nottingham.

Barrell, J. (1972) *The Idea of Landscape and the Sense of Place: An Approach to the Poetry of John Clare*, Cambridge University Press, Cambridge.

Barrell, J. (1980) *The Dark Side of the Landscape: The Rural Poor in English Painting 1730–1840*, Cambridge University Press, Cambridge.

Barrell, J. (1988) Being is perceiving; James Thomson and John Clare. In Barrell, J. (ed.) *Poetry, Language and Politics*, Manchester University Press, Manchester, 100–36.

Barrell, J. (1990) The public prospect and the private view: the politics of taste in eighteenth-century Birtain. In Pugh, S. (ed.) *Reading Landscape: Country – City – Capital*, Manchester University Press, Manchester, 19–40.

Bender, B. (1992) Introduction: landscape – meaning and action. In Bender, B. (ed.) *Landscape: Politics and Perspectives*, Berg, Providence and Oxford, 1–18.

Berger, J. (1972) *Ways of Seeing*, British Broadcasting Corporation and Penguin Books, Harmondsworth.

Bermingham, A. (1993) The aesthetics of ignorance. *The Oxford Art Journal*, **16**, 3–20.

Blunt, A. and Rose, G. (eds) (1994) *Writing Women and Space: Colonial and Postcolonial Geographies*, Guilford, New York.

Brewer, J. (1997) *The Pleasures of the Imagination: English Culture in the Eighteenth Century*, HarperCollins, London.

Cherry, D. (1993) *Painting Women: Victorian Women Artists*, Routledge, London.

Cosgrove, D. (1985) Prospect, perspective and the evolution of the landscape idea. *Transactions of the Institute of British Geographers*, **10**, 45–62.

Cosgrove, D. (1993) *The Palladian Landscape: Geographical Change and its Cultural Representations in Sixteenth-Century Italy*, Leicester University Press, London.

Cosgrove, D. and Daniels, S. (eds) (1988) *The Iconography of Landscape: Essays on the Symbolic Design and Use of Past Environments*, Cambridge University Press, Cambridge.

Cowell, B. (1997) The politics of park management in Nottinghamshire, *c.* 1750–1850. *Transactions of the Thoroton Society*, **101**, 133–43.

Cowper, W. (1985) *The Task, A Poem in Six Books*, J. Johnson, London.

Craton, M. (1982) *Testing the Chains: Resistance to Slavery in the British West Indies*, Cornell University Press, Ithaca.

Daniels, S. (1989) Marxism, culture, and the duplicity of landscape. In Thrift, N. and Peet, R. (eds) *New Models in Geography*, Unwin Hyman, London, 196–220.

Daniels, S. (1993) *Fields of Vision: Landscape Imagery and National Identity in England and the United States*, Polity Press, Cambridge.

Daniels, S. (1998) The political iconography of woodland. In Cosgrove, D. and Daniels, S. (eds) *The Iconography of Landscape: Essays on the Symbolic Design and Use of Past Environments*, Cambridge University Press, Cambridge, 43–82.

Daniels, S. and Cosgrove, D. (1993) Spectacle and text: landscape metaphors in cultural geography. In Duncan, J. S. and Ley, D. (eds) *Place/Culture/Representation*, Routledge, London, 57–77.

Daniels, S. and Seymour, S. (1990) Landscape design and the idea of improvement 1730–1900. In Dodgshon, R. A. and Butlin, R. A. (eds) *An Historical Geography of England and Wales*, Academic Press, London.

Drescher, S. (1977) *Econocide: British Slavery in the Era of Abolition*, University of Pittsburgh Press, Pittsburgh.

Drescher, S. (1982) Public opinion and the destruction of British colonial slavery. In Walvin, J. (ed.) *Slavery and British Society, 1776–1846*, Macmillan, London, 22–48.

Duckworth, A. (1971) *The Improvement of the Estate: A Study of Jane Austen's Novels*, Johns Hopkins Press, Baltimore and London.

Duncan, J. (1995) Landscape Geography, 1993–94. *Progress in Human Geography*, **19**, 414–22.

Duncan, J. and Duncan, N. (1988) (Re)reading the landscape. *Environment and Planning D: Society and Space*, **6**, 117–26.

Fabricant, C. (1979) Binding and dressing nature's loose tresses. *Studies in Eighteenth-Century Culture*, **8**, 109–35.

Fabricant, C. (1982) *Swift's Landscape*, Johns Hopkins Press, Baltimore.

Goveia, E. V. (1965) *Slave Society in the British Leeward Islands at the End of the Eighteenth Century*, Yale University Press, New Haven and London.

Gow, P. (1995) Land, people and paper in Western Amazonia. In Hirsch, E. and O'Hanlon, M. (eds) (1995) *The Anthropology of Landscape: Perspectives on Place and Space*, Clarendon Press, Oxford, 121–38.

Hamshere, C. (1972) *The British in the Caribbean*, Harvard University Press, Cambridge, Mass.

Harley, B. (1988) Maps, knowledge and power. In Cosgrove, D. and Daniels, S. (eds) *The Iconography of Landscape: Essays on the Symbolic Design and Use of Past Environments*, Cambridge University Press, Cambridge, 277–312.

Helsinger, E. (1987) Clare and the place of the peasant poet. *Critical Inquiry*, **13**, 509–31.

Helsinger, E. (1989) *Constable: The Making of a National Painter Critical Inquiry*, **15**, 253–79.

Helsinger, E. (1994) Turner and the Representation of England. In Mitchell, W. J. T. (ed.) *Landscape and Power*, Chicago University Press, Chicago and London, 103–26.

Hirsch, E. (1995) Introduction: landscape: between place and space. In Hirsch, E. and O'Hanlon, M. (eds) (1995) *The Anthropology of Landscape: Perspectives on Place and Space*, Clarendon Press, Oxford, 1–30.

Hirsch, E. and O'Hanlon, M. (eds) (1995) *The Anthropology of Landscape: Perspectives on Place and Space*, Clarendon Press, Oxford.

Hooke, D. (1998) *The Landscape of Anglo-Saxon England*, Leicester University Press, London.

Howkins, A. (1992) J. M. W. Turner at Petworth: agricultural improvement and the politics of landscape. In Barrell, J. (ed.) *Painting and the Politics of Culture: New Essays on British Art 1700–1850*, Oxford University Press, Oxford, 231–52.

Kingsley, M. (1982; first published 1897) *Travels in West Africa*, Virago Press, London.

Kinnaird, V., Morris, M., Nash, C., and Rose, G. (1997) Feminist geographies of environment, nature and landscape. In Women and Geography Research Group (eds) *Feminism and Geography: Diversity and Difference*, Addison Wesley Longman, Harlow, 146–89.

Langford, P. (1989) *A Polite and Commercial People: England 1727–1783*, Oxford University Press, Oxford and New York.

Lukacher, B. (1994) Nature historicized: Constable, Turner and romantic landscape painting. In Eisenman, S. F. (ed.) *Nineteenth-Century Art: A Critical History*, Thames and Hudson, London, 115–43.

Matless, D. (1992) An occasion for geography: landscape, representation and Foucault's corpus. *Environment and Planning D: Society and Space*, **10**, 41–56.

Matless, D. (1995) 'The art of right living': landscape and citizenship 1918–39. In Pile, S. and Thrift, N. (eds) *Mapping the Subject: Geographies of Cultural Transformation*, Routledge, London, 93–123.

Mellor, A. K. (1993) *Romanticism and Gender*, Routledge, London and New York.

Mitchell, W. J. T. (1994a) Imperial landscape. In Mitchell, W. J. T. (ed.), *Landscape and Power*, Chicago University Press, Chicago and London, 5–34.

Mitchell, W. J. T. (ed.) (1994b) *Landscape and Power*, University of Chicago Press, Chicago and London.

Mitchell, W. J. T. (1994c) Introduction. In Mitchell, W. T. J. (ed.) *Landscape and Power*, University of Chicago Press, Chicago and London, 1–4.

Morphy, H. (1992) Colonialism, history and the construction of place: the politics of landscape in Northern Australia. In Bender, B. (ed.) (1992) *Landscape, Politics and Perspectives*, Berg, Providence and Oxford, 205–4.

Morris, M. (1994) Home and Garden: Gender, Landscape and English Culture 1880–1930. Unpublished PhD thesis: University of Nottingham.

Muir, R. (1998) Landscape: a wasted legacy. *Area*, **30**, 263–71.

Nash, C. (1994) Remapping the body/land: new cartographies of identity, gender, and landscape in Ireland. In Blunt, A. and Rose, B. (eds) *Writing Women and Space: Colonial and Postcolonial Geographies*, Guilford, New York, 227–50.

Nash, C. (1996) Reclaiming vision: looking at landscape and the body. *Gender, Place and Culture*, **3**, 149–69.

Payne, C. (1993) *Toil and Plenty: Images of the Agricultural Landscape in England, 1780–1890*, Yale University Press, New Haven and London.

Pratt, M. L. (1992) *Imperial Eyes: Travel Writing and Transculturation*, Routledge, London.

Prince, H. (1988) Art and agrarian change, 1710–1815. In Cosgrove, D. and Daniels, S. (eds) *The Iconography of Landscape: Essays on the Symbolic Design and Use of Past Environments*, Cambridge University Press, Cambridge, 98–118.

Rose, G. (1993) *Feminism and Geography: the Limits of Geographical Knowledge*, Polity Press, Cambridge.

Rosenthal, M. (1982) *British Landscape Painting*, Phaidon, Oxford.

Said, E. (1978) *Orientalism*, Columbia University Press, New York.

Sales, R. (1983) John Clare and the politics of the pastoral. In Sales R., *English Literature in History, 1780–1830*, Hutchinson, London, 88–109.

Seymour, S. (1993) The Dukeries estates: improving land and landscape in the later eighteenth century. *Transactions of the Thoroton Society of Nottinghamshire*, **97**, 117–28.

Seymour, S., Daniels, S. and Watkins, C. (1995) Picturesque views of the British 'West Indies'. *The Picturesque*, **10**, 22–8.

Seymour, S., Daniels, S. and Watkins, C. (1998) Estate and empire; Sir George Cornewall's management of Moccas, Herefordshire and La Taste, Grenada, 1771–1819. *Journal of Historical Geography*, **24**, 313–51.

Sheridan, R. B. (1974) *Sugar and Slavery: An Economic History of the British West Indies, 1623–1775*, Eagle Hall, Barbados.

Smith, B. (1985) *European Vision and the South Pacific*, Yale University Press, New Haven and London.

Ward, J. R. (1988) *British West Indian Slavery, 1750–1834*, Clarendon Press, Oxford.

Watkins, C. (1998) The ancient oaks of Sherwood. In Watkins, C. (ed.) *European Woods and Forests: Studies in Cultural History*, CAB International, Wallingford, 93–113.

Williams, M. (1989) Historical geography and the concept of landscape. *Journal of Historical Geography*, **15**, 92–104.

Williams, R. (1973) *The Country and the City*, Chatto & Windus, London.

Chapter 9

Historical geographies of urbanism

Richard Dennis

Introduction

My aim in this chapter is to explore the significance of debates about modernity for studies of the historical geographies of cities. I begin with a brief history of recent urban historical geography, examining how different conceptions of the modern have left their mark. Then I focus on some selected 'arenas' of urban modernity, principally in nineteenth- and early twentieth-century cities in Britain and North America. But I also wish to explore the ways in which both contemporaries (then) and geographers (now) have written about these spaces – how spaces were socially and politically constructed and reconstructed in new forms of representation, from novels about city life and 'modern' art to ostensibly objective maps and social surveys.

'Modernity' and urban historical geography

Urban historical geographers have been concerned with 'modernity' for a long time. In 1975, for example, David Ward wrote an important paper with the title: 'Victorian cities: how modern?'. He was interested in questions of class formation and class consciousness which were then absorbing radical social historians, but – as a geographer – Ward was especially concerned to trace links between social structure and spatial structure, including the extent of residential segregation between different occupational groups and, by extension, 'classes' in nineteenth-century cities (Figure 9.1). Other geographers were interested in questions of mobility and 'community' in cities, situating their analyses along a sociological continuum from 'traditional' to 'modern' (Radford, 1981; Dennis, 1984). The former was associated with face-to-face, multi-stranded personal relationships and a high degree of continuity and stability, as indicated by low rates of residential mobility, high levels of support for local organisations such as churches, and intra-community marriage (as an indicator of more general locally based patterns of social interaction); while 'modern' was equated with impersonal, mechanistic relationships, anomie and 'communities of interest' extending over wider geographical areas. This was also the period of the quantitative revolution. The availability of quantifiable sources such as census enumerators' books, ratebooks (tax assessment rolls in North America) and city directories, and the sheer quantity of information available for rapidly growing industrial and commercial cities, encouraged urban historical geographers to

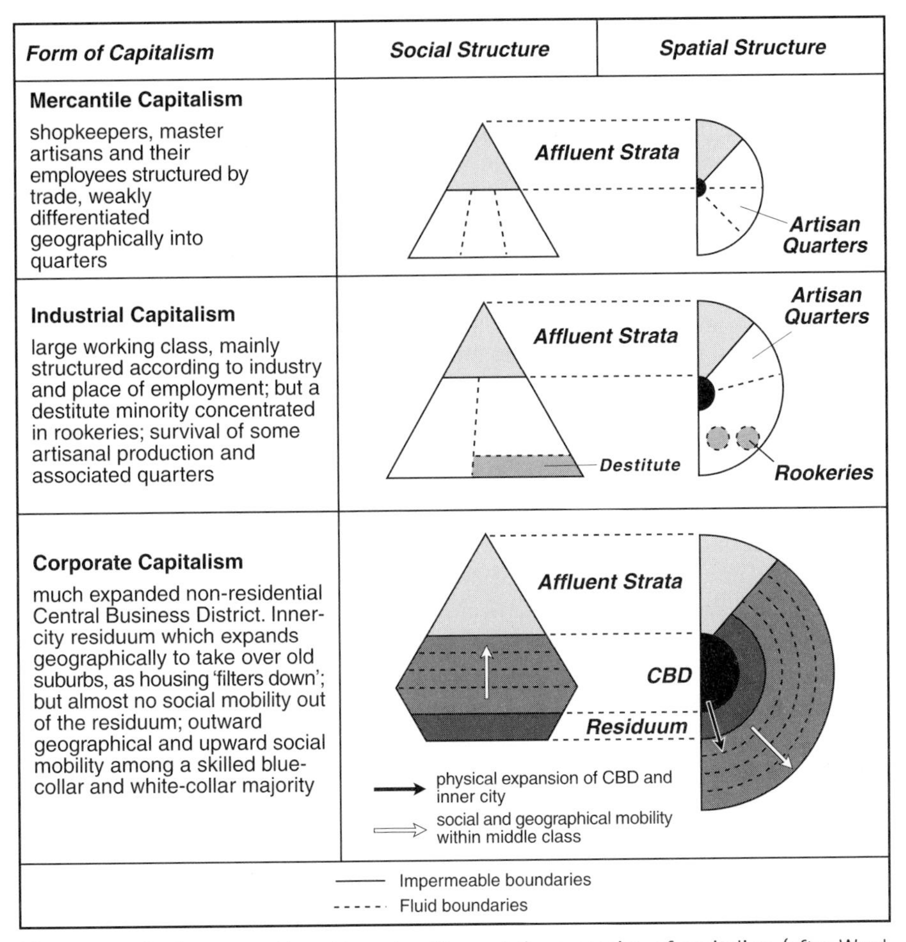

Figure 9.1 Socio-geographic patterns in cities and the expansion of capitalism (after Ward and Radford, 1983).

undertake quantitative analyses of segregation, mobility and community in cities. In its crudest form, 'modern' was equated with a particular spatial structure – the zones and sectors associated with Burgess's and Hoyt's models of twentieth-century American cities, especially Chicago. So, nineteenth-century cities, such as Liverpool and Toronto (Goheen, 1970; Pooley, 1984), ante-bellum American cities such as Charleston (Radford, 1979), early modern cities including Newcastle and Dublin, and occasionally even medieval cities such as Gloucester (Langton, 1975, 1977), were all compared to zonal and sectoral patterns to determine their modernity. What differentiated among these studies, and makes them of continuing value in a different theoretical climate, was the attention they devoted to *process* and *meaning*. Were the patterns underlain by processes that we think of as 'modern' – a capitalist land market with rents determined by location and accessibility; a socially mobile society in which people expressed their status through their geographical location; an ordered society in which, for reasons of 'efficiency' and profit maximisation, different functions were assigned different spaces, thereby necessitating lengthy journeys to work, to shop, to play, and a dependence on new forms of transportation and information circulation? Quantitative studies also examined questions of 'ethnic' segregation, attempting to differentiate between 'ghettoes' based on discrimination and exclusion and 'communities' based on choice (Pooley, 1977; Ward, 1989). Less attention was paid to geographies of gender in past cities. Nor was sufficient interest shown in the *meaning* of patterns for the everyday experience of different citizens: why did it matter and how did it feel to live in a 'modern' or a 'pre-modern' city?

This 'urban ecological' version of 'modern' was soon challenged by a political-economy approach, which placed more emphasis on processes of capital accumulation and investment, on the urbanisation and suburbanisation of capital, and on ideas of 'spatial fix' and 'creative destruction' (Walker, 1981; Harvey, 1985a; 1989). For David Harvey, the historical geography of modernity centred on the urbanisation of capital, and the diversion of profits from industrial production into secondary and tertiary circuits of capital: investment in the built environment, services, social infrastructure and means of social reproduction, such as education, health and social services. In various ways, crises of 'over-accumulation' in the primary, industrial circuit of capital could be circumvented. Profits from old industries, such as textiles, could be invested in new industries, such as those producing consumer durables. Production of clothes and furniture within the home was replaced by the purchase of these commodities ready-made. But for this to happen, people had to both want the new goods and have the means to acquire them and the place to accommodate them. The promotion of suburban home ownership thus fulfilled several 'needs' in modern society. Creating a new form of land-hungry built environment sustained the land market and the building industry; it provided 'dwellings' which could be personalised into 'homes' by the addition of appropriate furnishings; and by encouraging people to remain in the same 'home' for long periods (compared to moving frequently between rented rooms), it restored a degree of social stability that in turn facilitated stable government. The marketing and advertising of new products and associated lifestyles became big business. Moreover, large-scale capital investment in the built environment, and the encouragement of home ownership among people with little in the way of savings, stimulated new financial agencies: the evolution of credit finance from money-lenders and pawnbrokers to modern building societies, and the growth of clearing banks and insurance companies. Even if the *rate* of profit tended to decline over time as competition intensified, capital could still be used more efficiently by speeding-up its circulation, partly facilitated by faster forms of communication, such as the electric telegraph and the telephone, but also by the easier availability of credit.

Suburban development constituted one kind of 'spatial fix'; opening up new markets overseas and investing in colonial urban development was another. Closer to home, 'creative destruction' involved devaluing the existing built environment (and, in the cultural sphere, current fashions, tastes and practices) in order to justify a round of demolition and rebuilding, thereby increasing values and raising rents (or stimulating a demand for new cultural forms). Redevelopment might be privately initiated, associated with the expansion of business districts into old residential areas, or publicly funded, as in slum clearance projects in Victorian and interwar London (Yelling, 1986; 1992), or – on a grander scale – where state planning both required and served the interests of private capital, as in Haussmann's reconstruction of central Paris (Olsen, 1986). But whatever the catalyst, 'creative destruction' reflected the truth that 'For the developer, to stop moving . . . is death' (Berman, 1982: 69). Indeed, one of the most persistent dilemmas in modern cities involves the tension between vested interests who wish to protect their investments by resisting neighbourhood change – not on *our* street – and the necessity of change to sustain further rounds of economic growth. But change, initiated by the middle classes in their desire for economic development, necessarily engendered transience, uncertainty and insecurity among those same middle classes: 'all that is solid melts into air' (Marx and Engels, 1973: 46). A further consequence of modernisation was, therefore, a concern for individual identity, finding one's place in a changing world.

Another phrase often attributed to Marx, 'the annihilation of space by time', might seem to imply that space became less important. Certainly, the friction of distance was diminished by each speeding up of communications; instead, people 'began to pay more attention to smaller distinctions in time' (Thrift, 1994: 199). Complex transport systems required precise timetabling, new production processes necessitated scientific management to use time efficiently. However, the shrinking of space by speed was matched by a widening of opportunities to go further, faster. Within cities, separation of residence from workplace, or warehouse from office from retail store, each at its 'optimum' location, became possible. At a regional or even global scale, places could more easily specialise, exporting what they were good at, importing everything else. So annihilation of space by time led to both homogenisation, as more places accessed the same information and the same goods, and specialisation. It also facilitated the migration of capital: an ability to reinvest profits at a distance from where they were made. Slum landlords invested in suburban or out-of-town estates, profits from colonial trading, mining and plantation enterprises went to aggrandise imperial centres. In sum, there were processes of time–space compression, time–space distanciation and globalisation (Gregory, 1994).

Meanwhile, David Harvey had extended his analysis to incorporate a more humanistic conceptualisation of modernity, including issues of experience and identity. This returns us to the point made earlier, that as change became rapid enough to impinge on individuals' consciousness, they needed ways of coping with it, and reflecting on the fixity or fluidity of their own identity. The pioneer German sociologist, Georg Simmel, argued that city life gave people freedom, but led them to treat others – and to be treated in return – impersonally and instrumentally. For Simmel (1903), we became more conscious of our own identity – and potentially of our isolation from others – when we were one in a crowd. Another inspiration for Marxist humanist geographers was

Henri Lefebvre's writing on 'the production of space', and the circuit connecting 'representations of space' – ideologies of planning, surveillance and the creation of spectacle – and 'spaces of representation' – how space was contested and reclaimed in everyday life (Lefebvre, 1991). Harvey (1985b) explored this most fully in his essay on 'Paris, 1850–1870', where he began with the connections between finance capital, property ownership and the state, and their collaboration in the creation of a more differentiated, segregated, 'rational' cityscape, including the restructuring and geographical redistribution of labour. This led on to questions of how labour was reproduced, including discussions of housing and education. Some of the contradictions of restructuring become apparent: despite the oft-quoted claim that Haussmann's new boulevards made it easier to police the city, by carving through old centres of insurrection, and creating arteries too wide to be blocked by barricades but ideal for the rapid movement of troops, the displacement of inner-city populations to suburbs like Belleville and the consequent residential segregation of rich and poor actually made surveillance of working-class behaviour harder. Later stages in Harvey's essay explored questions of community, class, consciousness and identity, in part through an engagement with contemporary literature – Zola's novels, Baudelaire's poetry – leading finally to the expression of identity in action, in the geography of resistance of the Paris Commune. Gregory (1994: 221) interprets Harvey's aim as 'to illuminate the totality of capitalist modernization in Second Empire Paris'. This was a totalising metanarrative, modernist in its structure as much as in its subject.

Rationalism and pluralism in modern cities

While Harvey may be the leading practitioner of Marxist humanism *within* human geography, historical geographers have also been influenced by related approaches in art history and cultural studies, and especially the ideas of Walter Benjamin, whose autobiographical fragments such as *A Berlin Chronicle* illustrate the tension in his own experience between Marxist socialist and bourgeois *flâneur*. As Berman (1982: 146) noted:

> His heart and his sensibility draw him irresistibly towards the city's bright lights, beautiful women, fashion, luxury, its play of dazzling surfaces and radiant scenes; meanwhile his Marxist conscience wrenches him insistently away from these temptations, instructs him that this whole glittering world is decadent, hollow, vicious, spiritually empty, oppressive to the proletariat, condemned by history.

Benjamin's brief outline of 'Paris: Capital of the Nineteenth Century', a kind of research agenda for his unfinished project on the Paris Arcades, foreshadowed current interests in technology, panoramas and panoramic literature, exhibitions and city-as-spectacle (Benjamin, 1989). Savage (1995) has explored what was particularly *urban* about Benjamin's writing, emphasising his interest in 'the relationship between history, experience, memory, and the built environment'. For Benjamin, the city was a labyrinth, a place for 'shock experience' but holding out the possibility of 'redemption' as past dreams and hopes were laid bare. Modernity was more about discovery (and self-discovery) than progress. Important work in similar vein by geographers includes Pred's interpretations of the experience of modernity in the streets, the public space and the international exhibition grounds of nineteenth- and twentieth-century Stockholm (Pred, 1990; 1995) and Gregory's (1994) comparison of Harvey's, Benjamin's and Pred's writings on Paris and Stockholm.

The tension between increasingly self-aware and self-determining individuals and an ever more planned, rational and controlled environment was most creatively explored by the literary scholar, Marshall Berman, who traced the history of writing about modernity from Goethe's *Faust*, interpreted as the archetypal developer for whom no obstacles, human, material or cultural, could stand in the way of 'progress', through Marx and Baudelaire, the prototypical *flâneur*, and ending with Robert Moses, the twentieth-century New York planner, whose attitude to development replicated Faust's, especially – for Berman – by causing the destruction of the Grand Concourse in the Bronx, the stylish Jewish apartment-house district in which Berman grew up. Berman is central to the revival of a 'dynamic and dialectical modernism' concerned with the 'intimate unity of the modern self and the modern environment' (Berman; 1982: 35, 132). What Berman achieves, as does Harvey, rather differently, in *The Condition of Postmodernity* (1989), is the incorporation of postmodernity into the modern. For Harvey, postmodern urbanism – its emphasis on different forms of consumption, the ephemerality of fashion, flexibility in production, pastiche and historical quotation in architecture – is a response to the demands of advanced capitalism. For Berman, a world of regulation, discipline and order is a necessary precondition for acts of difference, display, spectacle and contestation.

Berman outlined some defining moments in the making of modern life, from the Renaissance through the Enlightenment and Darwinian evolutionary theory to the double-edged development of popular democracy and nationalism. He identified three phases in the history of modernity: between 1500 and 1800 when there were the beginnings of modern life, but no vocabulary to describe what was happening and no past shared experience to differentiate between 'now' and 'then'; from the late eighteenth to the beginning of the twentieth century, for Berman the pivotal period in which change was rapid, but the memory of the 'pre-modern' was cultivated – the 'invention of tradition', the beginnings of preservation and heritage movements, the awareness of differences between rural and urban, centre and periphery; and the period since 1900, when modernisation takes in the whole world, but the *idea* of modernity loses its vividness or its capacity to give meaning to our lives. Many commentators would challenge Berman's emphasis on the nineteenth century, suggesting that his lens was too narrowly focused on the western world, and on a masculine world, noting that the tension between order and freedom may be exhilarating for middle-class men but less relevant to the experiences of the poor and the marginalised, including most women; and arguing that we have always (or never) been 'modern' (Latour, 1993). Nonetheless, 'modernity' has particular value in understanding the geography of cities from at least the

Enlightenment onwards, because the various scientific, technological, political and demographic revolutions all had the effect of *both* promoting order (and constraint) *and* widening opportunities. Hence, an evolving self-awareness on the part of individuals for whom alternative ways to self-improvement, geographical and social mobility, were increasingly possible; but at the same time their choices were hedged around by supervision and regulation, and demands for the classification and specialisation of both people and places.

In their introduction to *The Landscape of Modernity*, a collection of essays on early twentieth-century New York, Ward and Zunz (1992: 3) described the modern city as 'between rationalism and pluralism'. They argued that the introduction of zoning ordinances in 1916, specifying which land uses were permitted in each part of the city in order to create specialised use districts for different activities, and regulating the height of buildings in each district, followed in the 1920s by more comprehensive and visionary planning agencies, created appropriate spaces for the perpetuation of ethnic and cultural diversity. It is easier for society to tolerate minority groups and nonconforming activities if they are assigned their own separate spaces. Likewise, it is important to recognise the symbiosis between agents of capital and social control in Napoleon III's Paris and apparent free spirits like Baudelaire engaged in their search for self-knowledge. Where would *flâneurs* have strolled or observed if there had been no boulevards and no cafés exploiting the opportunities created by a society wedded to conspicuous consumption? New art movements – Paris impressionists such as Manet, Monet, Pissarro and Caillebotte, who painted everyday scenes of urban life, in bars, streets and public gardens (Clark, 1984), or American 'ash-can' artists like Robert Henri, John Sloan and George Bellows, who specialised in realist scenes of dancehalls, boxing, and tenement-house life (Zurier *et al.*, 1995) – both celebrated and criticised the modern city, but their existence depended on the development of an art *market* – bourgeois patrons and clients purchasing works of art through commercial galleries, rather than gentry commissioning paintings in advance. Moreover, many new artists, especially in America, learnt their trade as commercial artists, producing advertising copy, or as newspaper illustrators, sketching court scenes or dramatic incidents such as fires and accidents in the days before photographs could be economically reproduced in newspapers.

There was also a close association between the rise of modern cities and modern novels. Keating (1984) identified three types of city novel, all relevant to geographical ways of thinking: the *portrait*, where the city was revealed through the experiences of the main character, often a youth from the country (for example, *Oliver Twist*) or a new immigrant (for example, Upton Sinclair's *The Jungle*, 1906, recounting the experiences of a Lithuanian immigrant new to Chicago); the *ecological*, centred on one small part of the city (for example, George Gissing's *The Nether World*, 1889, set in Clerkenwell); and the *synoptic*, connecting with the Victorian enthusiasm for panoramas, seeing the city laid out beneath one's feet, the novelist either playing God, orchestrating the connections and coincidences among apparently disparate characters (for example, many of Dickens's or Zola's novels) or, in more secular times, less moralising narrator and more agnostic reporter, but just as artfully recording city life like a series of randomly juxtaposed newspaper stories (for example, Dos Passos, *Manhattan Transfer*; 1925). Gilbert (1994) has examined the parallels between naturalist and organic metaphors in writing about Chicago among both novelists, like Sinclair and Dreiser, and sociologists, like Park and Burgess. To Malcolm Bradbury (1991: 97), the city is both the natural habitat of writers and critics and the natural setting for the modern novel:

> Here . . . are the intensities of cultural friction, and the frontiers of experience: the pressures, the novelties, the debates, the leisure, the money, the rapid change of personnel, the influx of visitors, the noise of many languages, the vivid trade in ideas and styles, the chance for artistic specialization.

More mundanely, here were the publishers, the printers, the bookshops and the new commercial lending libraries for whom authors like Dickens and Gissing supplied their stories in serial form, like today's soap-operas. So our evidence for new and diverse attitudes to and experiences of urban life – the *texts* that have become the stuff of *cultural* geography – emerges from the *economic* modernisation of cities.

Despite Berman's special pleading for the unique modernity of *nineteenth*-century cities, at least parts of his argument can be applied to earlier periods. Glennie and Thrift (1996) have discussed the growth of modern consumption practices in early modern England, using Pepys's diaries of Restoration London to depict the sociability of shopping among middle-class Londoners, and uncovering the roles of women as both shopkeepers and independent shoppers; and Ogborn (1998) has investigated a series of 'spaces of modernity' in eighteenth-century London – the Magdalen Hospital (for penitent prostitutes), the streets (newly paved and systematically lit), the pleasure gardens (as sites of commodification), and the Universal Register Office (a centralised exchange for commercial transactions and the circulation of information).

Space and identity

Ogborn's discussion of the Vauxhall Gardens 'macaronis' – fashionably dressed, effeminate young men – and of prostitutes subject to reform in Magdalen Hospital, contributes to an emerging spatialised history of gender and sexuality. Among cultural historians, Stansell (1986) and Gilfoyle (1992) have each discussed spaces of prostitution in nineteenth-century New York, interpreting the changing content and social geography of the city's sex industry as a response to both changes in spatial patterns of class and ethnicity, as the city expanded from Lower Manhattan towards and beyond Central Park (Figure 9.2), and innovations in technology: electrification accentuated the contrast between light and dark, moral and immoral landscapes, though an excess of 'bright lights', as around Times Square, was just as conducive to trade in sex as the darkest shadows; and the combination of telephony with the private apartment reversed the nineteenth-century trend, which had so worried moral reformers, of prostitution becoming more public, moving out of a concentration in a few bawdy houses in poor districts, and onto major public streets like Broadway.

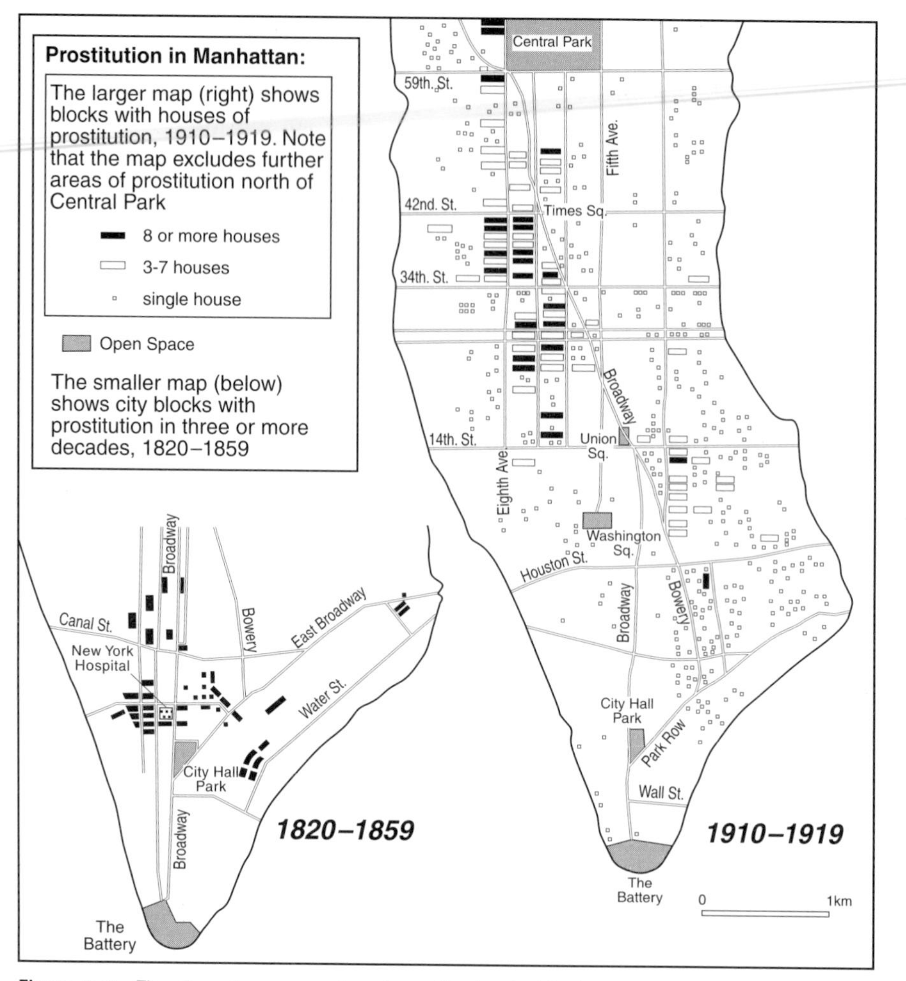

Figure 9.2 The changing geography of prostitution in Manhattan, 1820–1920 (after Gilfoyle, 1992).

Geographies of prostitution mostly accept the convention that modernity in the public spaces of cities involved male *flâneurs* objectifying unaccompanied women, thereby denying the possibility of *flâneuses*; women in public were to be looked at, not to do the looking. This interpretation connects with the notion of 'separate spheres', women being classified/segregated into the private/domestic/residential sphere, while men could be 'at home' in the public world of exchange and circulation. However, increasing doubt has been cast on the notion of 'separate spheres' and the supposed exclusion of women from the central districts of cities, except in the company of men (Walkowitz, 1992; Nord, 1995). By the 1880s, if not before, women were on the streets and public transport, on their own, as workers, tourists, shoppers and observers, with the potential to destabilise the position of male *flâneurs*.

Cultural geographers have recently devoted attention to 'tactical' *transgressions* of space, private and personal gestures which might not merit attention as acts of *resistance*, but which cumulatively prepare the way for more dramatic changes in the 'natural' spatial order. For example, Domosh (1998) discusses three illustrations which appeared in the *New York Illustrated News* during the 1860s. In one, a black family occupy the sidewalk on Broadway, decked out in their Sunday best, while the 'natural' occupants of the space – white, middle-class families – are displaced into the street. In another, a group of effete young men lounging in a gentlemen's club look down on women busily passing by on Fifth Avenue. Not only do the women ignore the male gaze (though it could be argued that their fashionable dress is itself a response), but the men are in the interior, private space while the women are in the public space of the street. However, while their roles are the reverse of expected, the picture's critique of the men depends on their being portrayed as weak and effeminate; it is criticising the place of these particular men, not the 'natural' positions of men and women as a whole. Domosh's reading of these images could no doubt be challenged. Nonetheless, she provides a valuable example of how we can use illustrative materials in a close and contextualised reading, equivalent to a contemporary urban ethnography.

Her third image, of crowds on Broadway at different times of day and night, emphasises the importance of *time* as well as *place*, the diurnal and seasonal rhythms of the city, the segregation of people and activities according to the hour of the day, or the day of the week. There was no reason to doubt the respectability of an unaccompanied woman on Broadway or Regent Street during daylight hours, but in the eyes of the law the same woman on her own in the same place after dark might be assumed to be a prostitute (Winter, 1993). When retailers on Fifth Avenue lobbied in 1913 for the introduction of restrictions on the height of new buildings, they wanted to exclude manufacturing 'lofts', from which garment workers were spilling out onto the street in their lunch hour at the same time as well-to-do shoppers were about (Zunz, 1990). The only way to eliminate this uncomfortable temporal clash was through spatial segregation.

Of course, crowds – whether organised demonstrations or spontaneous and accidental gatherings – could constitute more direct and immediately challenging transgressions than the examples in Domosh's illustrations. In *The Nether World*, in a chapter entitled 'Io Saturnalia!', George Gissing described a Bank Holiday Monday excursion to Crystal Palace – the south London park and exhibition site – by rowdy, happy-go-lucky inner Londoners. Their day comprised a combination of cheap music and dancing, vulgar sport, a fun-fair, and alcohol in abundance, and officially ended with one of the Crystal Palace's famous firework displays. In practice, it ended in fighting and a drunken train ride back to central London. In sum, 'A great review of the People. Since man came into being did the world ever exhibit a sadder spectacle?' (Gissing, 1889: 110). Yet our usual image of the Crystal Palace's clientèle is of the respectable, polite middle classes: the family groups, with mothers or governesses pushing

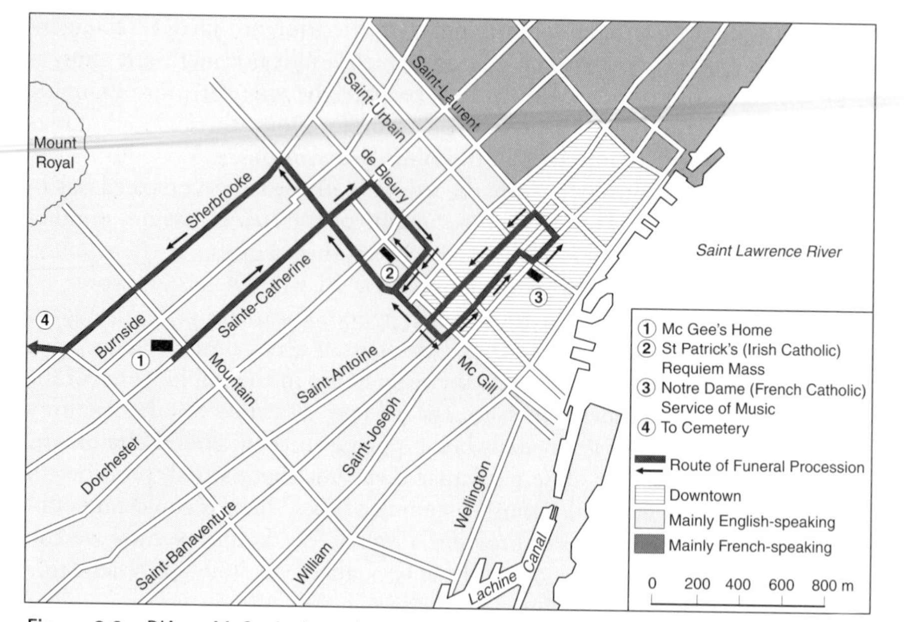

Figure 9.3 D'Arcy McGee's funeral procession and the social geography of Montreal, 1868 (after Goheen, 1993).

perambulators, who people Camille Pissarro's paintings of the district in 1870–71. What were *their* reactions to the Bank Holiday invasion?

Yet a further stage in the dialectic between Lefebvre's 'representations of space' and 'spaces of representation' is to consider how particular groups *deliberately* laid claim to public space. Goheen (1990; 1994) has investigated different kinds of street parade – funeral processions, trade union demonstrations, ethnic-group parades – in nineteenth-century Canadian cities, examining the selection or negotiation of routes, and the composition and order of parades. For example, the funeral of Thomas D'Arcy McGee, a prominent, pro-Confederation, Irish Catholic politician in Montreal in 1868 (the year after Canadian confederation was achieved), following his assassination, involved some delicate planning. It was appropriate for a service to be held in the city's main Catholic church, but this was in the majority Francophone part of the city, so the procession also had to visit the principal Irish Catholic church, where the funeral mass was conducted. But, as an Anglo-Irish politician, McGee was disliked by many working-class Québecois. If the procession passed too deeply through Francophone residential areas, it might be subject to abuse or simply ignored (Figure 9.3). The order in which different groups processed was also significant. In McGee's case, members of the municipal, provincial and federal governments led the way. The coffin was preceded by magistrates and judges, then came clergy, notaries, and members of the academic and medical professions, followed by ethnic and cultural benefit societies and associations, with workingmen's societies and 'citizens' (that is the general public) bringing up the rear. In Goheen's words (1990: 238), '[t]he organisers of this procession sought to portray a well articulated, hierarchical view of society resting on constituted authority and operating on a broad consensual base'.

By contrast, the organisers of a march in Hamilton, Ontario in 1872, to lobby for the establishment of a nine-hour working day, deliberately juxtaposed groups who had already been granted the concession alongside those whose employers had refused it. And the route had to visit factory sites where the concession had been granted, to raise morale and offer thanks, those where it had not, to provide a threatening show of strength, and the city centre, to advertise the cause to the rest of the community.

In essence, parades involved an appropriation of space and a display of power, as did other forms of spatial behaviour – demarcating territory with graffiti, setting up barricades, organising mass meetings in public spaces like Trafalgar Square and Hyde Park in London. They had to be choreographed as carefully as any theatrical performance, especially as their impact on society at large depended on how they were reported or illustrated in the press. In vertical cities, like New York, parades were even more theatrical, the audience looking down from office and apartment windows as if in a theatre balcony. Even on an everyday basis, then as now, the street was a stage, the people simultaneously performers, audience and critics.

Appropriations of residential space unsurprisingly evoked resistance, comparable to the 'not in my backyard (nimby)' attitudes of home owners today. Where a ground landlord retained control of a residential area – as in Mayfair and Belgravia in London, owned by the Duke of Westminster, or Bloomsbury, owned by the Duke of Bedford – restrictive covenants might be built into building leases, specifying what could and could not be built, and what kinds of land use were prohibited (Olsen, 1982). Such restrictions were not confined to eighteenth- and nineteenth-century London. For example, Paterson (1989) and Weiss (1987) record the proliferation of racial and anti-semitic covenants in North American cities, especially among self-styled 'community builders' developing suburban tracts in the first half of the twentieth century. While they excelled at creating middle-class, white, family suburbs, they denied the possibility of interclass or interracial areas.

Another form of self-defence was the erection of gates and barriers, or the imposition of tolls, to exclude 'undesirables' and through traffic from residential estates, the forerunners of today's 'gated communities'. In late Victorian London, there were more than 200 privately erected gates (Figure 9.4) (Atkins, 1993). A final stage in the process was to lobby local government to introduce 'residential restrictions' or 'zoning', supposedly on the modernist argument that segregation was both efficient and maximised property values, and was therefore in the 'public interest'. Often, the real objective was a form of apartheid. The first residential restrictions were directed against Chinese laundries (and, by extension, Chinese immigrants) in Californian cities in the 1880s. Immediately prior to the First World War, many North American cities passed anti-apartment by-laws, prohibiting the construction of multi-family apartment buildings in single-family residential suburbs. In these cases, the fear was that single-family dwellings would be devalued by the presence of neighbouring

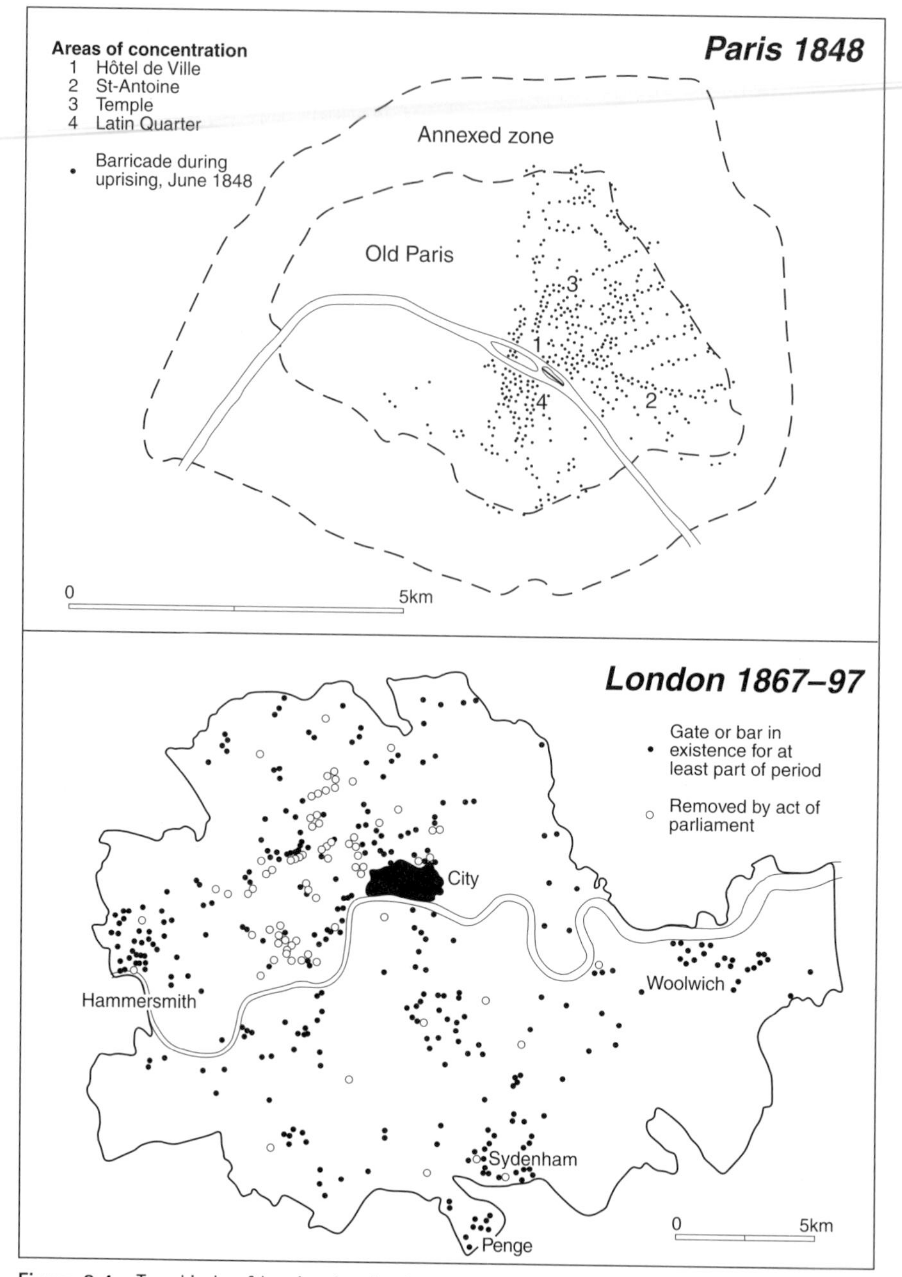

Figure 9.4 Two kinds of barrier: barricades in Paris, 1848 (after Harvey, 1985b) and gates and barriers in late nineteenth-century London (after Atkins, 1993).

apartment blocks, but objections were often cloaked in moral and public health arguments (Dennis, 1994). In Winnipeg, at least 59 anti-apartment by-laws, each relating to a particular street or city block, were passed between 1913 and 1926; in Toronto, an almost city-wide ban on apartment houses was introduced in 1912. Thereafter, developers wishing to erect such buildings had to seek exemptions from the original law. In commercially oriented or less prestigious districts, exemptions were easy to obtain, but in wealthier upper middle-class neighbourhoods, opposition was invariably fierce.

In all these cases, the character of space was being actively negotiated and contested, groups and functions were being marginalised. A modern meta-narrative of what was acceptable development was being expressed in the city's built environment. In the remainder of this chapter, I will examine some of the specialist landscapes that resulted, but also the contrasting forms of representation of those landscapes, by novelists, painters, photographers, film-makers, census takers, social reformers and journalists. I will focus principally on residential landscapes, but my examples (and my method) could be replicated among a huge range of new and specialist environments (Relph, 1987; Ward and Zunz, 1992): department stores and skyscrapers (Domosh, 1996a, b); museums and international exhibition buildings (Greenhalgh, 1988; Pred, 1995); railway stations, hotels and restaurants, music halls and theatres (Olsen, 1986); parks and cemeteries (Rosenzweig and Blackmar, 1992).

New residential spaces

Historical research on residential spaces in modern cities has tended to focus on the clearance and redevelopment of old slums (Gaskell, 1990) and the creation of new middle-class suburbs (Fishman, 1987). The assumption has been that suburbs, at least prior to twentieth-century social housing schemes, were invariably middle-class, conforming to the classic concentric-zone model of city structure. But since, historically, the vast majority of city-dwellers were working-class, it seems likely that at least some suburbanites were working-class too.

We can approach this question in at least two ways. One is to make our own analyses, based on the best evidence available to us, in census and property records and city directories. For example, Richard Harris (1996) has shown how suburban Toronto was more working-class than the central city until the onset of the Great Depression, when blue-collar families were hardest hit, forced to give up homes on which they could no longer afford mortgage repayments and return to rental housing, more often in inner urban areas. Their places were taken by middle-income, white-collar workers, whose jobs were generally more secure, thereby making the suburbs increasingly middle-class (Figure 9.5). Harris also argues that much working-class suburban housing was originally self-built. There is some evidence from contemporary writers, artists and photographers to this effect: illustrations showing primitive shacks next door to improved brick or stone dwellings (Figure 9.6), and diaries and oral histories provided by pioneer self-builders and their children. But Harris

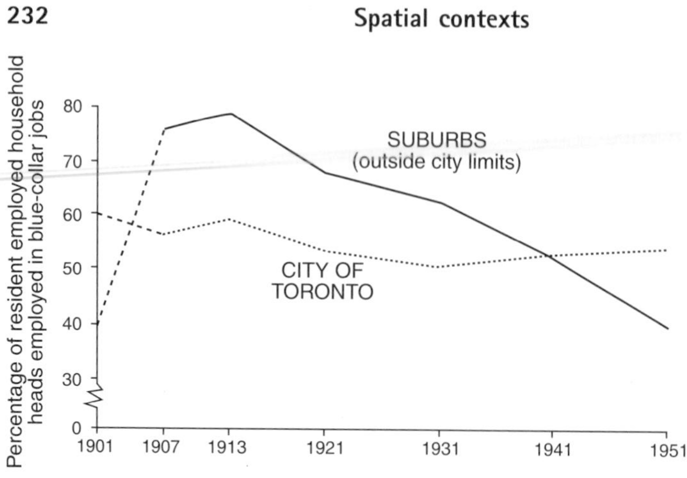

Figure 9.5 The rise and fall of Toronto's blue-collar suburbs, 1901–1951 (after Harris, 1996).

Figure 9.6 The two stages of development in Earlscourt (Toronto): painting by Henry Copping, 1911 (after Harris, 1996).

also infers the extent and location of self-built suburbs by some imaginative use of quantitative sources. Speculative builders tended to build rows of identical houses, so where adjacent properties had identical assessed values, it is likely that they had been built speculatively, unless they were *very* low-value properties. Comparing values in assessment records with field evidence indicated just how cheaply speculative builders could go. Houses valued below this threshold were almost certainly owner-built. Drawing samples of entries from assessment records for different years, Harris estimated that at least a third of all dwellings erected in Toronto between 1901 and 1913 were self-built. In districts newly annexed into the City of Toronto, more than half of new dwellings were self-built, and in suburbs still outside city limits in 1913, more than 90 per cent were self-built. Nor was Toronto's experience unique among North American cities.

A second way of studying suburbia is to examine the range of contemporary representations of suburban life in newspapers, magazines and novels. Roger Miller (1991) explored how innovations in domestic technology such as washing machines and refrigerators were advertised to American housewives between 1910 and 1930. With the move to more remote suburbs, middle-class women found it harder to recruit live-in domestic servants. There were more job opportunities for young women as shop assistants, typists, telephonists, nurses and on assembly lines in new manufacturing plants, so the supply of servants was drying up. Before the First World War, domestic equipment was quite often advertised with the assumption that it would be used by servants, but by the 1920s it was clear that housewives would be the principal users. Advertisements for vacuum cleaners depicted fashionably dressed women getting the home ready for a dinner party. Those for washing machines stressed the modern science behind the washing process, but also the ease with which machines could be used: no heavy, physical work which would be demeaning or supposedly beyond housewives' physical capabilities. An intermediate solution to the servant problem was to employ a charwoman for a few hours a day. One washing machine advertisement depicted a delicate young wife in negotiation with a coarse-featured and drably dressed charlady. The caption – '$4.00 a Day – and Lunch – and Car Fare?' – implied that the charwoman was driving a hard bargain; the solution to this dilemma was not to employ her at all but buy a Maytag Electric Washer (Figure 9.7). But, as Miller (1983) discussed in an earlier paper, the reality was more usually the employer exploiting the servants, paying by the hour rather than the day. And the demand for 'car fare' (that is tram/streetcar) was a consequence of the increasing scale of residential segregation: potential servants no longer lived within walking distance.

Marketing was also critical in attracting residents to settle in different kinds of suburb (Paterson, 1989). In pre-First World War Toronto, advertisements for 'Cedarvale' appeared in a delicate typeface accompanying an illustration of well-dressed people, standing beside elaborate, ornamental gates. In the background, a modern automobile and horses pass on a tree-lined avenue. The message was clear: this was to be an exclusive estate for the well-off, a 1910s gated community. In the same newspaper, 'Silverthorn Park Addition' was promoted quite differently: fewer words, a heavier typeface, a working-class family plonking down their model home 'right in the heart of the factory district'. With no concerns about pollution, the district's attractiveness was denoted by thick smoke belching from the factory chimneys (Figure 9.8). A third example, for the 'Nairn & Parsons' Estate', was directed at the poorest self-builders:

Figure 9.7 Advertisement for Maytag Cabinet Electric Washer, *The Saturday Evening Post*, 4 September 1920 (after Miller, 1991).

> Choose a Home Lot – Pay Down $10.00 and then get possession immediately – Build a little place until you are able (with saved rent money) to enlarge it. If you can't build, put up a tent for the summer. What you save in rent builds you a little home in the fall. Then build larger next year.

But it also pandered to their desire to emulate the rich: plots were laid out 'on the Rosedale Plan' (curved streets and irregularly shaped plots just like Rosedale, the home of the city's élite); and all three advertisements assumed that home ownership was an unmitigated blessing. Many North American cities promoted themselves as 'cities of homes', by which they meant cities of home owners rather than tenants, and cities of single-family houses rather than apartment buildings (Daunton, 1988).

The apartment house was another new and controversial form of residence in the nineteenth century, prominent in another medium of representation, the novel. In Britain and North America, living in flats was perceived as 'foreign'. Promoters advertised apartments as 'Parisian Buildings', implying a degree of continental sophistication; critics called them 'French Flats', hinting

Figure 9.8 Advertisement for Silverthorn Park Addition, *Toronto Evening Telegram*, 23 August 1912 (after Paterson, 1989).

at their bohemian squalor and immorality (Tarn, 1974; Cromley, 1990). They were especially suspicious of dwellings arranged on one level with no vertical segregation of 'sleeping' and 'living' areas (bungalows were equally suspect!). Edith Wharton satirised this view in *The Age of Innocence*, where she described the home of elderly matriarch, Mrs Mingott. Hers was not a real flat; she lived on one floor simply because she could no longer manage the stairs. But the result was the same; visitors could see through to the bedroom from the sitting room. They 'were startled and fascinated by the foreignness of this arrangement, which recalled scenes in French fiction, and architectural incentives to immorality such as the simple American had never dreamed of' (Wharton, 1974: 27). Perhaps Wharton had in mind Emile Zola's *Pot Bouille* (*Restless House*) which graphically detailed the intrigues and affairs of families and their servants living on different floors of a Parisian apartment building. The building's internal geography included servants' quarters concentrated in the attic (so servants from different families could freely gossip about one another's masters and mistresses), separate back stairs connecting the attic with the kitchens of each flat (allowing servants or secret visitors to pass through the building undetected), and a central 'area' or ventilation shaft, onto which the kitchens faced, and across which servants could exchange news. When 'French flats' were proposed in London, the architectural press was full of earnest debate on how to improve these arrangements: servants had to be supervised, but they also had to be segregated. In microcosm, this was the spatial dilemma that had to be resolved in all nineteenth-century cities: the rich wanted to be separate from the poor, but close enough for the poor to meet their every need.

Figure 9.9 Oxford and Cambridge Mansions, photograph in advertising brochure, no date, c. 1910 (City of Westminster Archives).

Zola and Wharton wrote to entertain as much as comment, and we may doubt their reliability as witnesses. How do novelists' observations compare with the bald facts of censuses and tax returns? In *The Whirlpool* (Gissing, 1897, but set in 1889), the footloose, childless Hugh and Sibyl Carnaby return to London after an extended world tour to a flat in Oxford & Cambridge Mansions, a real block erected in 1883 near Edgware Road Station. The block's awkward situation on an acute-angled corner meant that, when viewed from a low vantage point opposite the corner, it could be advertised as 'progressive', like the bow of an ocean liner (Figure 9.9). This was how numerous American skyscrapers were photographed. Behind this progressive facade, the interior was more modest. There was no lift and the angled site made for a very confined courtyard at the rear, little better than the backyard of a working-class tenement block. Compared to many apartment buildings, there were few servants resident in the building on census day, 1891: 69 servants out of a total population of 204 persons living in 62 households. The Carnabys had returned from Queensland. Real-life residents of Oxford & Cambridge Mansions included two single women born in Jamaica and one born in India, a banker's clerk whose wife was from the West Indies, a colonial broker's agent from Tasmania, an elderly widower born in Calcutta, and a stock-exchange dealer from the Punjab. Flats were convenient for people temporarily or newly returned from the colonies, as yet uncertain where to settle permanently. There was an 'inner colonial perimeter' around London's West End, including Westminster, Kensington, Bayswater and Maida Vale, equivalent to the 'inner industrial perimeter' to the north and east of the City. Also notable was the large number of female-headed households: thirteen households in Oxford & Cambridge Mansions were headed by spinsters, eight by widows, four by married women (with no husband present), and six by female servants (with no master or mistress on the premises). In 1890s London most women living independently in flats would have been prosperous spinsters and widows reliant on investments, rents from property, or pensions. A few would have been 'odd women', the contemporary epithet for independently minded, working women, mainly teachers, nurses and typists. But flats were ideal accommodation for the explosion of single working women during the twentieth century. In Toronto, the proportion of female-headed apartment households increased from 15 per cent in 1909 to 39 per cent in 1930. Whereas almost two-thirds of these women were widows in 1909, by 1930 the majority (of a much larger total number) were single women.

Mapping the city

We can see how suburbs and apartments fitted into the new social geography of cities, by making our own analyses using assessment rolls (as in Harris's and Dennis's work on Toronto) or census enumerators' books (as in the example of flats in London), but sampling dwellings and households from throughout cities, not just particular buildings or estates. Alternatively, we can draw on surveys undertaken by contemporary social explorers. Most such surveys focused on slum districts, and terms like 'explorer' should alert us to the values behind their research. They were mounting expeditions into London's East End or New York's Lower East Side, searching for sources of poverty and depravity, much as African explorers were tracing the source of the Nile: 'darkest England' paralleled 'darkest Africa'. But some surveys attempted to situate poverty in larger metropolitan contexts (Bulmer *et al.*, 1992). By the early twentieth century, the survey movement in North America was characterised by unemotional, scientific analyses, such as the multi-volume Pittsburgh Survey (1907–9). In London, Charles Booth's *Life and Labour of the People of London* (1889–1902) drew on reports by school-board officers, local government employees whose job was to ensure that children were attending the new board schools. Armed with their information, Booth calculated rates of poverty in every part of London, and mapped poverty street-by-street, from streets coloured gold, which signified the wealthiest areas, to black, which denoted 'Lowest class. Vicious, semi-criminal'. While there was an overall pattern of zones and sectors – higher levels of poverty in the East End than the West End, and in the inner city than the suburbs – there were still huge differences among neighbouring streets: the gold of Oxford & Cambridge Mansions was only a stone's throw from the blue–black of Bowman's Buildings (Figure 9.10).

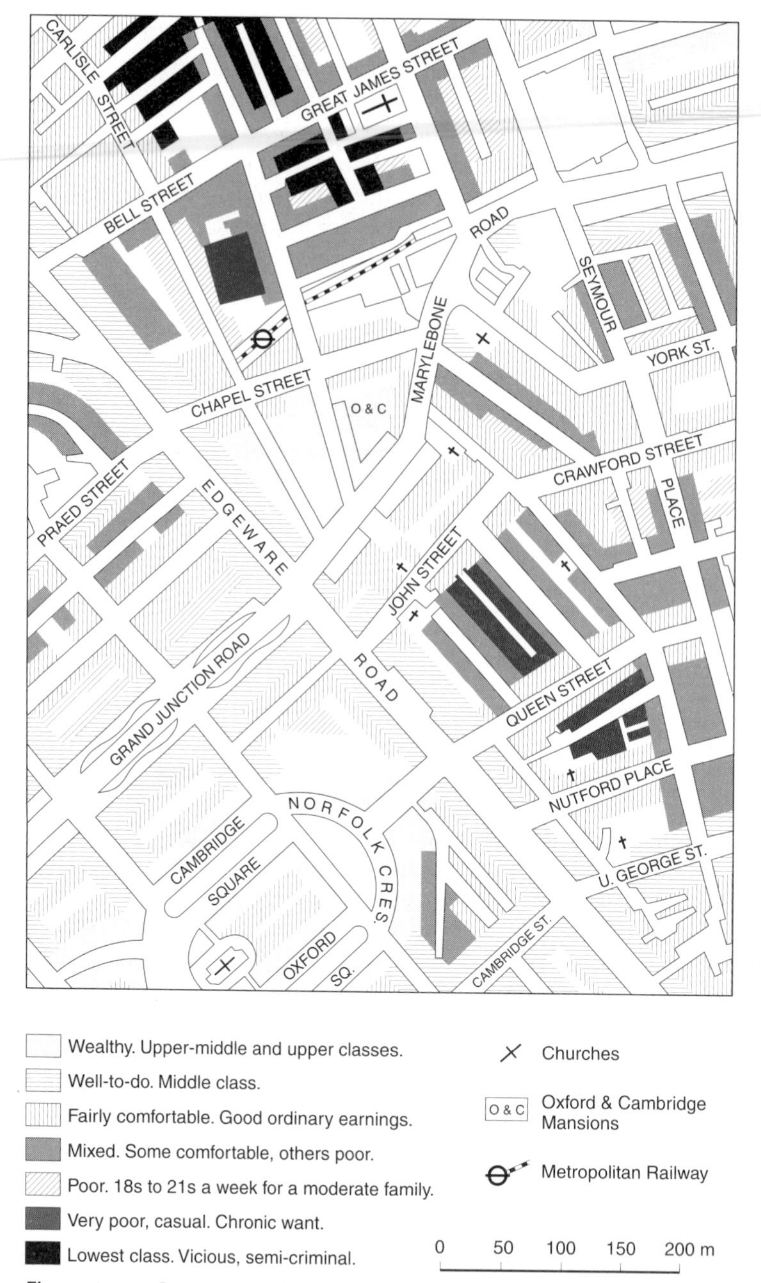

Figure 9.10 Degrees of poverty and wealth in the area around Oxford and Cambridge Mansions, from Charles Booth's *Descriptive Map of London Poverty*, 1889 (after Reeder, 1987).

Figure 9.11 *The Railway Station*, W. P. Frith (1862) (reproduced by kind permission of Royal Holloway, University of London).

Surveys like Booth's followed a Foucauldian logic, connecting knowledge and power. To map was to legitimate authority, to classify was to impose order. Booth was not just exploring London; by classifying and mapping he was engaged in a work of colonisation, effectively if coincidentally legitimating the social policies of the London County Council which came into existence in 1889, the same year that Booth published his *Descriptive Map of London Poverty*.

But for all their social scientific objectivity, Booth's maps were just another means of visual representation, more factually detailed but just as artfully constructed as a 'bird's eye view' or 'balloon view' of the city from above, or even a panormaic painting like Frith's 'The Railway Station', which paraded a cast of characters from different backgrounds and social classes (Figure 9.11) (Daniels, 1985). For example, Booth's 'vicious, semi-criminal' is a different kind of coding from 'wealthy'! His colour scheme may seem 'natural', but it reinforced the idea that slums were 'black spots', cancers that could be removed surgically by 'slum clearance'. The act of mapping implied that social problems could be contained within spatial boundaries: poverty was associated with particular places – 'slums', not with the economic system or the form of government.

The word 'slum' is a classic example of a social construction imposed by outsiders to make sense of the 'Other' (Mayne, 1993). Note its similarity to a variety of sl- words, almost all unsavoury. Slums went along with slime, sloth, slush, sludge, slop. They were inhabited by slovenly, slatternly, sluts! The 'otherness' of slums was often compounded by their association with immigrants, reflected in the labelling of areas (again, by outsiders) as 'Chinatowns' or 'The Ghetto' (Ward, 1989; Anderson, 1991). Other terms, which predated the language of 'slum', included 'rookery' and 'den', indicating the sub-human, animal-like character attributed to their 'migratory' residents, people who were 'savages' or a 'race apart', living in an 'urban jungle'. More apocalyptic language referred to poor districts as a 'nether world', a 'city of dreadful night', an 'inferno' or an 'abyss' from which it was impossible to escape.

To summarise, modernisation was taking place on the ground, in the creation of new spaces, new scales and new patterns of segregation and specialisation, and new forms of technology, including the technology to make a 'networked city', interconnected by water pipes, sewers, gas mains, electricity cables, telegraph and telephone wires, buses, trams and trains (Tarr and Dupuy, 1988). But modernisation was also taking place in the mind, in how cities were spoken and written about, how they were visualised, mapped, painted and photographed (Sutcliffe, 1984).

Integrating the city, integrating representations

A final example should demonstrate the interplay of culture, economy and technology in the context of alternative constructions of city life. The railway was an obvious symbol of progress. The plume of steam trailing behind a locomotive indicated the single-minded direction of improvement. Later, the headlight of the electric train fulfilled the same role. But the urban railway, whether raised on a masonry viaduct on its approach to a city terminus, or an intra-urban elevated railway, suspended overhead on an iron or steel frame, as in many American cities, was associated with modernity in multiple ways. It allowed development, literally and symbolically, to leapfrog over the dilapidation and disorder of the old city. Commuters and middle-class shoppers could be carried to and from city centres, avoiding the congestion of inner-city streets and thereby enhancing the idea of the city as a circulatory system. Seen from above, the elevated appeared to weave its way effortlessly and efficiently through the complexity of the city. Even after the advent of subway systems, elevated railways remained popular with film-makers as evidence of the *life* of the city, from King Vidor's *The Crowd* (1926) to Woody Allen's *Manhattan* (1979) or Wayne Wang's *Smoke* (1995).

Elevated railways exposed the slums to view. In the 1840s Engels (1969: 79) noted how, because the major radial streets in Manchester were all lined with shops, wealthy suburbanites could 'go in and out daily without coming into contact with a working-people's quarter or even with workers'. Yet in the same year that Engels' account was first published, Reynolds was describing the view of London's East End from the new Eastern Counties Railway:

> The traveller upon this line may catch, from the windows of the carriage in which he journeys, a hasty but alas! too comprehensive glance of the wretchedness and squalor of that portion of London. He may actually obtain a view of the interior and domestic misery peculiar to the neighbourhood; he may penetrate with his eyes into the secrets of those abodes of sorrow, vice and destitution. In summertime the poor always have their windows open, and thus the hideous poverty of their rooms can be readily descried from the summit of the arches on which the railroad is constructed. (Reynolds, 1845, quoted in Allen, 1998: 120)

Figure 9.12 *Rails Over London*, Gustave Doré (from Doré and Jerrold, 1872).

Reynolds was the archetypal outsider, intent on revealing 'the mysteries of London'. To him, the poverty was 'hideous', the dwellings were 'abodes of sorrow, vice and destitution'. He went on to describe women preparing 'the sorry meal', 'scolding, swearing and quarrelling', and men who were 'all the day long smoking'. Truly, a remarkable detail of observation from a 'hasty glance'! Nonetheless, he foreshadowed later observers who were inspired or appalled by the view from the viaduct. More than half a century later, in New York, John Sloan made paintings and sketches looking through tenement windows, a perspective made possible by the elevated railway, and itself elevated artistically by Edward Hopper in the generation after Sloan (Zurier *et al.*, 1995). These pictures may be considered disturbingly voyeuristic, but they are a form of local panopticism, a way in which middle-class commuters could claim a knowledge of 'how the other half lives'.

Gustave Doré's illustration of *Rails Over London* (1872) is often used as evidence of the appalling housing conditions revealed by the railways (Figure 9.12). However, his picture is not so simply interpreted. It is less a depiction of terrible slums as a warning of the dehumanising consequences of progress. The houses were not slums by the standards of the 1870s: each was self-contained with its own backyard and outdoor toilet, exactly the kinds of houses required by local by-laws. The speeding train could be indicating the solution to the

Figure 9.13 *Under the 'L'*, Philip Reisman (1928) (print in possession of Strang Print Room, University College London).

problem, not just for the middle classes but for everybody. More likely, however, Doré was commenting on the regimentation and sheer density of life in an underworld, hitherto unknown to the rich.

In New York, too, there was a very different life 'under the el': robust working-class communities of mothers and children and market stalls, as depicted by Philip Reisman (Figure 9.13) (Hemingway, 1996), or seedier scene of prostitutes, petty criminals and a skid-row of cheap rooming houses, a demi-monde in the shadows and half-light, a perspective favoured by Reginald Marsh in paintings of 'The Bowery' in the depression years of the 1930s. Either way, the implication was that the underworld was an inevitable and necessary complement to the life above. The attraction of cities is that they are full of paradoxes, contradictions, margins and boundaries, what Victorian observers called 'the attraction of repulsion'.

I have argued through a succession of examples that modern urban historical geographers should be eclectic in the sources they use and the approaches they adopt, that both quantitative and qualitative sources are essential, that *every* source, literary or numerical, is a socially constructed representation, and that we need a form of 'triangulation' whereby different kinds of source inform one another. Literary accounts of city life should provoke us to examine the statistical record, while quantitative sources require the contextualisation and the embodiment that more discursive narratives can provide. Hence, the recent history of urban historical geography should not be read as ecology supplanted by Marxism supplanted by iconography, but as an accumulation of additional layers of understanding. That archaeological metaphor is particularly apposite as urban historians increasingly link their work to that of urban archaeologists. Almost all literary, visual and numerical representations are 'from above', but urban archaeology can at least uncover the material lives of those who left no textual records, as major excavations in 'Five Points' in Manhattan, 'The Rocks' in Sydney, and 'Little Lon' in Melbourne – themselves facilitated by current rounds of 'creative destruction' – are demonstrating (Mayne, 1999).

I have concentrated in this chapter on culture and society, on the representation of technology and economy more than their material consequences. Yet modern cities necessarily entailed modern economies: new ways of raising capital, new scales of capitalist enterprise, new ways of accessing credit, new forms of industrial production. For British cities in the nineteenth century these modern economies were also imperial. For example, one of the earliest property development companies in the City of London was established in the 1860s by rum importers who invested profits from the sale of property in Jamaica in purchasing City buildings ripe for improvement or redevelopment; and among the City's grandest new buildings were East India Company offices, erected in 1866, which also provided space for the Hudson's Bay Company (Thorne, 1984). London's Tate Gallery (1897) and Horniman Museum (1896–1901) were built on the profits of sugar and tea businesses, respectively (King, 1990a). Geographically less central, but hugely significant in the history of British town planning, pioneering garden suburbs at Bournville, near Birmingham, and Port Sunlight, near Liverpool, were developed from the proceeds of Cadbury's cocoa and Lever's coconut oil businesses. King (1990b: 78) posed the question 'whether the real development of London or Manchester can be understood without reference to India, Africa, and Latin America'. Clearly, the answer is that they cannot.

Finally, I should emphasise that 'modern cities' are, and were, not confined to Europe and North America. Studies of colonial cities, such as Calcutta (Hornsby, 1997) show how the centre was reconfigured and re-established on the periphery. Colonial architecture was 'largely a reproduction of metropolitan forms' (King, 1990b: 60), sometimes imitative, as in the erection of classical or gothic banks, offices and municipal buildings in 'Victorian cities' overseas, such as Melbourne and Toronto, but also, often, an embodiment of colonial power, designed to reproduce metropolitan power structures and institutions. Bombay, Delhi and Lucknow were all redeveloped from the 1860s 'using the urban technologies and knowledge from industrial Britain' (King, 1990b: 39). More dramatically, European planning ideals informed plans for colonial capitals: New Delhi, Pretoria, Canberra, Algiers. But at the same time, European 'imperial cities' were incorporating empire into the metropolitan centre. Empire was being performed in imperial exhibitions, such as the 1851 Great Exhibition and the British Empire Exhibition at Wembley in 1924–5; in the naming of countless pubs, office buildings and streets (this book is being written in a suburban semi-detached only yards from a 1920s 'Imperial Drive'); in moments of imperial fervour like the City of London's celebration of the relief of Mafeking; and it was being physically reproduced in 'orientalist' architecture, and in the pagodas and exotica of Kew Gardens and London Zoo (Crinson, 1996; Driver and Gilbert, 1998). There was a 'colonial urban system' whose very existence, tied together by steamships and telegraphy, was witness to the spreading net of modernity.

References

Allen, R. (1998) *The Moving Pageant: A Literary Sourcebook on London Street-Life, 1700–1914*, Routledge, London.

Anderson, K. (1991) *Vancouver's Chinatown: Racial Discourse in Canada, 1875–1980*, McGill-Queen's University Press, Montreal.

Atkins, P. (1993) How the West End was won: the struggle to remove street barriers in Victorian London. *Journal of Historical Geography*, **19**, 265–77.

Benjamin, W. (1989) Paris, capital of the nineteenth century. In *Reflections: Essays, Aphorisms, Autobiographical Writings*, Schocken Books, New York, 146–62.

Berman, M. (1982) *All That Is Solid Melts Into Air: The Experience of Modernity*, Verso, London.

Bradbury, M. (1991) The cities of modernism. In Bradbury, M. and McFarlane, J. (eds) *Modernism: A Guide to European Literature 1890–1930*, Penguin, Harmondsworth, 96–104.

Bulmer, M., Bales, K. and Sklar, K. K. (eds) (1992) *The Social Survey in Historical Perspective, 1880–1940*, Cambridge University Press, Cambridge.

Clark, T. J. (1984) *The Painting of Modern Life: Paris in the Art of Manet and His Followers*, Princeton University Press, Princeton, N.J.

Crinson, M. (1996) *Empire Building: Orientalism and Victorian Architecture*, Routledge, London.

Cromley, E. (1990) *Alone Together: A History of New York's Early Apartments*, Cornell University Press, Ithaca, N.Y.

Daniels, S. (1985) Images of the railway in nineteenth-century paintings and prints. In Nottingham Castle Museum *Train Spotting: Images of the Railway in Art*, NCM, Nottingham, 5–19.

Daunton, M. (1988) Cities of homes and cities of tenements. *Journal of Urban History*, **14**, 283–319.

Dennis, R. (1984) *English Industrial Cities of the Nineteenth Century: A Social Geography*, Cambridge University Press, Cambridge.

Dennis, R. (1994) Interpreting the apartment house: modernity and metropolitanism in Toronto, 1900–1930. *Journal of Historical Geography*, **20**, 305–22.

Domosh, M. (1996a) The feminized retail landscape: gender, ideology and consumer culture in 19th-century New York City. In Wrigley, N. and Lowe, M. (eds) *Retailing, Consumption and Capital: Towards the New Retail Geography*, Longman, Harlow, 257–70.

Domosh, M. (1996b) *Invented Cities: The Creation of Landscape in Nineteenth-Century New York and Boston*, Yale University Press, New Haven, Conn.

Domosh, M. (1998) Those 'gorgeous incongruities': polite politics and public space on the streets of nineteenth-century New York City. *Annals of the Association of American Geographers*, **88**, 209–26.

Doré, G. and Jerrold, D. (1872) *London: A Pilgrimage*, Grant, London.

Driver, F. and Gilbert, D. (1998) Heart of empire? Landscape, space and performance in imperial London. *Environment and Planning D: Society and Space*, **16**, 11–28.

Engels, F. (1969) *The Condition of the Working Class in England*, Panther, London (originally 1845).

Fishman, R. (1987) *Bourgeois Utopias: The Rise and Fall of Suburbia*, Basic Books, New York.

Gaskell, M. (ed.) (1990) *Slums*, Leicester University Press, Leicester.

Gilbert, E. (1994) Naturalist metaphors in the literature of Chicago, 1893–1925. *Journal of Historical Geography*, **20**, 283–304.

Gilfoyle, T. (1992) *City of Eros: New York City, Prostitution, and the Commercialization of Sex, 1790–1920*, Norton, New York.

Gissing, G. (1889) *The Nether World*, (1973 edn), Dent, London.

Gissing, G. (1897) *The Whirlpool*, Lawrence & Bullen, London.

Glennie, P. and Thrift, N. (1996) Consumers, identities, and consumption spaces in early-modern England. *Environment and Planning A: Society and Space*, **28**, 25–45.

Goheen, P. (1970) *Victorian Toronto 1850–1900*, University of Chicago Department of Geography Research Paper No. 127, Chicago.

Goheen, P. (1990) Symbols in the streets: parades in Victorian urban Canada. *Urban History Review/Revue d'histoire urbaine*, **18**, 237–43.

Goheen, P. (1993) Parades and processions. In Gentilcore, R. L. (ed.) *Historical Atlas of Canada, Volume II: The Land Transformed 1800–1891*, University of Toronto Press, Toronto, 150–1.

Goheen, P. (1994) Negotiating access to public space in mid-nineteenth-century Toronto. *Journal of Historical Geography*, **20**, 430–49.

Greenhalgh, P. (1988) *Ephemeral Vistas: The Expositions Universelles, Great Exhibitions and World's Fairs, 1851–1939*, Manchester University Press, Manchester.

Gregory, D. (1994) *Geographical Imaginations*, Blackwell, Oxford.

Harris, R. (1996) *Unplanned Suburbs: Toronto's American Tragedy, 1900 to 1950*, Johns Hopkins University Press, Baltimore.

Harvey, D. (1985a) *The Urbanization of Capital*, Blackwell, Oxford.

Harvey, D. (1985b) *Consciousness and the Urban Experience*, Blackwell, Oxford.

Harvey, D. (1989) *The Condition of Postmodernity*, Blackwell, Oxford.

Hemingway, A. (1996) *Philip Reisman's Etchings: Printmaking and Politics in New York 1926–33*, College Art Collections, University College London, London.

Hornsby, S. (1997) Discovering the mercantile city in South Asia: the example of early nineteenth-century Calcutta. *Journal of Historical Geography*, **23**, 135–50.

Keating, P. (1984) The metropolis in literature. In Sutcliffe, A. (ed.) *Metropolis 1890–1940*, Mansell, London, 129–45.

King, A. (1990a) *Global Cities: Post-imperialism and the Internationalization of London*, Routledge, London.

King, A. (1990b) *Urbanism, Colonialism, and the World-Economy: Cultural and Spatial Foundations of the World Urban System*, Routledge, London.

Langton, J. (1975) Residential patterns in pre-industrial cities: some case studies from seventeenth-century Britain. *Transactions of the Institute of British Geographers*, **65**, 1–28.

Langton, J. (1977) Late medieval Gloucester: some data from a rental of 1455. *Transactions of the Institute of British Geographers*, **NS 2**, 259–77.

Latour, B. (1993) *We Have Never Been Modern*, Harvester Wheatsheaf, London.

Lefebvre, H. (1991) *The Production of Space*, Blackwell, Oxford.

Marx, K. and Engels, F. (1973) *Manifesto of the Communist Party*, Progress Publishers, Moscow (originally 1848).

Mayne, A. (1993) *The Imagined Slum: Newspaper Representations in Three Cities, 1870–1914*, Leicester University Press, Leicester.

Mayne, A. (1999) On the edge of history. *Journal of Urban History*, **25**, forthcoming.

Miller, R. (1983) The Hoover in the garden: middle-class women and suburbanization, 1850–1920. *Environment and Planning D: Society and Space*, **1**, 73–87.

Miller, R. (1991) Selling Mrs Consumer: advertising and the creation of suburban socio-spatial relations, 1910–1930. *Antipode*, **23**, 263–301.

Nord, D. E. (1995) *Walking the Victorian Streets: Women, Representation and the City*, Cornell University Press, Ithaca, N.Y.

Ogborn, M. (1998) *Spaces of Modernity: London's Geographies 1680–1780*, Guilford Press, New York.

Olsen, D. J. (1982) *Town Planning in London: The Eighteenth and Nineteenth Centuries*, Yale University Press, New Haven, Conn.

Olsen, D. J. (1986) *The City as a Work of Art: London, Paris, Vienna*, Yale University Press, New Haven, Conn.

Paterson, R. (1989) Creating the packaged suburb: the evolution of planning and business practices in the early Canadian land development industry, 1900–1914. In Kelly, B. M. (ed.) *Suburbia Re-examined*, Greenwood Press, Westport, Conn., 119–32.

Pooley, C. (1977) The residential segregation of migrant communities in mid-Victorian Liverpool. *Transactions of the Institute of British Geographers*, **NS 2**, 364–82.

Pooley, C. (1984) Residential differentiation in Victorian cities: a reassessment. *Transactions of the Institute of British Geographers*, **NS 9**, 131–44.

Pred, A. (1990) *Lost Words and Lost Worlds: Modernity and the Language of Everyday Life in Late Nineteenth-Century Stockholm*, Cambridge University Press, Cambridge.

Pred, A. (1995) *Recognizing European Modernities*, Routledge, London.

Radford, J. (1979) Testing the model of the pre-industrial city: the case of ante-bellum Charleston, South Carolina. *Transactions of the Institute of British Geographers*, **NS 4**, 392–410.

Radford, J. (1981) The social geography of the nineteenth-century US city. In Herbert, D. T. and Johnston, R. J. (eds) *Geography and the Urban Environment: Progress in Research and Applications, Volume IV*, John Wiley, Chichester, 257–93.

Reeder, D. (1987) *Charles Booth's Descriptive Map of London Poverty 1889*, London Topographical Society, London.

Relph, E. (1987) *The Modern Urban Landscape*, Johns Hopkins University Press, Baltimore.

Rosenzweig, R. and Blackmar, E. (1992) *The Park and the People: A History of Central Park*, Cornell University Press, Ithaca, N.Y.

Savage, M. (1995) Walter Benjamin's urban thought: a critical analysis. *Environment and Planning D: Society and Space*, **13**, 201–16.

Simmel, G. (1903) The metropolis and mental life. In Levine, D. (ed.) (1971) *On Individuality and Social Form*, University of Chicago Press, Chicago, 324–39.

Stansell, C. (1986) *City of Women: Sex and Class in New York, 1789–1860*, Alfred A. Knopf, New York.

Sutcliffe, A. (ed.) (1984) *Metropolis 1890–1940*, Mansell, London.

Tarn, J. (1974) French flats for the English in nineteenth-century London. In Sutcliffe, A. (ed.) *Multi-Storey Living*, Croom Helm, London, 19–40.

Tarr, J. and Dupuy, G. (eds) (1988) *Technology and the Rise of the Networked City in Europe and America*, Temple University Press, Philadelphia.

Thorne, R. (1984) Office building in the City of London 1830–1880. Paper presented to the Urban History Group Colloquium, London.

Thrift, N. (1994) Inhuman geographies: landscapes of speed, light and power. In Cloke, P., Doel, M., Matless, D., Phillips, M. and Thrift, N. *Writing the Rural: Five Cultural Geographies*, Paul Chapman Publishing, London, 191–248.

Walker, R. A. (1981) A theory of suburbanization: capitalism and the construction of urban space in the United States. In Dear, M. and Scott, A. (eds) *Urbanization and Urban Planning in Capitalist Society*, Methuen, London, 383–429.
Walkowitz, J. (1992) *City of Dreadful Delight: Narratives of Sexual Danger in Late-Victorian London*, Virago, London.
Ward, D. (1975) Victorian cities: how modern? *Journal of Historical Geography*, **1**, 135–51.
Ward, D. (1989) *Poverty, Ethnicity, and the American City, 1840–1925: Changing Conceptions of the Slum and the Ghetto*, Cambridge University Press, Cambridge.
Ward, D. and Radford, P. (1983) *North American Cities in the Victorian Age*, Historical Geography Research Series No. 12, Geo Books, Norwich.
Ward, D. and Zunz, O. (eds) (1992) *The Landscape of Modernity: New York City, 1900–1940*, Russell Sage Foundation, New York.
Weiss, M. (1987) *The Rise of the Community Builders: The American Real Estate Industry and Urban Land Planning*, Columbia University Press, New York.
Wharton, E. (1974) *The Age of Innocence*, Penguin, Harmondsworth (originally 1920).
Winter, J. (1993) *London's Teeming Streets 1830–1914*, Routledge, London.
Yelling, J. (1986) *Slums and Slum Clearance in Victorian London*, Allen & Unwin, London.
Yelling, J. (1992) *Slums and Redevelopment: Policy and Practice in England, 1918–45*, UCL Press, London.
Zunz, O. (1990) *Making America Corporate 1870–1920*, University of Chicago Press, Chicago.
Zurier, R., Snyder, R. W. and Mecklenburg, V. M. (1995) *Metropolitan Lives: The Ashcan Artists and Their New York*, National Museum of American Art/W.W. Norton, New York.

PART IV

PAST AND PRESENT

Chapter 10

Historical geographies of the present

Nuala C. Johnson

Introduction

> Now is the winter of our discontent
> Made glorious summer by this sun of York
> (Gloster, *Richard III*)

This opening phrase from William Shakespeare's *Richard III* forms the interpretative stepping stone of Al Pacino's late twentieth-century exploration of an historical drama. What does the phrase mean? Making sense of a play – written four centuries ago about an English king who reigned for two years in the late fifteenth century – is the central preoccupation of Pacino's documentary movie, *Looking for Richard*. As an exercise in translation, Pacino's treatment of the play brings into sharp relief the possibilities and challenges presented by attempting to bring the past into the present. The performance and interpretation of the drama by an American cast, the filming of the play in New York City, and the conversations between actors, directors, camera crew, Shakespearean specialists, taxi-drivers, high-school students, and construction workers, all underpin the queries that Pacino raises about translating meaning across space (from the United Kingdom to the United States) and through time (from the past to the present).

The adverb of time – 'Now' – which dramatically introduces the opening speech by Gloster, immediately unfetters the temporal chain conventionally employed to locate moments of the past. Each rehearsal of *Richard III* arises from the perspective of 'Now' and Pacino's search for Richard is one moment in that quest for meaning. From discussions of iambic pentameters, the War of the Roses, the internecine intrigue of the English court, to the psycho-political and sexual motives underpinning the actions of the principal characters, Pacino's documentary film seeks to translate both the *process* of interpretation and the *interpretation* itself to a popular audience. In so doing it makes translucent the relationship between the text and context in any rendition of the past. Similarly it melts the distinction between historical fact and creative fiction by historicising the fiction and dramatising the facts.

An Al Pacino movie, based on a Shakespearean tragedy, may seem a far cry from the subject matter of this chapter, yet our translation of the past into the present shares many of the same conceptual and cultural problems that Pacino's cast experience when dealing with Shakespeare. How do we render extremely complex familial, political and geographical relationships of the past comprehensible? To what extent can the language and syntax of a different period be translated to audiences schooled in new modes of communication? How visible should the lens of interpretation be in any recounting of the past? All these questions impinge on the approaches taken by the monument-maker, the artist, the museum curator, the historian or the geographer making sense of and translating the past to popular audiences.

In this chapter I seek to identify a selection of the theoretical, moral and practical questions raised by bringing the past into the present. My focus will specifically rotate around two themes. First, the articulation of a public memory and the creation of landscapes of remembrance in the aftermath of the First World War is examined. Secondly, the debates surrounding the representation of the past through heritage tourism are addressed and some of the opportunities and challenges encountered at heritage sites will be elucidated through an Irish case study. Combining these themes, I suggest that the historical geographies of the present, like each rehearsal of Shakespeare's Gloster, are fundamentally conjugated in the present tense, a continuous present calibrated by the shifting sands of interpretation. In the following section we begin to unravel some of the meanings in popular consciousness attached to the First World War.

War and public memory

Although in his writings James Joyce made only one direct reference to the Great War, literary historians have contended that *Ulysses*:

> . . . constitutes a response in content and form, not only to World War I, the Easter Rising, and other upheavals, but to the preceding quarter of a century – a period of intensified imperial and national rivalries, of technological innovation, of social change. (Fairhall, 1993: 164)

The novel's principal character, Stephen Dedalus, '. . . complains that history is a nightmare from which he is trying to escape' (Joyce, 1993: 55). For European society, the years, 1914–18, can also be seen as a nightmare out of which it was seeking escape. The release, however, was never complete and fragments of the nightmare persisted in the memory of both the individual soldier and the larger society. The structuring of this post-war memory, both private and public, entails some discussion of the relationship between history as past events, and history as a narrative account of past events. For the historical geographer the written account is central but, as Frederic Jameson (1981: 35) contends, the past itself 'is not a text, not a narrative'.

For nineteenth-century historians, the text may have been construed as a straightforward, simple presentation of what actually happened. In this century, however, it has been more fully recognised that the evidence of history cannot be so easily separated from the interpretation built upon it (Collingwood, 1946). This is particularly true of efforts to situate the First World War in social, economic and intellectual history. For instance, feminist historians – in beginning to address the impact of the War on gender relations – have drawn quite varied conclusions. Some have viewed the War as a deciding moment in the re-articulation of gender roles through documenting the extension of female social, economic and sexual freedoms during the conflict (Gilbert and Gubar, 1989). Others, however, have interpreted the evidence in a different manner. Using the image of the 'double helix', with its structure of two intertwined strands, Margaret and Patrice Higonnet (1987: 39) have attempted:

> . . . to trace the continuity behind the wartime material changes in women's lives. That continuity lies in the subordination of women's new roles to those of men, in their symbolic function, and more generally in the integrative ideology through which their work is perceived.

This example illustrates that our account of past events cannot rely on the robustness of the evidence alone; it is also dependent on the theoretical framework guiding it.

Representations of the Great War, and the construction of a collective memory of the conflict, have also been subject to multiple analyses. From an examination of the visual arts and fiction, literary historians have argued that the War represented a critical juncture in the evolution of an ironic modernism (Fussell, 1975). These types of study have focused attention on élite manifestations of the War. Alternative views of commemoration stress the linkages between post-war memory and the cultivation of nationalist politics, especially in Germany and Italy (Mossé, 1990). One historian claims that:

> Modern memory was born not just from the sense of a break with the past, but from an intense awareness of the conflicting representations of the past and the effort of each group to make its version the basis of national identity. (Gillis, 1994: 8)

The links between memory and national identity are complex. A number of studies have stressed the need for a contextual approach to commemoration, which integrates into the analysis the voices of a variety of different actors: soldiers, veterans' organisations, the public and the state (Whalen, 1984; Gregory, 1994).

The distinction between modern and traditional memory is perhaps best represented in the writings of Pierre Nora who has suggested that modern memory emerged out of the economic and political revolutions of the late eighteenth century and replaced 'true' memory '. . . which has taken refuge in gestures and habits, in skills passed down by unspoken traditions, in the body's inherent self-knowledge, in unstudied reflexes and ingrained memories' (Nora, 1989: 26). In contrast, modern memory is self-conscious, historical, individual and archival. In terms of war commemoration, the validity of the distinction between traditional and modern forms of mourning has been recently challenged. Winter (1995: 5) has suggested that traditional forms of mourning persisted in post-war commemoration, precisely because such practices had healing powers in ways that modern ironic responses to the Great War did not. He claims that modern memory's 'multi-faceted sense of dislocation, paradox, and the ironic, could express anger and despair, and did so in enduring ways; it was melancholic, but it could not heal'.

The spectacle of memory

Unlike formal academic histories, in which an account of the past is conventionally structured around the linking together of episodes into a narrative, public memory may be more suitably articulated as a spatial arrangement of objects around a spectacle. Leersen (1996: 7) puts it as follows: '. . . one way of unifying history [is] to rearrange its consecutive events from a narrative order into a spectacle, a conspectus of juxtaposed "freeze-frame" images'. The collapsing of time into space through the annual rehearsal and repetition of a spectacle provides a framework, not only for understanding remembrance, but also for the public enactment of forgetfulness. Geographers have begun to theorise the extent to which spectacle has become the total lens through which modern society is experienced and controlled. In this vein, Ley and Olds (1988: 194) suggest, that '. . . spectacle is the manifestation of the power of commodity relations, and the instrument of hegemonic consciousness', where the masses of spectators are rendered passive and duplicitous in their own impotence. This monolithic sense of control of the spectator by those creating the spectacle has been challenged, however, through the analysis of parody and other subversive uses of spectacle (Bonnett, 1989).

The genealogy of the spectacle metaphor has also been explored and the different meanings associated with the term outlined. These range from spectacle as ordinary display to spectacle as '. . . the sense of a mirror through which truth which cannot be stated directly may been seen reflected and perhaps distorted' (Daniels and Cosgrove, 1993: 58). This latter view of spectacle borrows from Roland Barthes who drew on parallels with ancient theatre in his discussion of the 'spectacle of excess' witnessed in wrestling. He claims (1972: 23) that what 'is thus displayed for the public is the great spectacle of Suffering, Defeat, and Justice'. Analysing the cultural meaning of spectacle, Barthes stressed the significance, not only of words and actions, but also of objects themselves (the bodies of the wrestlers) as signifiers in the production of meaning. In so doing he moved beyond the linguistic study of signs initially developed by Saussure (Barthes, 1967), extending the analysis from an interpretation of an individual image (for example, a photograph), to the investigation of an entire event or series of events.

The strength of this approach to the study of remembrance of the Great War is that it was popularly represented, precisely through large-scale drama or theatre. The construction of a spectacle of remembrance translated individual responses to loss and victory into a collective response, where the relationship between the 'actors' in the spectacle, the audience viewing it, and the geographical setting which framed it, all created the context for interpretation. In his discussion of wrestling, Barthes stressed these precise types of connections. The exaggerated antics of the wrestlers, the moral expectations of the audience, and the arenas in which the meaning was adjudicated, were all interrelated. For Barthes, wrestling was not a sport, viewed to see who would win or lose; it was a spectacle where the ethics of the physical encounter were negotiated. While modern-day wrestling may seem remote to the slaughter of the First World

War, the question of the intelligibility of death, and in this case the prodigious loss of life, is germane, as each death was simultaneously a private moral matter (for family and friends) and a public one (for the state and the army). The response of a civilian audience to that which they themselves did not experience directly, raised questions about the moral and political meaning of modern warfare. European society, in the aftermath of the War, attempted to present and reconcile these questions through staging annual parades and creating commemorative landscapes. By treating these as ritual spectacles, albeit considerably different in kind to more orthodox spectacular events, we begin to unravel the ways in which large-scale death could be culturally and morally harmonised in a peacetime environment.

Landscapes of death and remembrance

An account of the past relayed through public spectacle, like narrative history, is partly mediated through the lens of current political preoccupation. In the context of the British Empire's response to its war dead, a heated public debate took place which was anchored around the twin issues of where and how these soldiers could be best officially commemorated (Heffernan, 1995). Before the Great War, the construction of official memorials and burial sites to the rank and file war dead was rare. But the sheer volume of casualties, and the fact that the army was largely comprised of volunteer recruits rather than professional soldiers, altered the entire terms of reference. The geographical scale of fighting, the number of soldiers being killed, and the manner in which they were being killed, quickly necessitated action on the registration and burial of the dead. Under Fabian Ware, the Graves Registration Commission began this task and by May 1916, 50,000 graves had been registered and 200 battlefield cemeteries were under construction in France and Belgium. While it was initially thought that corpses would be repatriated to Britain once the War ended, the horrendous volume of casualties pointed to the immense logistical and moral dilemmas that would be raised by such a strategy. Consequently, as the War continued, debates began to take place on how to remember and represent the dead.

In a complex exchange of views, the Imperial War Graves Commission favoured the construction of simple uniform headstones (instead of crosses which were seen as too Catholic a symbol to embrace all the religious and non-religious participants in the War), while opponents to this proposal objected to the uniformity, impersonality and secularism implied by the design (Heffernan, 1995). After heated parliamentary debates, and vitriolic exchanges in the letters pages of the national press, the central issue to emerge concerned the ownership of the bodies and the legitimacy of the state to manage their burial and remembrance. By 1920 this issue was resolved: 'The war dead were henceforth public property, and their commemoration was to be organised not by individuals in private burial places but by an official bureaucracy' (Heffernan, 1995: 305). An official landscape of remembrance was to be inaugurated along the Western Front and, to this end, thousands of Portland stone headstones were transported across the English Channel and erected in a series of specially

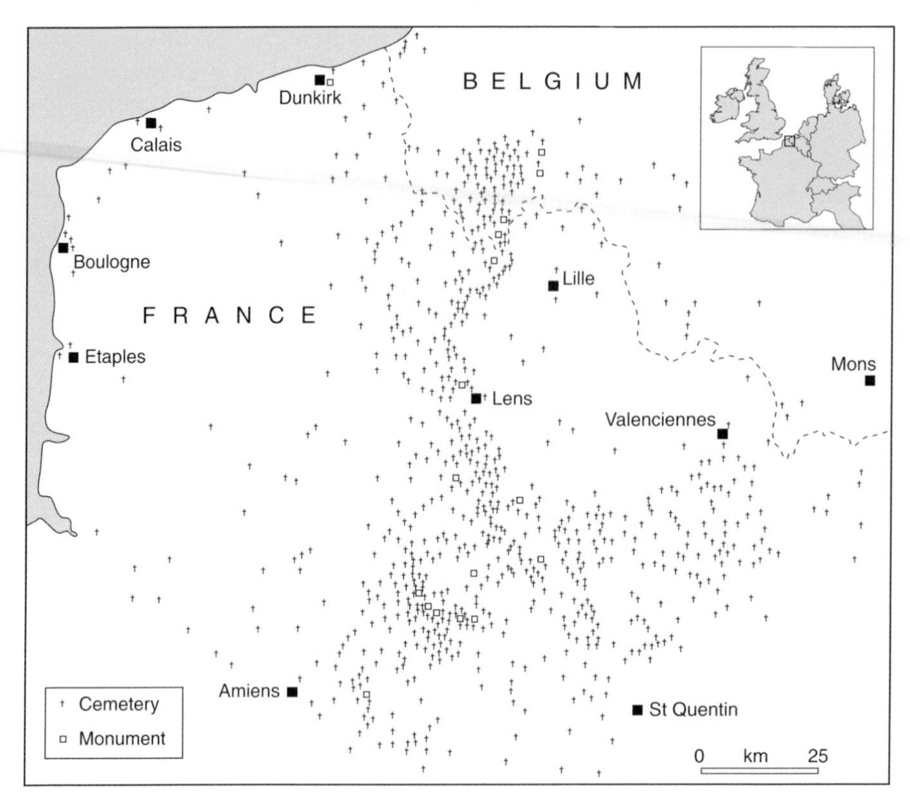

Figure 10.1 Imperial (now Commonwealth) War Graves Commission cemeteries in France and Belgium (after Heffernan, 1995).

commissioned war cemeteries in Belgium and France. By 1930 there were over 540,000 headstones established in 891 cemeteries designed to replicate many elements of a quintessential English style of landscape architecture (Morris, 1997) (Figure 10.1). Although buried abroad then, there was an attempt to domesticate the landscape in a way in keeping with the familiar tastes of those at home (Figures 10.2 and 10.3).

The symbolic keystone in the calendar of remembrance, inaugurated after the First World War, corresponds with the Sunday closest to the signing of the Armistice on November 11. An official spectacle of commemoration at the Cenotaph in London is matched by smaller rituals of memory in towns and villages across the United Kingdom. If the memorial sites at home represent the abstract and communal articulation of loss, the war cemeteries along the Western Front represent the physical and more individualised evidence of the large-scale destruction of soldiers in the First World War. As Heffernan (1995: 295) comments:

> The official commemoration of the war dead is . . . articulated around a complex geography, combining domestic ceremonials from which the dead are excluded and

Figure 10.2 Courcelette British Cemetery, 1916 Somme battlefield: the Portland stone headstone marking the grave of a Canadian soldier killed on 15 September 1916.

> a vast network of overseas memorials and cemeteries where individual soldiers are recalled and their actual remains interred.

Pilgrimages to these sites remain as popular today as they did for the families of the dead in the years immediately following the conflict. Despite the fact that most of that generation have now passed away, the collective memory of the War is maintained through these landscapes of memory and the tour companies which organise excursions to these cemeteries. The emotional impact of witnessing (either first hand or through the media) the thousands of war graves and memorials to the missing in France and Belgium cannot be underestimated and their role in popular understandings of the British past remains important, albeit now modified in meaning. The commitment by the state to maintain these cemeteries in perpetuity ensures that the past will remain in the consciousness of those living in the present. Although the meanings attached to these sites will vary along a suite of different axes (for example,

Figure 10.3 The Commonwealth War Graves Commission cemetery at 'Y' Ravine, 1916 Somme battlefield.

social or gender lines), each encounter with these shrines of remembrance will be an exchange between the 'Now' of the Great War and the 'Now' of the viewer. Indeed Remembrance Sunday also includes those killed in the Second World War and subsequent conflicts. Since the Vietnam War, the conduct of war, public attitudes towards conscript armies and casualties has altered radically, yet the sacrifice of a generation of young men and women in the second decade of this century continues to capture the public imagination, not least because the very magnitude and circumstances of their deaths now seems to stand as *the* symbol of the utter futility of war.

Heritage and the making of history

If the historical geography of military conflict and widespread death is brought to contemporary audiences through ritual, spectacle and monument, the past is similarly brought into focus through the representational practices and philosophies underlying the heritage industry. The considerable expansion, over the last several decades, in the number of heritage tourism sites available for public consumption has heralded important discussions about the translation of the past through spaces of historic interest. It is towards these questions that the remainder of the chapter now turns.

The relationship between heritage, history and memory has been subject to much debate recently among geographers, historians and cultural critics (Lowenthal, 1996; Tunbridge and Ashworth, 1996). Conventionally, a rigid line of demarcation ran between the past as narrated by professional historians on the one hand, and by the heritage industry on the other. Heritage, as a concept, begins with the highly individualised notion of personal inheritance or bequest (for example, through family wills and legacies). We are more concerned here, however, with the collective notions of heritage, which link a group to a shared inheritance. The basis of this group identification varies in time and in space and can be based on allegiance derived from a communal religious tradition, a class formation, geographical propinquity, or a national grouping. Indeed, it is with respect to the 'imagined community' of nationhood that heritage is often most frequently connected (Anderson, 1991).

While the origins of the nation-state may be relatively recent, nations themselves are often based on the assumption that group identity derives from a collective inheritance that spans centuries and at times millennia. National states attempt to maintain this identity by highlighting the historical trajectory of the cultural group through preservation of elements of the built environment, through spectacle and parade, through art and craft, through museum and monument (Hobsbawm and Ranger, 1983; see Chapter 3). The heritage industry, then, has often been viewed as a mechanism for re-inscribing nationalist narratives in the popular imagination (Wright, 1985). Lowenthal (1994: 43) claims that 'heritage distils the past into icons of identity, bonding us with precursors and progenitors, with our own earlier selves, and with promised successors'. As such, the historical narratives transmitted through heritage are seen to be selective, partial and distorting. They offer a 'bogus' history which ignores complex historical processes and relationships, and sanitises or forgets the less savoury dimensions of the past. This contrasts with the work of professional historians where 'testable truth is [the] chief hallmark [and] . . . historians' credibility depends on their sources being open to scrutiny' (Lowenthal, 1997: 120).

The distinction between true history and false heritage, however, may be more illusory than actual when viewed from the perspective of deconstructionism within the social sciences. Making the claim that all historical narration is interpretative, postmodern accounts make problematic the distinction between representation and reality, between fake heritage and genuine history. Postmodernism involves a 'dissolving of boundaries, not only between high and low cultures, but also between different cultural forms, such as tourism, art, music, sport, shopping and architecture' (Urry, 1990: 82). Drawing from the insights of semiotics, it suggests that signs are all that we consume and that we do so knowingly (Baudrillard, 1988). The signs which represent episodes from the past can be found in historians' scholarly texts as well as at heritage sites. The treatment of time and historical explanation though may be significantly different between heritage sites and works of professional 'scientific' history. In the latter, historical sequence (for example, ancient, medieval, early modern, modern) structures the narration. Without this chronology, history would be rendered a series of random events taking place outside of the strictures of time or context. Thus although the past that is mediated through heritage may be only one element in a whole suite of historical representations, its handling of time is of significance in the conjugation of the historical narrative.

Putting the past to work: heritage tourism

The globalisation of tourism destinations over the past few decades and the large revenues it generates for individual states is now well-documented (for example, Pearce, 1995). With an annual worldwide growth rate of five to six per cent per annum, it is estimated that tourism is now the world's largest employer. This new pattern of tourist activity has been linked to a number of factors: to the changing circumstances of the workplace, where conventional distinctions between work time and leisure time are increasingly blurred (Urry, 1990); to the rise of a new service class rich in cultural capital; and to rapidly changing technologies which have accelerated a sense of time–space compression (Bourdieu, 1977; Soja, 1988). Coupled with these changes is the socially and geographically selective nature of sites of tourism production and consumption. While access to leisure time and the capital resources necessary to travel are unevenly distributed at a global scale, the pattern of fashionable destinations is also differentially distributed and subject to rapid transformations. MacCannell (1992: 1) notes, however, that tourism cannot be reduced solely to commercial transactions since 'it is an ideological framing of history, nature and tradition; a framing that has the power to reshape culture and nature to its own needs'. The framing of history and its relationship with narratives of national identity have assumed increased importance with the appreciable expansion of heritage tourism (or 'gazing on the past', Urry, 1990).

While heritage tourism forms a distinct niche market, Ashworth (1994: 21) suggests that heritage is intrinsically a place-based activity: 'whether or not heritage is deliberately designed to achieve pre-set spatio-political goals, place identities at various spatial scales are likely to be shaped or reinforced by heritage planning'. Sheilds (1991) outlines the manner in which specific spatial arrangements and cultural practices become appropriate for particular types of activities, and together constitute a place-myth which is undergirded by a suite of core place-images, both symbolic and material. Geographers have long recognised the importance of place promotion in evoking and disseminating powerful place icons (see, for example, Burgess and Gold, 1985; Kearns and Philo, 1993; Gold, 1994; Tunbridge and Ashworth, 1996). Particular rural and urban landscapes, for instance, can play a central role in the heritage industry's 'recovery' of the past. The 'regal' landscape of the West End of London, the Mall in Washington D.C., and the sharecroppers' houses in Tuscany, all somehow come to represent the quintessential mirrors of a culture's collective past, while their reinvention for tourist consumption fixes them spatially in the historical imagination and helps ensure their future protection.

Analyses of postcards, tourist brochures and advertising literature have in various ways begun to deconstruct the influential images of place traded to tourists (Selwyn, 1990, 1996; Cohen, 1995; Crang, 1996). The significance of these representations, however, does not reside solely in identifying whether they are effective or authentic expositions of place. More particularly, it is their role as part of a larger network of circulated ideas about the nature of place and the past which is of import. In this context, Britton (1991: 475) suggests that the tourism imaginary is a 'lesson in the political economy of the social construction of "reality" and the social construction of place and people, whether from the point of view of the visitors, the host communities, or the state'. It has been suggested that tourism sites '. . . are centres of physical and emotional sensation from which temporal and spatial continuities have been abolished' (Selwyn, 1990: 24). This may be to overstate the case, however, as there is no necessary imperative for the eradication of the historical imagination at tourism sites. (Samuel, 1994, provides a host of useful examples where popular representations of the past are effectively executed.) The intellectual focus on the imaginary of place, and the consequent reading of the text of place using the insights of semiotics, has tended to overshadow the demarcation and understanding of time in the representational practices of sites of tourist consumption. While heritage is grounded in particular spaces, it is the relationship between space and time – the awarding of space a past – that is central to heritage tourism planning. If, as Sorenson (1989: 65) suggests, theme parks are 'visits to time's past', questions about the representation of historical knowledge is of equal importance to the representation of space. In the following discussion, an Irish example is used to illuminate some of these relationships.

Selling pasts: travel to Ireland

The last decade has witnessed an increased push towards transforming the Irish countryside from a predominantly working agricultural one to a tourism landscape (for a discussion of the distinction between 'working country' and landscape, see Williams, 1973). The expansion in the number of heritage centres, historic trails, nature reserves and interpretative centres has hastened the remapping of Ireland from a peripheral European state specialising in the export of agricultural products (see for example, Ó Gráda, 1994), to a 'pleasure periphery' designed to retail and retell historical narratives to an ever-expanding volume of travellers visiting the Republic of Ireland (O'Connor and Cronin, 1993; Breathnach, 1994; McManus, 1997). Overseas tourist numbers increased from 3.5 million in 1991 to 5.0 million in 1997.

While historically it has been the rural environment which has been most vigorously promoted, Irish cities have increasingly become part of the tourism package, frequently acting as entrepôts for overseas tourists. Dublin, in particular, has marketed itself effectively in a drive to increase tourism revenue especially since it's designation as European City of Culture, 1991 (Clohessy, 1994). Even though Dublin houses some of the most popular Irish heritage attractions (for example, the Book of Kells), the influence of modernist developers has resulted in much of the historic fabric of the city being razed (Lincoln, 1993). In the 1990s, therefore, the city has been marketed not so much as a place to see, but as a place to experience, combining its historical features with more contemporary ones.

Travelling to Ireland and associated travel writing, however, is not a recent activity. Indeed Cronin (1993: 52) observes that '[for centuries] travellers in search of salvation, instruction or the godsend of novelty have either left the island or landed on it, tracking the signs of specificity'. What is relatively new is the increased volume of visitors to Ireland, the expanding number of travel books published on Ireland since 1945, and the increased role of the state in the promotion of Ireland as a tourist destination. Through advertising, dominant images of place are represented and analyses of tourist brochures suggest that Ireland offers the promise of 'empty space' – space that is uninhabited and in this respect reminiscent of colonial accounts of overseas territories ripe for European settlement (Pratt, 1992). Visual representations picture Irish 'natives' as predominantly working in agriculture and suggest implicitly that an organic relationship between people and their natural environment is to be found in Ireland (Gibbons, 1988; Gallagher, 1989). Concomitant with depictions which evoke 'empty space', tourist literature similarly conveys Ireland as occupying 'empty time', where today is like yesterday and yesterday is like tomorrow.

In his analysis of academic anthropology, Fabian (1983) suggests that the discipline emerged as an allochronic discourse: the science of studying other people in another time. This practice, he claims, was inevitably political as anthropologists positioned themselves in relation to the society under study along spatial and temporal co-ordinates. Remote places at the 'uttermost ends of the earth' frequently proved to be fertile ground for the advancement of evolutionary theories of human societies (Gamble, 1992). Paradoxically, the proximity of Ireland to the centre of Western intellectual thought, coupled with a pattern of colonial relationships between itself and its nearest neighbour, has rendered it a transitional space, neither exclusively traditional nor modern *vis-à-vis* progressivist frameworks of the past (Graham, 1997). Tourist images and travel writing about Ireland have adopted similar representational practices where Ireland is placed 'behind' modern time. The tourist text mimics its colonial antecedents, which underwrote a particular historical narration of Ireland (Kiberd, 1995). In his analysis of travel writing Cronin (1993: 61) observes that 'the hegemony of linear, unidirectional time in the post-Renaissance West, is subverted by the digressive, anarchic disrespect for its imperatives in daily life in Ireland'. Over the past decade the Irish state has published a variety of strategy documents, outlining its approach to the development of heritage tourism attractions and I now focus on the most substantial initiative.

Planning a heritage landscape

In 1992 Bord Fáilte, the Irish Tourist Board, published a document which would form the basis for the future development of heritage attractions in Ireland. The introductory section of the report claimed that;

> History and culture are fundamental to the core Irish tourism product, as perceived by potential (overseas) tourists . . . Irish history, due to the influence of many peoples, cultures and conflicts, is not easily understood by visitors. (Bord Fáilte, 1992: 1–2)

To overcome this difficulty, it was proposed that a chronological approach be discarded in favour of a thematic one, because chronologies are problematic to represent and are poorly understood by visitors (Bord Fáilte, 1992). Cronin (1993) noted a similar absence of chronology in travel accounts and he connected this with the discursive strategies of the genre itself. Travel writers comment on time being ordered differently in Ireland, running counter to the 'chronocracy' of Western conceptions of time, and this facilitates a strategy which attempts to convey a unique sense of time and place. The travel writer digresses where 'the peripeteia of incident are paralleled by the wanderings of narrative, shifting between description, comment and speculation' (Cronin, 1993: 62). Similarly, Ashworth and Larkham (1994) have observed a reciprocal link between heritage and place, where place is sacralised by its general historical associations. Bord Fáilte's desire to structure the Irish past, not around chronological time but around 'themed spaces' borrows some of the techniques employed by travel writers. In tourism planning of this type the plot of the past is loosely arranged around a series of themes acted out in spaces but the sequencing of events, which may inform academic historians' structuring of the past, is displaced. The marketing initiative of Bord Fáilte broadly reflects this diagnosis where 'emotional experience [is at] its core positioning' (Bord Fáilte, 1997: 3).

Specifically, it was proposed that the Irish past be mediated through a series of 'interpretative gateways'. These are arranged under five broad themes, each explored through a series of 'storylines'. The five themes – each supported by specific subthemes – comprise: live landscapes; making a living; saints and religion; building a nation; and the spirit of Ireland (Figure 10.4). Live landscapes, for instance, include what we might conventionally call nature, with storylines consisting of land and sea, mountain and moorland. This themed strategy seeks to avoid replication of the product and to enable the tourism package to be *regionalised* (fixed in space) in a coherent manner. This approach is not guided by a desire to explicate the interconnections between times and spaces; rather, it is based on a principle of sound economic management and the political requirement to spread income from tourism around the country. Under the theme, making a living, for instance, is the storyline of emigration and famine. The explicit omission of any reference to time masks the reality that Ireland has experienced periodic emigration and famines over the past two millennia. Implicitly, this category is making reference to the nineteenth century, and while famine and emigration were important processes in accounting for changing patterns of land tenureship and settlement in that century, and with cultural transformations associated with the decline of the Irish language and the re-organisation of the Catholic Church (Ó Gráda, 1988), their incorporation under the theme, making a living, appears unintentionally ironic. The regionalisation of the famine theme to ports of departure (for example, Cobh, Co. Cork), or to workhouses in the west of Ireland, disguises its effects on political, economic and social processes elsewhere in the country.

Ideally, for the Tourist Board, each heritage site in the state should be accommodated under one of these themes and should market itself accordingly. Moreover, each site must avoid 'copying the storyline of an existing attraction' (Bord Fáilte, 1992: 1–4). While the immediate experience of the Great Famine

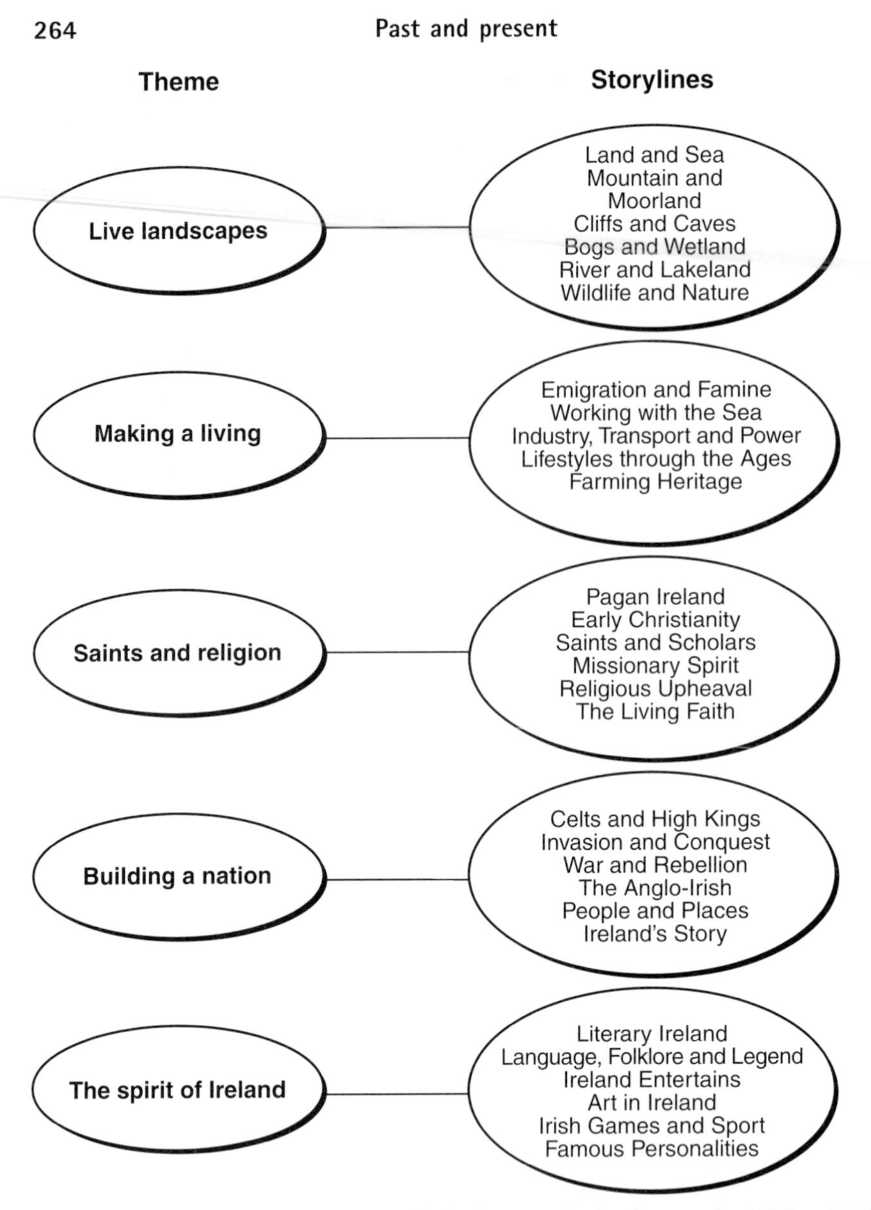

Figure 10.4 Themes and storylines in Irish heritage marketing (*source*: Bord Fáilte, 1992).

in Ireland during the 1840s was socially and spatially selective (Whelan, 1986), this policy suggests that two adjacent heritage sites should not publicly represent their history of the Famine because to do so would be copying one another's storyline. The static conception of place implied here allows none of the spatial dynamics of the Famine to be represented. Tourism planning then can be likened to a literary text – a series of short stories – which can be read independently of each other by the tourist. While the text metaphor has been popular in landscape interpretation, especially with its emphasis on intertextuality (the relationships between different texts), the Tourist Board's approach is underwritten by a narrative of place exceptionality. Time is obliterated by place as heritage mapping becomes a reference guide to spatialised storylines rather than to a series of localised yet interdependent histories. Consequently, as Agnew (1996: 28) has noted, time and space have suffered by 'by expressing one in the reductionist terms of the other'. Moreover, for the consumer this strategy conveys a random sense of time where people's 'lives . . . are experienced as a succession of discontinuous events' (Urry, 1990: 92).

In Ireland, where the contestation of what has been regarded as a nationalist historiographical canon has occasioned heated debate (see, for example, Boyce and O'Day, 1996), the interpretative gateway approach to heritage planning has relevance to the wider debate in academic history (Graham, 1994, 1997). The overlapping of different interpretations of the past, suggested by historical revisionism, is hindered by a themed framework where avoidance of repetition or overlap on a site-basis constrains the possibility for multivocal representations. Unlike some analyses of the British heritage industry, which link nostalgia for the past to economic decline (Wright, 1985; Hewison, 1987), in Ireland it is precisely the pressures to reject older interpretations of the past which generate a crisis of representation in the heritage sector. In the late 1990s, for instance, visits to Buckingham Palace or the Eiffel Tower could be construed as apolitical activities in neutral spaces – sites of entertainment – unconnected with current political issues: in Dublin, a visit to Kilmainham Gaol is difficult to divorce from the independence movement of the early twentieth century and thus to the constitutional legitimacy of the state itself. The themed framework adopted by Bord Fáilte may enhance the marketing of Ireland as a tourism destination, and may render it comprehensible for European Union grant-aid, but the privileging of a themed framework of heritage management underestimates the potential for heritage sites to mediate the past in different but by no means antithetical ways to academic history.

A National Heritage Park?

The Irish National Heritage Park in County Wexford, designed and opened in 1987 under the auspices of the local authority, Wexford County Council, falls within Bord Fáilte's theme of building a nation, the storyline being 'Ireland's story'. The site was developed to achieve the twin aims of attracting tourists to south-eastern Ireland and educating the public in field monuments. Historic theme parks have a long history, and Sorenson (1989) draws a distinction between those that have evolved *in-situ* from existing archaeological sites and those that are purpose-built. While the distinction may be best thought of as a continuum, the Irish National Heritage Park is largely the latter. Drawing upon professional archaeological expertise, the Park presents examples of field monuments from the Mesolithic (7000 BC) to the Anglo-Norman period (*c.* 1200–1500 AD). The site is comprised of exact replicas of field monuments based on academic research. Using Urry's (1990) tripartite division of heritage sites – Wexford is designed for the collective gaze, while it is historic

Figure 10.5 A purpose-built heritage: the Irish National Heritage Park.

and unauthentic. The Park is not specifically designed to offer a history of settlement in Ireland over several thousand years. In contrast, it is devised to convey something of 'everyday life' – the ways in which earlier peoples produced, used and consumed objects and tools of manufacture. The tourist has the opportunity to view, for example, a Stone Age house, a Viking ship, or a medieval castle (Figure 10.5).

Although the map of the site suggests that the visitor tours the Park in a roughly chronological order, beginning with the pre-historic monuments, there is no necessity to do so. In a carefully landscaped setting, the tourist can pick and mix at a micro-scale (Figure 10.6). There is little explanation of how location, context, social and economic parameters informed the development of specific aspects of the material culture represented in the Park. Material culture just is, it does not become. As in other theme parks 'death and decay are, it seems, denied' (Sorenson, 1989: 65). Hoyau's claim (1988: 29) of French heritage sites that '. . . dead labour is restaged, with the violence done

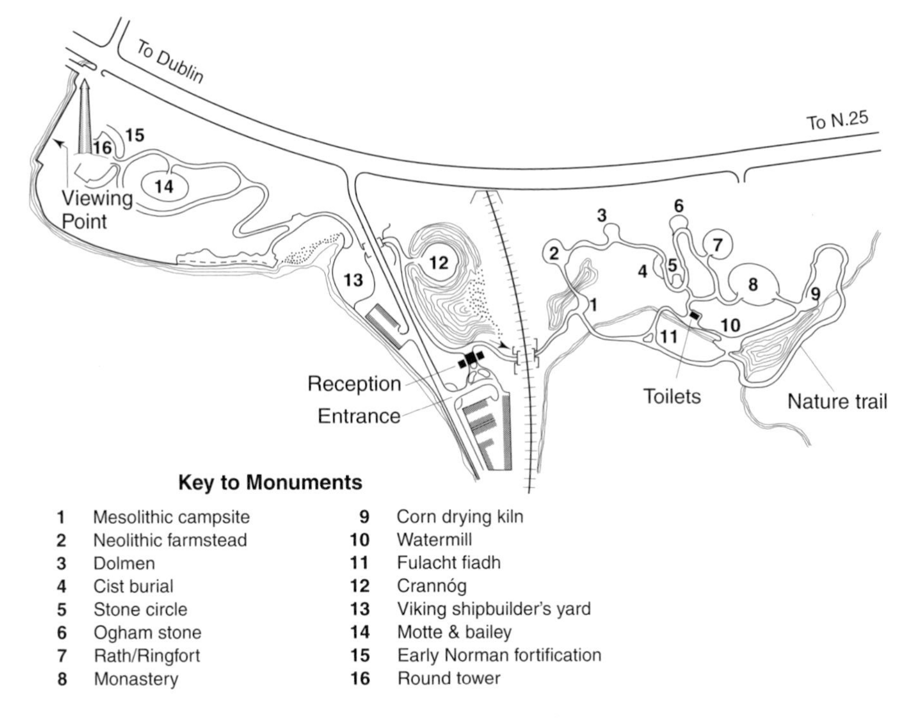

Figure 10.6 The layout of the Irish National Heritage Park.

to the producers and the environment spirited away in a search for lived experience and past forms of social life' also applies at Wexford. The focus on material culture alone creates a static experience of life in the past. While Ashworth (1994: 20) argues that it is the interpretation which 'is traded [at heritage sites], not its various physical resources', in Wexford the role of the latter cannot be underestimated, as visitors regularly climb into the Viking ship, operate an early milling machine or touch a standing stone. But much attention is focused on the visual representation and the interpretation is consistently implicit rather than explicit. The beholding eye of the tourist is naturalised as a cogent observer of the world, making sense of a visual display which is logical and rational. Crang (1997) has recently reminded us, however, of the necessity of querying these conventional assumptions about the practices of visualising.

In addition the Park's cut-off point in the sixteenth century suggests that the Irish 'nation' has produced little of value since then, or that the history of the island becomes messier and less amenable to the neat categorisations employed in the Park. While historians have generally located the emergence of ideas of nationhood in the last two centuries (Anderson, 1991), the Park's displays suggest that nationhood can be situated in the early antiquities of the island. In line with nationalist discourse, nations are held to exist from time immemorial and thus are constituted outside the usual conventions of time.

While a case can be made for the popular representation of field monuments, particularly where original examples no longer exist, in Wexford the irony is that even as a fake motte-and-bailey was under construction there, not far away, the real Anglo-Norman motte-and-bailey at Kells, Co. Kilkenny, was being bulldozed by a farmer. The commitment of the state to fund a simulacrum of archaeological monuments to attract tourists can be contrasted with a notably flimsier commitment to preserving archaeological remains.

In the context of any museum display, Lumley (1988: 13) contends that '. . . the museum text needs also to manifest the metatext, so that the very ability to read and make sense, as well as the choices leading to a particular display, are visible to the public'. In Wexford, the metatext is obscured in an effort to produce accurate replicas of field monuments. The fact that the monuments represented at the site would rarely be found in the field at such close proximity to each other is not sufficiently exposed. The functional and ideological links between the periods represented (for example, Mesolithic and Neolithic) at the Park, and the rationale for adopting Bord Fáilte's category – building a nation – are not adequately developed. The round tower erected to commemorate Wexfordmen killed during the Crimean War of 1854–55, for instance, is a nineteenth-century reproduction of an Early Christian tower. The reasons why there was a revival of interest in Early Christian architectural forms in the middle of the nineteenth century and how this is connected with a nationalist discourse is not elucidated (Sheehy, 1980). Consequently, the reader is given the impression that there is no difference whatsoever between the original and the reproduction, and thus the consumer has no context from which to understand the cultural politics underlying the use of the round tower in commemorative activity in the nineteenth century. While the Park provides students and visitors with some information on the fundamental characteristics of field monuments, and the reproductions are well-executed, the claim to be a 'National' Heritage Park is over-ambitious. The historical geography exposed at this site is of an archaeological and architectural variety, giving priority to the built environment over questions of process.

Conclusion

As I watched the annual Belfast Orange parade on 12 July 1998, I was participating in a ritual of social memory which was not simply an exercise in recollections of times past but a symbolic spectacle connecting past events with the present. The parade not only carved out a geography of city through its routing along specific thoroughfares, but the dress code, iconography of banners, music performed and relationship between the spectators and performers all contributed to the conveyance of meaning. Even though the parade takes place every year and commemorates the same event – the victory of William of Orange at the Battle of the Boyne in 1690 – I was spectating this ritual on a gloomy, wet July day in 1998. The signing of the 'Good Friday' Belfast Peace Agreement, the stand-off at Drumcree, Co. Armagh, where Orangemen were insisting on their right to march through a predominantly Catholic, nationalist area, and the apparently related murder of three children in a fire-bombing of a Catholic home in Ballymoney, Co. Antrim, all informed the performative spectacle witnessed at that specific moment. The significance of this event in conjugating diverse contemporary political identities cannot be underestimated. Like Shakespeare's Gloster, the past was brought into the present through the juxtaposition of the historic event being recalled to public memory, and the particular circumstances in which the recollection was situated.

The meaning attached to the commemoration of the Great War in the years immediately after 1918 differs from popular understandings of war today although the poignancy of the iconography – the grave of the unknown soldier, the Poppy, the Portland headstones – continues to move those whose connections with that conflict are distant and remote. The representational practices of the heritage industry in bringing the past into the present similarly evoke and provoke diverse responses in our reading of our relationship with the past. While Shakespeare's Gloster may ultimately have been prepared to substitute his kingdom for a horse to save his skin, the historic figure, Richard III, was killed by his enemies. While his tragedy is communicated to us through the play, our visits to times past – like Al Pacino's – are never so literal or so final. At the end of the performance Al Pacino 'rises' from the dead, takes his bow and departs, and we, his audience, do likewise, having been enriched or diminished by the process. In this chapter I have set out some of the key debates surrounding these acts of translation. Focusing in particular on the cultivation of a collective memory of the past through an analysis of landscapes of remembrance created after the First World War and the representational practices of the heritage industry, I have sought an explication, which highlights the complex, and at times contradictory, interconnections between the past and our interpretation of it.

References

Agnew, J. A. (1996) Time into space: the myth of 'backward' Italy in modern Europe. *Time and Society*, **5**, 27–45.

Anderson, B. (1991) *Imagined Communities: Reflections on the Origins and Spread of Nationalism*, (revised edition) Verso, London.

Ashworth, G. J. (1994) From history to heritage – from heritage to identity. In search of concepts and models. In Ashworth, G. J. and Larkham, P. (eds) *Building a New Heritage. Tourism, Culture and Identity in the New Europe*, Routledge, London, 13–30.

Ashworth, G. J. and Larkham, P. (eds) (1994) *Building a New Heritage. Tourism, Culture and Identity in the New Europe*, Routledge, London.

Barthes, R. (1967) *The Elements of Semiology*, Cape, London.

Barthes, R. (1972) *Mythologies*, Hill and Wang, New York.

Baudrillard, J. (1988) *America*, Verso, London.

Bonnett, A. (1989) Situationism, geography and poststructuralism. *Environment and Planning D: Society and Space*, **7**, 131–46.

Bourdieu, P. (1977) *Outline of a Theory of Practice*, Cambridge University Press, Cambridge.

Bord Fáilte (1992) *Heritage and Tourism*, Bord Fáilte, Dublin.

Bord Fáilte (1997) *The Fáilte Business*, Bord Fáilte, Dublin.

Boyce, D. G. and O'Day, A. (eds) (1996) *Modern Irish History: Revisionism and the Revisionist Controversy*, London, Routledge.

Breathnach, P. (1994) Employment creation in Irish tourism. In Breathnach, P. (ed.), *Irish Tourism Development*, Geographical Society of Ireland, Special Publications No. 9, Dublin, 41–60.

Britton, S. (1991) Tourism, capital and place: towards a critical geography of tourism *Environment and Planning D: Society and Space*, **9**, 451–78.

Burgess, J. and Gold, J. (eds) (1985) *Geography, the Media and Popular Culture*, Croom Helm, London.

Clohessy, L. (1994) Culture and urban tourism: 'Dublin 1991' – European city of culture. In Kockel, U. (ed.) *Culture, Tourism and Development: The Case of Ireland*, Liverpool University Press, Liverpool, 189–96.

Cohen, E. (1995) Contemporary tourism – trends and challenges: sustainable authenticity or contrived post-modernity. In Butler, R. W. and Pearce, D. G. (eds) *Change in Tourism: People, Places and Processes*, Routledge, London, 12–29.

Crang, M. (1996) Envisioning urban histories: Bristol and palimpsest, postcards and snapshots. *Environment and Planning A: Society and Space*, **28**, 429–52.

Crang, M. (1997) Picturing practices: research through the tourist gaze. *Progress in Human Geography*, **21**, 359–73.

Collingwood, R. (1946) *The Idea of History*, Oxford University Press, Oxford.

Cronin, M. (1993) Fellow travellers: contemporary travel writing and Ireland. In O'Connor, B. and Cronin, M. (eds) *Tourism in Ireland: A Critical Analysis*, Cork University Press, Cork, 51–67.

Daniels, S. and Cosgrove, D. (1993) Spectacle and text: landscape metaphors in cultural geography. In Duncan, J. and Ley, D. (eds) *Place/Culture/Representation*, Routledge, London, 57–77.

Fabian, J. (1983) *Time and Other: How Anthropology Makes its Object*, Columbia University Press, New York.

Fairhall, J. (1993) *James Joyce and the Question of History*, Cambridge University Press, Cambridge.

Fussell, P. (1975) *The Great War and Modern Memory*, Oxford University Press, Oxford.

Gallagher, M. (1989) Landscape and Bord Fáilte films. *Circa*, **43**, 24–8.

Gamble, C. (1992) Archaeology, history and the uttermost ends of the earth – Tasmania, Tierra del Fuego and the Cape. *Antiquity*, **66**, 712–20.

Gibbons, L. (1988) Coming out of hibernation? The myth of modernity in Irish culture. In Kearney, R. (ed.) *Across the Frontiers: Ireland in the 1990s*, Wolfhound, Dublin, 205–18.

Gilbert, S. and Gubar, S. (1989) *No Man's Land: The Place of the Woman Writer in the Twentieth Century*, Yale University Press, New Haven, Conn.

Gillis, J. R. (1994) Memory and identity: the history of a relationship. In Gillis, J. R. (ed.) *Commemorations: The Politics of National Identity*, Princeton University Press, Princeton, N.J., 3–24.

Gold, J. (1994) Locating the message: place promotion as image communication. In Gold, J. and Ward, S. (eds) *Place Promotion: The Use of Publicity and Marketing to Sell Towns and Regions*, John Wiley, Chichester, 19–38.

Graham, B. (1994) Heritage conservation and revisionist nationalism in Ireland. In Ashworth, G. J. and Larkham, P. (eds) *Building a New Heritage. Tourism, Culture and Identity in the New Europe*, Routledge, London, 135–58.

Graham, B. (ed.) (1997) *In Search of Ireland: A Cultural Geography*, Routledge, London.

Graham, G. (1997) *The Shaping of the Past*, Oxford University Press, Oxford.

Gregory, A. (1994) *The Silence of Memory*, Berg, Oxford.

Heffernan, M. (1995) For ever England: the Western Front and the politics of remembrance in Britain. *Ecumene*, **2**, 293–324.
Hewison, R. (1987) *The Heritage Industry*, Routledge, London.
Higonnet, M. and Higonnet, R. (1987) The double helix. In Higonnet, M., Jensen, J., Michel, S. and Weistz, M. (eds) *Behind the Lines: Gender and Two World Wars*, Yale University Press, New Haven, Conn., 31–50.
Hobsbawm, E. and Ranger, T. (1983) *The Invention of Tradition*, Cambridge University Press, Cambridge.
Hoyau, P. (1988) Heritage and the 'conserver society': the French case. In Lumley, R. (ed.) *The Museum Time Machine*, Routledge, London, 27–35.
Jameson, F. (1981) *The Political Unconscious*, Cornell University Press, Ithaca.
Joyce, J. (1993 edn) *Ulysses*, Oxford University Press, Oxford (originally published in 1922).
Kearns, G. and Philo, C. (1993) *Selling Places*, Pergamon, Oxford.
Kiberd, D. (1995) *Inventing Ireland: The Literature of the Modern Nation*, Jonathan Cape, London.
Leersen, J. (1996) *Remembrance and Imagination: Patterns in the Historical and Literary Representation of Ireland in the Nineteenth Century*, Cork University Press, Cork.
Ley, D. and Olds, K. (1988) Landscape as spectacle: world's fairs and the culture of heroic consumption. *Environment and Planning D: Society and Space*, **6**, 191–201.
Lincoln, C. (1993) City of culture: Dublin and the discovery of urban heritage. In O'Connor, B. and Cronin, M. (eds) *Tourism in Ireland: A Critical Analysis*, Cork University Press, Cork, 203–32.
Lowenthal, D. (1994) Identity, heritage and history. In Gillis, J. R. (ed.) *Commemorations: The Politics of National Identity*, Princeton University Press, Princeton, N.J., 41–57.
Lowenthal, D. (1996) *The Heritage Crusade and the Spoils of History*, Viking, London.
Lumley, R. (1988) Introduction. In Lumley, R. (ed.) *The Museum Time Machine*, Routledge, London, 1–24.
MacCannell, D. (1992) *Empty Meeting Grounds: The Tourist Papers*, Routledge, London.
McManus, R. (1997) Heritage and tourism in Ireland – an unholy alliance? *Irish Geography*, **30**, 90–8.
Morris, M. S. (1997) Gardens 'for ever England': landscape, identity and the First World War British cemeteries on the Western Front. *Ecumene*, **4**, 410–34.
Mossé, G. (1990) *Fallen Soldiers: Shaping the Memory of Two World Wars*, Oxford University Press, Oxford.
Nora, P. (1989) Between memory and history: les lieux de memoire. *Representations*, **26**, 10–18.
O'Connor, B. and Cronin, M. (eds) (1993) *Tourism in Ireland: a Critical Analysis*, Cork University Press.
Ó Gráda, C. (1988) *Ireland Before and After the Famine: Explorations in Economic History 1800–1930*, Manchester University Press, Manchester.
Ó Gráda, C. (1994) *Ireland: A New Economic History 1780–1939*, Clarendon Press, Oxford.
Pearce, D. (1995) *Tourism Today: A Geographical Analysis* (second edn), Longman, Harlow.
Pratt, M. L. (1992) *Imperial Eyes: Travel Writing and Transculturation*, Routledge, London.
Samuel, R. (1994) *Theatres of Memory: Past and Present in Contemporary Culture*, Verso, London.
Selwyn, T. (1990) Tourist brochures as postmodern myth. *Problems of Tourism*, **13**, 13–26.
Selwyn, T. (ed.) (1996) *The Tourist Image: Myths and Myth Making in Tourism*, John Wiley, Chichester.
Sheehy, J. (1980) *The Rediscovery of Ireland's Past: The Celtic Revival 1830–1930*, Thames and Hudson, London.
Shields, R. (1991) *Places on the Margin: Alternative Geographies of Modernity*, London: Routledge.

Soja, E. (1988) *Postmodern Geographies: The Reassertion of Space in Social Thought*, Verso, London.

Sorenson, C. (1989) Theme parks and time machines. In Vergo, P. (ed.) *The New Museology*, Reaktion Books, London, 60–73.

Tunbridge, J. and Ashworth, G. J. (1996) *Dissonant Heritage: The Management of the Past as Resource in Conflict*, John Wiley, Chichester.

Urry, J. (1990) *The Tourist Gaze: Leisure and Travel in Contemporary Societies*, Sage, London.

Whalen, R. W. (1984) *Bitter Wounds: German Victims of the Great War*, Cornell University Press, Ithaca.

Whelan, K. (1986) The Famine and post-Famine readjustment. In Nolan, W. (ed.) *The Shaping of Ireland*, Mercier Press, Dublin, 121–64.

Williams, R. (1973) *The Country and the City*, Hogarth, London.

Winter, J. (1995) *Sites of Memory: Sites of Mourning*, Cambridge University Press, Cambridge.

Wright, P. (1985) *On Living in an Old Country*, Verso, London.

Index

Pages on which illustrations appear are shown in **bold**.